Corynebacterium diphtheriae and Related Toxigenic Species

Andreas Burkovski
Editor

Corynebacterium diphtheriae and Related Toxigenic Species

Genomics, Pathogenicity and Applications

Springer

Editor
Andreas Burkovski
Friedrich-Alexander-Universität
Erlangen
Bayern
Germany

ISBN 978-94-007-7623-4 ISBN 978-94-007-7624-1 (eBook)
DOI 10.1007/978-94-007-7624-1
Springer Dordrecht Heidelberg London New York

Library of Congress Control Number: 2013951785

Printed on acid-free paper

Springer is part of Springer Science+Business Media (www.springer.com)

Preface

Diphtheria, the strangling angel of children, has lost its threatening potential with the development of anti toxin, toxoid vaccine and antibiotics in the last century. However, even today several thousand cases per year are reported to the World Health organization and especially the outbreak of diphtheria in the former states of the Soviet Union demonstrated impressively that diphtheria is not completely eradicated.

The outbreak in the 1990s, the development of new molecular biology tools and especially the availability of genome sequence information gave new impetus to research in this field. Several strains of *Corynebacterium diphtheriae*, the etiological agent of diphtheria and the type species of the genus *Corynebacterium* have been sequenced and genome data are available for two closely related pathogenic species, *Corynebacterium ulcerans* and *Corynebacterium pseudotuberculosis*.

The book summarizes the latest advances made in understanding physiology and host pathogen interaction of *C. diphtheriae* and its relatives. Topics addressed are genomics of toxigenic corynebacteria, host-pathogen interactions, diagnosis, surveillance and treatment strategies as well as application aspects.

Contents

Contributors

Gayatri Amirthalingam Centre for Infectious Disease Surveillance and Control, Public Health England, London, UK

Vasco Ariston de Carvalho Azevedo Departmento de Biologia, Instituto de Ciências Biológicas, Universidade Federal de Minas Gerais, Belo Horizonte, MG, Brazil

Muhammet Bektaş Istanbul Faculty of Medicine, Department of Biophysics, Istanbul University, Istanbul, Turkey

Anja Berger Bayerisches Landesamt für Gesundheit und Lebensmittelsicherheit (LGL), Oberschleißheim, Germany

Andreas Burkovski Lehrstuhl für Mikrobiologie, Friedrich-Alexander-Universität Erlangen-Nürnberg, Erlangen, Germany

Natasha S. Crowcroft Infectious Diseases, Public Health Ontario, Toronto, Ontario, Canada

Bilge Özerman Edis Istanbul Faculty of Medicine, Department of Biophysics, Istanbul University, Istanbul, Turkey

Androulla Efstratiou Public Health England, Microbiology Services Division: Colindale, London, UK

Ana Luíza de Mattos Guaraldi Laboratório de Difteria e Corinebactérias de Importância Clínica-LDCIC, Disciplina de Microbiologia e Imunologia, Universidade do Estado do Rio de Janeiro. Rio de Janeiro, RJ, Brazil

Michael Hogardt Bayerisches Landesamt für Gesundheit und Lebensmittelsicherheit (LGL), Oberschleißheim, Germany

Paul A. Hoskisson Strathclyde Institute of Pharmacy and Biomedical Science, University of Strathclyde, Glasgow, UK

Raphael Hirata Júnior Laboratório de Difteria e Corinebactérias de Importância Clínica-LDCIC, Disciplina de Microbiologia e Imunologia, Universidade do Estado do Rio de Janeiro. Rio de Janeiro, RJ, Brazil

Seonghun Kim Jeonbuk Branch Institute, Korea Research Institute of Bioscience and Biotechnology (KRIBB), Jeongeup, Korea

Regina Konrad Bayerisches Landesamt für Gesundheit und Lebensmittelsicherheit (LGL), Oberschleißheim, Germany

Ohsuk Kwon Biochemicals and Synthetic Biology Research Center, Korea Research Institute of Bioscience and Biotechnology (KRIBB), Yuseong-gu, Daejeon, Korea

Enrico Malito Novartis Vaccines and Diagnostics, Siena, Italy

Doo-Byoung Oh Biochemicals and Synthetic Biology Research Center, Korea Research Institute of Bioscience and Biotechnology (KRIBB), Yuseong-gu, Daejeon, Korea

Diana M. Oram Department of Microbial Pathogenesis, School of Dentistry, University of Maryland, Baltimore, MD, USA

Laboratory of Respiratory and Special Pathogens, Division of Bacterial, Parasitic, and Allergenic Products, Center for Biologics Evaluation and Research, Food and Drug Administration, Bethesda, MD, USA

Lisa Ott Lehrstuhl für Mikrobiologie, Friedrich-Alexander-Universität Erlangen-Nürnberg, Erlangen, Germany

Rino Rappuoli Novartis Vaccines and Diagnostics, Siena, Italy

Melissa E. Reardon-Robinson Department of Microbiology & Molecular Genetics, University of Texas Health Science Center, Houston, TX, USA

Vartul Sangal Strathclyde Institute of Pharmacy and Biomedical Science, University of Strathclyde, Glasgow, UK

Michael P. Schmitt Laboratory of Respiratory and Special Pathogens, Division of Bacterial, Parasitic, and Allergenic Products, Center for Biologics Evaluation and Research Food and Drug Administration, Bethesda, MD, USA

Ingo Schubert Department of Biology, Chair of Microbiology, University of Erlangen-Nuremberg, Erlangen, Germany

Andreas Sing Bayerisches Landesamt für Gesundheit und Lebensmittelsicherheit (LGL), Oberschleißheim, Germany

Andreas Tauch Institut für Genomforschung und Systembiologie, Centrum für Biotechnologie, Universität Bielefeld, Bielefeld, Germany

Hung Ton-That Department of Microbiology and Molecular Genetics, The University of Texas Medical School at Houston, Houston, TX, USA

Department of Microbiology & Molecular Genetics, University of Texas Health Science Center, Houston, TX, USA

Eva Trost Institut für Genomforschung und Systembiologie, Centrum für Biotechnologie, Universität Bielefeld, Bielefeld, Germany

Başak Varol Istanbul Faculty of Medicine, Department of Biophysics, Istanbul University, Istanbul, Turkey

Karen S. Wagner Centre for Infectious Disease Surveillance and Control, Public Health England, London, UK

Joanne M. White Centre for Infectious Disease Surveillance and Control, Public Health England, London, UK

Aleksandra Anna Zasada National Institute of Public Health—National Institute of Hygiene, Department of Bacteriology, Warsaw, Poland

Katherina Zakikhany Swedish Institute for Communicable Disease Control, Solna, Sweden

The European Programme for Public Health Microbiology Training (EUPHEM), European Centre for Disease Prevention and Control (ECDC), Stockholm, Sweden

Chapter 1
Diphtheria and its Etiological Agents

Andreas Burkovski

Abstract Diphtheria, the 'strangling angel of children', plagued mankind for thousands of years. With the discovery of its etiological agent, *Corynebacterium diphtheriae*, it became a paradigm of an infectious disease. According to Koch's postulates *C. diphtheriae* was isolated by Klebs and Loeffler from infected patients, grown in pure culture and used to re-infect guinea pigs as test animals. Loeffler also recognized that the bacterium predominantly colonizes the nasopharyngeal cavity and, based on this observation, postulated that the secretion of a toxin might cause the often fatal damage of distant organs, a hypothesis, which was further supported by Roux and Yersin. While toxin production is the most dangerous aspect of diphtheria infection, the diphtheria toxin has also been the basis for effective diphtheria treatment and control. Already in 1890, von Behring suggested antitoxin application as a means of diphtheria treatment; in 1913 he developed a first vaccine against diphtheria toxin and in 1920 first mass vaccinations started. Mass immunization is still the most efficient means to prevent and control diphtheria, while antibiotics are effective to eradicate the bacteria from infected patients.

Keywords *Corynebacterium diphtheriae* · *Corynebacterium pseudotuberculosis* · *Corynebacterium ulcerans* · Toxoid

1.1 Introduction

Diphtheria of the upper respiratory tract is characterized by pseudomembrane, which renders breathing difficult. Eventually, coughing can remove parts of the pseudomembrane, easing the situation of the patient temporarily, and after several fits of coughing the pseudomembrane might even be removed and healing might be achieved. More often obstruction of airways results in suffocation, agony and death.

A. Burkovski (✉)
Friedrich-Alexander-Universität Erlangen-Nürnberg, Erlangen, Germany
e-mail: andreas.burkovski@fau.de

A. Burkovski (ed.), *Corynebacterium diphtheriae and Related Toxigenic Species*,
DOI 10.1007/978-94-007-7624-1_1,

In tropical and subtropical regions respiratory tract diphtheria is outnumbered by cutanous diphtheria, which is characterized by skin lesions located predominantly on feet, lower legs and hands. Typically this infection is persistent for months.

Treatment of diphtheria is unproblematic today since *C. diphtheriae* can be eliminated easily using antibiotics, and the diphtheria toxin can be neutralized by antitoxin application. Nevertheless, mass immunization is the means of choice for diphtheria control. A highly effective toxoid vaccine is available which made diphtheria an extremely rare disease in industrialized countries. However, local outbreaks and even a full scale epidemic have been observed during the last three decades. Several thousand diphtheria cases are reported to the World Health organization each year, showing that diphtheria is not completely eradicated and that reservoirs still exist. Putative carriers are people with insufficient access to medical care including vaccination programs, for example poor parts of the population in developing countries. Consequently, for citizens of industrialized countries a major risk factor for infection with *C. diphtheriae* might be travel to an endemic country; however, animals also seem to play an increasing role in transmission of the infection. Isolation of *C. diphtheriae* strains was reported from domestic cats and horses. Animal reservoirs are even more common for other toxigenic *Corynebacterium* species, which might also be connected to diphtheria-like illness. *Corynebacterium ulcerans* has been primarily recognized as a commensal bacterium in domestic and wild animals; however, an increasing frequency and severity of human infections associated with *C. ulcerans* is observed and *C. ulcerans* strains producing diphtheria toxin are reported with rising frequency from industrialized countries.

1.2 A brief History of Diphtheria

Diphtheria has been well known since Babylonian and Sumerian times and documents describing this disease can be found in the Talmud and in writings by Hippocrates. Its alternative designation 'strangling angel of children' refers to the dreadful death by suffocation especially of infants infected by classical diphtheria of the upper respiratory tract. Based on the appraisal of written reports, diphtheria seems to originate from the Middle East and was most likely disseminated from there first to Europe and later to all other continents. Numerous waves of epidemic outbreaks of diphtheria occurred in Europe, often connected to times of war. The cycles include gaps of 100 years and more (English 1985) and outbreaks were reported, e.g. in the sixteenth century in France and Germany, in the seventeenth century in Italy and Spain, and in the eighteenth century in Germany and Sweden. During the eighteenth century, diphtheria also reached the East coast of North America, where in 1799 George Washington became maybe the most prominent victim of this disease. At the end of the eighteenth and the beginning of the nineteenth century, diphtheria became more prevalent and developed into a leading cause of infant mortality. Up to four fifth of children infected with diphtheria died. The reason for this increase in infection numbers and mortality might be the concomitant rise of urban population with beginning industrialization coupled with inferior housing and nutrition conditions of working class people.

The start of industrialization at the end of the nineteenth century brought not only an increase of diphtheria and other diseases connected with poverty, overcrowding and malnutrition, such as tuberculosis, but also the rise of bacteriology. In fact, diphtheria became a paradigm of an infectious disease and provided key evidence for Koch's postulates on the germ theory (Dolman 1973; Levy 1975; English 1985). In 1883 Klebs delivered the first description of *C. diphtheriae* isolated from a patient's swab (Klebs 1883) and based on Klebs' work, Loeffler was the first to grow *C. diphtheriae* in pure culture. In consistence with Koch's postulates, this pure culture was able to evoke diphtheria in guinea pigs used as test animals. Furthermore, Loeffler passed the disease from guinea pig to guinea pig and found that even after more than twenty of these passes the animals still developed diphtheria-like symptoms. In his groundbreaking work, Loeffler described that *C. diphtheriae* colonizes the membranes of the nasopharyngeal cavity but not deeper parts of the body. Based on this observation, he postulated the secretion of a toxin being responsible for the often fatal damage of internal organs connected with upper respiratory tract diphtheria (Loeffler 1884). His hypothesis was supported by Roux and Yersin (1888), who injected bacteria-free filtrates of *C. diphtheriae* cultures in animals and observed that these developed organ damages indistinguishable from those of human diphtheria cases.

While toxin production is the most dangerous aspect of diphtheria infection, leading to severe, often fatal damages of heart, nervous system and muscles (see below), the diphtheria toxin has also been the key to successful treatment, vaccination and control, based on scientific advances made in the late nineteenth century. Already in 1890, von Behring and Kitasato suggested antitoxin application as a means of choice to treat diphtheria (von Behring and Kitasato 1890; for overview, see Grundbacher 1992). In fact, von Behring was also the first who used antitoxin from a horse to successfully treat diphtheria and in 1913 he developed the vaccine against diphtheria toxin. The basis of the vaccine and the vaccination programs starting in the 1920s is the diphtheria toxoid, a formaldehyde-inactivated diphtheria toxin, produced and secreted in vast amounts by strain Park-Williams no. 8 (PW8), originally isolated from a mild case of diphtheria (Park and Williams 1896; Iwaki et al. 2010). Immunization with the inactivated toxin that remains antigenetically intact is extremely effective and changed diphtheria from a former main cause of infant mortality to an extremely rare disease, which will never be seen by most pediatricians or other physicians (English 1985).

Despite this extraordinary success story, diphtheria is not eradicated today. From 2000 to 2008 between three to eleven thousand cases were reported to the World Health Organization per year. Furthermore, especially the epidemic outbreak starting after the breakdown of the former Soviet Union in 1989 showed impressively that diphtheria is still a serious health hazard.

This large scale epidemic started in 1990 with the Russian Federation and Ukraine being centers of the outbreak. 15,211 diphtheria cases were reported for Russia and further 2,987 cases for Ukraine in 1993 (Galazka et al. 1995; Eskola et al. 1998; Popovic et al. 2000), from where the disease spread quickly to neighboring countries. The most characteristic feature of the outbreak was the infection of adolescents and adults, instead of infants being victims of the disease. In addition to Belarus, Russia and Ukraine, this was especially pronounced in the Baltic States Estonia, Latvia and Lithuania, where diphtheria cases among people from 15 years and older ranged from two thirds to four fifths of total cases (Hardy et al. 1996; Dittmann et al. 2000). Consequently, mass immunization of adults, which was started in 1993, was one crucial step to eradicate the disease and stop the epidemic.

The 1990s epidemic showed impressively that socioeconomic instability, large-scale population movements and absence or break-down of health infrastructure favor the spread of diphtheria. It also indicated, together with constant numbers of case reports from various countries, that *C. diphtheriae* might persist in a population for a long time (for example see Marston et al. 2001). In many cases, carriers of infection are unknown, sometimes these are groups of people with insufficient access to medical care including vaccination programs (John 2008) or drug abusers (Lowe et al. 2011). For inhabitants of industrialized countries, a major risk factor for infection with *C. diphtheriae* might be travel to a country where diphtheria is endemic (Connell et al. 2005; Bonmarin et al. 2009; Wagner et al. 2010).

1.3 Transmission and Clinical Manifestations of Diphtheria

1.3.1 Classical Diphtheria of the Upper Respiratory Tract

Diphtheria is a classical infectious disease which typically spreads from person to person by respiratory droplets produced by an infected person by coughing or sneezing (Fig. 1.1). Establishment of the disease takes 2–5 days and patients are infectious for two to three weeks. Besides people showing diphtheria symptoms, asymptomatic carriers may also contribute to spreading of the disease. Loeffler showed before that about five percent of the school children in Berlin (Germany) were carriers of *C. diphtheriae* without showing diphtheria-like symptoms.

Classical diphtheria is an infection of the upper respiratory tract that causes sore throat and low fever. Symptoms range from mild pharyngitis to suffocation due to airway obstruction by pseudomembrane formation. In severe cases, the air passages might be completely blocked. Typically, infection starts with the colonization of epithelial cells of the upper respiratory tract by *C. diphtheriae*. The bacteria multiply on the surface of the mucous membrane, but do not advance into deeper parts of the body, as already described by Loeffler (1884). Tissues infected show inflammatory symptoms, later edema and necroses develop due to the detrimental action of the diphtheria toxin. The toxin is also responsible for inflammation of deeper lying capillaries which results in fibrin secretion into the damaged epithelium. The released fibrin protein, destroyed epithelial cells and bacteria together form the so-called pseudomembrane, a grayish or yellowish-white coating. Later during infection, the

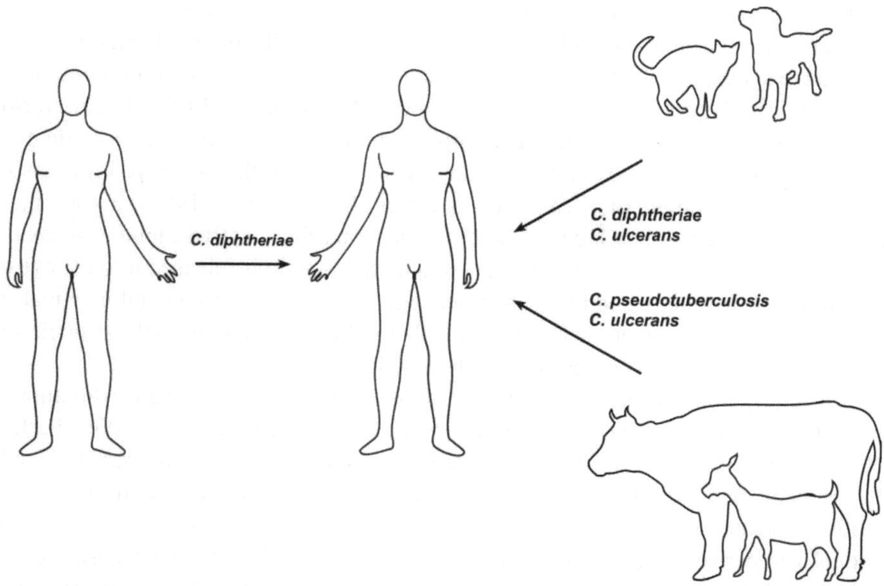

Fig. 1.1 Transmission of toxigenic corynebacteria. Transmission of *Corynebacterium diphtheriae* from person to person occurs by close physical contact. Respiratory diphtheria is mainly disseminated by respiratory droplets; however, smear infections are also possible, especially in case of cutaneous diphtheria. *C. ulcerans* infections occur via close physical contact with pets and farm animals or animal bites. Man-to-man transmission was not reported. *C. pseudotuberculosis* is transferred from infected cattle, especially sheep and goats, to humans by close physical contact

damage of the capillaries due to inflammatory processes causes bleedings, and due to decaying erythrocytes, the color of the pseudomembrane can turn into a dirty brown and the breath of the patients becomes sweetish-putrid, a characteristic indication of diphtheria. With progressive pseudomembrane formation reaching the larynx, a barking cough develops. Furthermore, the voice of the patients becomes affected and hoarseness develops which can even result in complete loss of the voice. Later, trachea and bronchi may be covered by firmly or loosely attached pseudomembranes. At this stage, breathing renders difficult for the patients and their lips turn blue. With increasing dyspnea the patients become restless, their pulse beats become faster and the look of their faces becomes timid and frightened. Sometimes coughing can remove parts of the pseudomembranes at this stage of the disease, easing the situation of the patient temporarily. After several of these fits of coughing, the pseudomembranes might even be completely removed and healing might be achieved in some cases. More often, suffocation results in death agony and, finally, the patients faint and heart failure results in death.

As mentioned above, the colonization of the human host by *C. diphtheriae* remains localized to the upper respiratory tract, although satellite infections may occur in the esophagus, stomach or lower airpassages. The severe complications observed in later stages of infection are the result of the diphtheria toxin which is secreted by *C. diphtheriae* and distributed by the circulatory system to remote parts of the body.

Diphtheria toxin is synthesized depending on the iron concentration in the environment. When iron becomes limiting, as it is the case in the human host, the bacteria start to synthesize the toxin which is then secreted into the extracellular medium as a single polypeptide chain (Pappenheimer 1977; Holmes 2000). In this form the toxin is inactive, can be absorbed into the circulatory system and disseminated to distant parts of the body. When binding to its receptor, the uncleaved precursor of the heparin-binding EGF-like growth factor (Naglich et al. 1992), it can enter the cell by endocytosis. Once inside the endosome, the inactive diphtheria toxin, consisting of an A and B chain, is cleaved, and the A chain is released into the cytoplasm, where it is activated by further cleavage into the active toxin and inactivates its cellular target elongation factor 2 (EF-2) by ADP-ribosylation (Pappenheimer 1977; Lord et al. 1999; Falnes and Sandvig 2000).

Especially myocardium and peripheral motor neurons are affected by diphtheria toxin (Harrison et al. 1972; Murray and Noble 1985; Hadfield et al. 2000; Perles et al. 2000). Up to two thirds of patients show some evidence of myocarditis and 10–25 % of cases develop clinical cardiac dysfunctions. Cardiac symptoms are directly correlated to the extent and severity of local *C. diphtheriae* infection in the patient's upper respiratory tract and often may prove fatal weeks later during convalescence (MacGregor 1995; Hadfield et al. 2000; Perles et al. 2000). Histological changes in the heart differ significantly from patient to patient and may include edema, congestion, infiltration by mononuclear cells and presence of fat drops. The myocardium may show areas of granular degradation, hyaline degradation, necrosis, inflammation and loss of cross striation (Kline and Kaplan 1998; Hadfield et al. 2000; Perles et al. 2000). The heart may be dilated, pale and flabby (Hadfield et al. 2000) and as a result of toxin damage to the cardiac conduction, muscle and nervous system electrical disturbances such as bradyarrhthmia, tachyarrythmia, artioventricular and bundle branch blocks are often found (Perles et al. 2000).

The nervous system is, besides the heart, a main target of the toxin. About three-fourths of patients with severe diphtheria infection develop neurologic complications. First symptoms of neuropathy are paralysis of the soft palate and posterior pharyngeal wall, resulting in regurgitation of swallowed fluids through the nose. Additionally, often a paralysis of the muscles of the eyes and dysfunction of facial, pharyngeal or laryngeal nerves are observed. In later stages, the nerves of trunk, neck, arms and hands might be affected leading to paralysis. Histological investigations to characterize the basis of the neuropathies led to the observation of paranodal and segmental demyelination, resulting in degeneration of myelin sheaths and axons (Baba et al. 1984; Hadfield et al. 2000).

1.3.2 Cutaneous Diphtheria

In tropical and subtropical regions, cutaneous diphtheria is more common and prevails over the respiratory tract form (Höfler 1991). It is still endemic in some African and Asian countries, where climate, overcrowding, poverty, poor hygiene and frequent, slowly healing wounds favor the infection. Cutaneous diphtheria is easily spread by contact with infected skin, respiratory droplets of a patient infected with respiratory tract diphtheria or exposure to dust or clothing contaminated with *C. diphtheriae* (Höfler 1991).

Common sites for cutaneous diphtheria are feet, lower legs and hands (for example see Hamour et al. 1995; Connell et al. 2005). Due to the various skin lesions that can be colonized by the bacteria and frequently occurring co-infections by different other bacterial pathogens the clinical manifestation of cutaneous diphtheria can be extremely variable.

The infection typically starts with a painful, fluid-filled pustule which breaks down later and progresses as a punched-out ulcer. The diameter of the ulcer might range between a few millimeters to centimeters. During the first weeks of infection lesions are covered by a smeary grayish-brown pseudomembrane, which might change color to a dirty or dark reddish brown over time. Later, the infection becomes anesthetic and the pseudomembrane falls off, leaving a hemorrhagic base with a surrounding of edematous grayish-white, pink or purple-colored tissue (Höfler 1991; Hadfield et al. 2000). Spontaneous healing takes several weeks to months, cases lasting one year have been observed (Höfler 1991). This long persistence might favor dissemination of the disease and might explain the extremely high infection rates of skin lesions with *C. diphtheriae* observed previously. Infection rates of up to 60 % were found in some African and Asian countries (Liebow et al. 1946; Livingood et al. 1946; Bezjak and Farsey 1970a, 1970b; Höfler 1991).

1.3.3 Systemic Infections

C. diphtheriae is not only the etiological agent of classical diphtheria of respiratory tract and skin, but can also cause, although generally rare, systemic infections. Cases of bacteraemia, endocarditis, hepatic and splenic abcesses, meningitis, mycotic aneurysm, osteomyelitis, pneumonia as well as septic arthritis caused by non-toxigenic and toxigenic *C. diphtheriae* were reported (Isaac-Renton et al. 1981; Puliti et al. 2006; Hirata et al. 2008; Honma et al. 2009; Muttaiyah et al. 2011; and references therein). The best documented systemic infections are related to *C. diphtheriae* endocarditis. Endocarditis as a result of *Corynebacterium* infection has been described as aggressive disease often requiring surgical intervention (Mishra et al. 2005). Typically, the left heart of adult males is infected and underlying valvular disease is frequently found. Up to 28 % of patients require valve replacements and more than 40 % die (Belmares et al. 2007).

1.4 Diagnosis, Treatment and Control

The diagnosis of respiratory tract diphtheria is, as in former times, still based on the classical symptoms of this disease: sore throat, formation of a pseudomembrane and the typical sweetish-putrid smell of the patients' breath. With the identification of *C. diphtheriae* as its etiological agent by Loeffler (1884) and the development of modern biochemistry and molecular biology, the diagnostic toolbox was constantly improved. Besides different screening and identification tests, the classical Elek test (Efstratiou et al. 1998) for toxicity testing is most commonly used (for a overview of tests and quality evaluation, see Neal and Efstratiou 2009). Furthermore, also a number of other toxigenic *Corynebacterium* species (see below) can also be reliably identified today (Schuhegger et al. 2008; Sing et al. 2011).

Before the introduction of antitoxin and antibiotics, physicians were restricted to treatments preventing suffocation such as tracheostomy introduced by Bretonneau in 1825 and intubation introduced by Bouchut in 1859 (English 1985). Despite some success, mortality stayed high, since, besides the severe side-effects of the proposed treatments, the detrimental action of the toxin could not be avoided by these treatments. The situation improved dramatically with the introduction of antitoxin, which is able to neutralize the toxin in the circulatory system. Even more effective for controlling diphtheria is the immunization with diphtheria toxoid, formaldehyde-inactivated diphtheria toxin that remains antigenetically intact. With increasing levels of immunity, the annual incidence of diphtheria dropped to 0.1–0.2 per million (Kwantes 1984; Höfler 1991; Murphy 1996; Vitek 2006; Roush and Murphy 2007). Diphtheria toxoid is widely used as a component of DPT (diphtheria, pertussis, tetanus) vaccine. Immunization typically starts with early childhood; after four doses of the vaccine within the first 2 years immunization against diphtheria is effective up to 97%. Since antibody titers wane over time, a large percentage of adults in the United States and Europe have antitoxin levels below the protective level (Murphy 1996; von Hunolstein et al. 2000). Therefore, booster immunization of adults is recommended.

The discovery of antibiotics was the next hallmark of effective diphtheria treatment. In contrast to other corynebacteria such as *Corynebacterium jeikeium*, which causes severe infections in intensive care units, multi-resistance against antibiotics is not the problem in *C. diphtheriae* and penicillin and erythromycin are first line antibiotics used for its eradication (Begg 1994; Kneen et al. 1998; Pereira et al. 2008; Zasada et al. 2010). In cases of cutaneous diphtheria, additional local application of drugs such as bacitracin or gentian violet is recommended (Höfler 1991). The therapy with antibiotics might become more difficult in future with the emergence of multidrug resistant strains in some countries. While a recent study on antimicrobial resistance found no multidrug resistant strains among isolates circulating in Poland (Zasada et al. 2010), a considerable number of isolates resistant against one or more antibiotics were observed among Brazilian *C. diphtheriae* strains and reservations about the use of penicillin were risen (Pereira et al. 2008).

To avoid complications due to action of diphtheria toxin, e.g. myocarditis, antitoxin is normally given. However, even if properly treated with antibiotics and antitoxin, five to ten percent of cases can end fatally. Therefore, mass immunization of the entire population is the best means to control diphtheria.

1.5 Diphtheria as a Zoonotic Infection

Besides humans, animals seem to play a role as a reservoir of the disease (Fig. 1.1). Isolations of *C. diphtheriae* strains were reported for example from domestic cats (Hall et al. 2010), cows (Corboz et al. 1996) and horses (Henricson et al. 2000; Leggett et al. 2010). The existence of animal reservoirs is even more common, when other toxigenic *Corynebacterium* species are taken into consideration (Bonmarin et al. 2009). *C. diphtheriae* is closely related to two further *Corynebacterium* species, *Corynebacterium pseudotuberculosis* and *Corynebacterium ulcerans* (Riegel et al. 1995). The three species can be lysogenized by similar corynebacteriophages and, if these carry a *tox* gene, all three species produce diphtheria toxin (Groman 1984; Buck et al. 1985; Cianciotto and Groman 1996), making them a potential health threat.

Infections due to *C. pseudotuberculosis*, which causes caseous lymphadenitis in sheep and goat populations worldwide (Dorella et al. 2006; Baird and Fontaine 2007), are rare in humans and classical diphtheria of the upper respiratory tract or skin connected with *C. pseudotuberculosis* have not been observed. However, *C. pseudotuberculosis* may serve as a reservoir of corynebacteriotoxphages. In contrast, during the last decade, the frequency and severity of human infections associated with *C. ulcerans* appear to be increasing in various countries. Cases of respiratory tract diphtheria caused by toxigenic *C. ulcerans* strains were reported from various industrialized countries (Wagner et al. 2001; Hatanaka et al. 2003; De Zoysa et al. 2005; Tiwari et al. 2008) and became even more common than *C. diphtheriae* infections in the United Kingdom (Wagner et al. 2010). Infections with toxigenic *C. ulcerans* can be fatal in unvaccinated patients and usually occur in adults. Besides upper respiratory tract diphtheria, *C. ulcerans* can also cause severe skin and pulmonary infections (Dessau et al. 1995; Nureki et al. 2007; Mattos-Guaraldi et al. 2008). Transmission by person to person contact was not reported up to now. In contrast, close contact with domestic animals (Wagner et al. 2010) and consumption of raw, unpasteurized milk (Bostock et al. 1984; Hart 1984) seem to be risk factors. This observation is in accordance with the identification of *C. ulcerans* strains commensals in various domestic and wild animals (Schuhegger et al. 2008; Dixon 2010; Sykes et al. 2010) , which may serve as a reservoir for zoonotic infections.

1.6 Development and Persistence of *C. diphtheriae* Populations

As described above, there seems to be a shift from *C. diphtheriae* to *C. ulcerans* as etiological agent of diphtheria in at least some Western countries. Interestingly, a shift within *C. diphtheriae* populations has been observed as well. With the introduction of diphtheria toxoid vaccines, not only the number of diphtheria cases but also the number of isolated toxigenic *C. diphtheriae* strains decreased, suggesting a protection by the vaccine not only against the fatal action of the toxin but, at least partially, against the bacteria itself. Interestingly, anti-parallel to this development, an increasing number of non-toxigenic strains has been isolated (Zuber et al. 1992; Gilbert 1997; Hadfield et al. 2000; Wagner et al. 2011). These non-toxigenic *C. diphtheriae* are persisting over years in different, often poor populations (Romney et al. 2006; Lowe et al. 2011; Shashikala et al. 2011), where they are connected especially to skin infections. Unfortunately, also an increasing number of systemic infections by non-toxigenic strains has also been observed.

Following the outbreak by a unique clonal group of *C. diphtheriae* in Russia in 1990 (Popovic et al. 2000), a rising heterogenicity of circulating strains after the epidemic, emergence of new toxigenic variants, and persistence of invasive non-toxigenic strains were observed (Mokrousov 2009).

All in all, the emergence of new diphtheria-causing corynebacteria such as *C. ulcerans* and the adaptation of *C. diphtheriae* populations to medical treatment are supporting the need of continuous surveillance of *C. diphtheriae* and its relatives and justify, besides the basic scientific interest, experimental efforts to characterize these pathogens.

Acknowledgements The author's work was supported by the Deutsche Forschungsgemeinschaft (SFB796, B5). Help with the manuscript and preparation of Fig. 1.1 by S. Morbach (Friedrich-Alexander-Universität Erlangen-Nürnberg) is gratefully acknowledged.

References

Baba M, Gilliatt RW, Harding AE, Reiners K (1984) Demyelination following diphtheria toxin in the presence of axonal atrophy. J Neurol Sci 64(2):199–211

Baird GJ, Fontaine MC (2007) *Corynebacterium pseudotuberculosis* and its role in ovine caseous lymphadenitis. J Comp Pathol 137(4):179–210

Begg N (1994) Manual of the management and control of diphtheria in the European region. Vol ICP/EPI 038 (B). World Health Organization, Copenhagen

Belmares J, Detterline S, Pak JB, Parada JP (2007) *Corynebacterium* endocarditis species-specific risk factors and outcomes. BMC Infect Dis 7:4

Bezjak V, Farsey SJ (1970a) *Corynebacterium diphtheriae* carriership in Ugandan children. J Trop Pediatr 16(1):12–16

Bezjak V, Farsey SJ (1970b) *Corynebacterium diphtheriae* in skin lesions in Ugandan children. Bull World Health Organ 43(5):643–650

Bonmarin I, Guiso N, Le Fleche-Mateos A, Patey O, Patrick AD, Levy-Bruhl D (2009) Diphtheria: a zoonotic disease in France? Vaccine 27(31):4196–4200

Bostock AD, Gilbert FR, Lewis D, Smith DC (1984) *Corynebacterium ulcerans* infection associated with untreated milk. J Infect 9(3):286–288

Buck GA, Cross RE, Wong TP, Loera J, Groman N (1985) DNA relationships among some tox-bearing corynebacteriophages. Infect Immun 49(3):679–684

Cianciotto NP, Groman NB (1996) Extended host range of a beta-related corynebacteriophage. FEMS Microbiol Lett 140(2–3):221–225

Connell TG, Rele M, Daley AJ, Curtis N (2005) Skin ulcers in a returned traveller. Lancet 365(9460):726

Corboz L, Thoma R, Braun U, Zbinden R (1996) Isolation of *Corynebacterium diphtheriae* subsp. *belfanti* from a cow with chronic active dermatitis. Schweiz Arch Tierheilkd 138(12):596–599

De Zoysa A, Hawkey PM, Engler K, George R, Mann G, Reilly W, Taylor D, Efstratiou A (2005) Characterization of toxigenic *Corynebacterium ulcerans* strains isolated from humans and domestic cats in the United Kingdom. J Clin Microbiol 43(9):4377–4381

Dessau RB, Brandt-Christensen M, Jensen OJ, Tonnesen P (1995) Pulmonary nodules due to *Corynebacterium ulcerans*. Eur Respir J 8(4):651–653

Dittmann S, Wharton M, Vitek C, Ciotti M, Galazka A, Guichard S, Hardy I, Kartoglu U, Koyama S, Kreysler J, Martin B, Mercer D, Ronne T, Roure C, Steinglass R, Strebel P, Sutter R, Trostle M (2000) Successful control of epidemic diphtheria in the states of the Former Union of Soviet Socialist Republics: lessons learned. J Infect Dis 181(Suppl 1):S10–22

Dixon B (2010) Sick as a dog. Lancet Infect Dis 10(2):73

Dolman CE (1973) The Donald T. Fraser Memorial Lecture, 1973. Landmarks and pioneers in the control of diphtheria. Can J Public Health 64(4):317–336

Dorella FA, Pacheco LG, Oliveira SC, Miyoshi A, Azevedo V (2006) *Corynebacterium pseudotuberculosis*: microbiology, biochemical properties, pathogenesis and molecular studies of virulence. Vet Res 37(2):201–218

Efstratiou A, Engler KH, Dawes CS, Sesardic D (1998) Comparison of phenotypic and genotypic methods for detection of diphtheria toxin among isolates of pathogenic corynebacteria. J Clin Microbiol 36(11):3173–3177

English PC (1985) Diphtheria and theories of infectious disease: centennial appreciation of the critical role of diphtheria in the history of medicine. Pediatrics 76(1):1–9

Eskola J, Lumio J, Vuopio-Varkila J (1998) Resurgent diphtheria-are we safe? Br Med Bull 54(3):635–645

Falnes PO, Sandvig K (2000) Penetration of protein toxins into cells. Curr Opin Cell Biol 12(4):407–413

Galazka AM, Robertson SE, Oblapenko GP (1995) Resurgence of diphtheria. Eur J Epidemiol 11(1):95–105

Gilbert L (1997) Infections with *Corynebacterium diphtheriae*—changing epidemiology and clinical manifestations. Report of the third international meeting of the European Laboratory Working Group on Diphtheria (ELWGD), Institute Pasteur, Paris 7–8 June 1996. Commun Dis Intell 21(12):161–164

Groman NB (1984) Conversion by corynephages and its role in the natural history of diphtheria. J Hyg (Lond) 93(3):405–417

Grundbacher FJ (1992) Behring's discovery of diphtheria and tetanus antitoxins. Immunol Today 13(5):188–190

Hadfield TL, McEvoy P, Polotsky Y, Tzinserling VA, Yakovlev AA (2000) The pathology of diphtheria. J Infect Dis 181(Suppl 1):S116–S120

Hall AJ, Cassiday PK, Bernard KA, Bolt F, Steigerwalt AG, Bixler D, Pawloski LC, Whitney AM, Iwaki M, Baldwin A, Dowson CG, Komiya T, Takahashi M, Hinrikson HP, Tondella ML (2010) Novel *Corynebacterium diphtheriae* in domestic cats. Emerg Infect Dis 16(4):688–691

Hamour AA, Efstratiou A, Neill R, Dunbar EM (1995) Epidemiology and molecular characterisation of toxigenic *Corynebacterium diphtheriae var mitis* from a case of cutaneous diphtheria in Manchester. J Infect 31(2):153–157

Hardy IR, Dittmann S, Sutter RW (1996) Current situation and control strategies for resurgence of diphtheria in newly independent states of the former Soviet Union. Lancet 347(9017):1739–1744

Harrison BM, McDonald WI, Ochoa J (1972) Remyelination in the central diphtheria toxin lesion. J Neurol Sci 17(3):293–302

Hart RJ (1984) *Corynebacterium ulcerans* in humans and cattle in North Devon. J Hyg (Lond) 92(2):161–164

Hatanaka A, Tsunoda A, Okamoto M, Ooe K, Nakamura A, Miyakoshi M, Komiya T, Takahashi M (2003) *Corynebacterium ulcerans* Diphtheria in Japan. Emerg Infect Dis 9(6):752–753

Henricson B, Segarra M, Garvin J, Burns J, Jenkins S, Kim C, Popovic T, Golaz A, Akey B (2000) Toxigenic *Corynebacterium diphtheriae* associated with an equine wound infection. J Vet Diagn Invest 12(3):253–257

Hirata R, Pereira GA, Filardy AA, Gomes DL, Damasco PV, Rosa AC, Nagao PE, Pimenta FP, Mattos-Guaraldi AL (2008) Potential pathogenic role of aggregative-adhering *Corynebacterium diphtheriae* of different clonal groups in endocarditis. Braz J Med Biol Res 41:986–991

Höfler W (1991) Cutaneous diphtheria. Int J Dermatol 30(12):845–847

Holmes RK (2000) Biology and molecular epidemiology of diphtheria toxin and the tox gene. J Infect Dis 181(Suppl 1):S156–S167

Honma Y, Yoshii Y, Watanabe Y, Aoki N, Komiya T, Iwaki M, Arai H, Arakawa Y, Takahashi M, Kimura H (2009) A case of afebrile pneumonia caused by non-toxigenic *Corynebacterium diphtheriae*. Jpn J Infect Dis 62(4):327–329

Isaac-Renton JL, Boyko WJ, Chan R, Crichton E (1981) *Corynebacterium diphtheriae* septicemia. Am J Clin Pathol 75(4):631–634

Iwaki M, Komiya T, Yamamoto A, Ishiwa A, Nagata N, Arakawa Y, Takahashi M (2010) Genome organization and pathogenicity of *Corynebacterium diphtheriae* C7(-) and PW8 strains. Infect Immun 78(9):3791–3800

John TJ (2008) Resurgence of diphtheria in India in the 21st century. J Med Res 128:668–670

Klebs E (1883) Über Diphtherie. Verh Cong Inn Med 2:139–154

Kline MW, Kaplan SL (1998) Infections, immunizations, and principles of antimicrobial therapy. In: Garon A, Bricker JT, Fisher DJ, Neish SR (eds) The Science and practice of pediatric cardiology. Williams and Wilkins, Baltimore, pp 2856–2857

Kneen R, Pham NG, Solomon T, Tran TM, Nguyen TT, Tran BL, Wain J, Day NP, Tran TH, Parry CM, White NJ (1998) Penicillin vs. erythromycin in the treatment of diphtheria. Clin Infect Dis 27(4):845–850

Kwantes W (1984) Diphtheria in Europe. J Hyg (Lond) 93(3):433–437

Leggett BA, De Zoysa A, Abbott YE, Leonard N, Markey B, Efstratiou A (2010) Toxigenic *Corynebacterium diphtheriae* isolated from a wound in a horse. Vet Rec 166(21):656–657

Levy FM (1975) The 50th anniversary of diphtheria and tetanus immunization. Prev Med 4(2):226–237

Liebow AA, Maclean PD, Bumstead JM (1946) Tropical ulcers and cutaneous diphtheria. Arch Intern Med (Chic) 78:255–295

Livingood CS, Perry DJ, Forrester JS (1946) Cutaneous diphtheria; a report of 140 cases. J Invest Dermatol 7(6):341–364

Loeffler F (1884) Untersuchungen über die Bedeutung der Mikroorganismen für die Entstehung der Diphtherie beim Menschen, bei der Taube und beim Kalbe. Mitt Klein Gesundheitsamte Berlin 2:421–499

Lord JM, Smith DC, Roberts LM (1999) Toxin entry: how bacterial proteins get into mammalian cells. Cell Microbiol 1(2):85–91

Lowe CF, Bernard KA, Romney MG (2011) Cutaneous diphtheria in the urban poor population of Vancouver, British Columbia, Canada: a 10-year review. J Clin Microbiol 49(7):2664–2666

MacGregor RR (1995) *Corynebacterium diphtheriae*. In: Mandell GL, Douglas JE, Dolin R (eds) Principles and practice of infectious diseases. 4th edn. Churchill Livingston, New York, pp 1866–1869

Marston CK, Jamieson F, Cahoon F, Lesiak G, Golaz A, Reeves M, Popovic T (2001) Persistence of a distinct *Corynebacterium diphtheriae* clonal group within two communities in the United States and Canada where diphtheria is endemic. J Clin Microbiol 39(4):1586–1590

Mattos-Guaraldi AL, Sampaio JL, Santos CS, Pimenta FP, Pereira GA, Pacheco LG, Miyoshi A, Azevedo V, Moreira LO, Gutierrez FL, Costa JL, Costa-Filho R, Damasco PV, Camello TC, Hirata Jr R (2008) First detection of *Corynebacterium ulcerans* producing a diphtheria-like toxin in a case of human with pulmonary infection in the Rio de Janeiro metropolitan area, Brazil. Mem Inst Oswaldo Cruz 103(4):396–400

Mishra B, Dignan RJ, Hughes CF, Hendel N (2005) *Corynebacterium diphtheriae* endocarditis—surgery for some but not all!. Asian Cardiovasc Thorac Ann 13(2):119–126

Mokrousov I (2009) *Corynebacterium diphtheriae*: genome diversity, population structure and genotyping perspectives. Infect Genet Evol 9(1):1–15

Murphy JR (1996) *Corynebacterium diphtheriae*. In: Baron S (ed) Medical Microbiology. 4th edn. University of Texas Medical Branch at Galveston, Galveston, pp 99–100
Murray K, Noble M (1985) In vitro studies on the comparative sensitivities of cells of the central nervous system to diphtheria toxin. J Neurol Sci 70(3):283–293
Muttaiyah S, Best EJ, Freeman JT, Taylor SL, Morris AJ, Roberts SA (2011) *Corynebacterium diphtheriae* endocarditis: a case series and review of the treatment approach. Int J Infect Dis 15(9):e584–e588
Naglich JG, Metherall JE, Russell DW, Eidels L (1992) Expression cloning of a diphtheria toxin receptor: identity with a heparin-binding EGF-like growth factor precursor. Cell 69(6):1051–1061
Neal SE, Efstratiou A (2009) International external quality assurance for laboratory diagnosis of diphtheria. J Clin Microbiol 47(12):4037–4042
Nureki S, Miyazaki E, Matsuno O, Takenaka R, Ando M, Kumamoto T, Nakano T, Ohkusu K, Ezaki T (2007) *Corynebacterium ulcerans* infection of the lung mimicking the histology of Churg-Strauss syndrome. Chest 131(4):1237–1239
Pappenheimer AM Jr (1977) Diphtheria toxin. Annu Rev Biochem 46:69–94
Park WH, Williams AW (1896) The production of diphtheria toxin. J Exp Med 1:164–185
Pereira GA, Pimenta FP, Santos FR, Damasco PV, Hirata Junior R, Mattos-Guaraldi AL (2008) Antimicrobial resistance among Brazilian *Corynebacterium diphtheriae* strains. Mem Inst Oswaldo Cruz 103(5):507–510
Perles Z, Nir A, Cohen E, Bashary A, Engelhard D (2000) Atrioventricular block in a toxic child: do not forget diphtheria. Pediatr Cardiol 21(3):282–283
Popovic T, Mazurova IK, Efstratiou A, Vuopio-Varkila J, Reeves MW, De Zoysa A, Glushkevich T, Grimont P (2000) Molecular epidemiology of diphtheria. J Infect Dis 181(Suppl 1):S168–177
Puliti M, von Hunolstein C, Marangi M, Bistoni F, Tissi L (2006) Experimental model of infection with non-toxigenic strains of *Corynebacterium diphtheriae* and development of septic arthritis. J Med Microbiol 55(Pt 2):229–235
Riegel P, Ruimy R, de Briel D, Prevost G, Jehl F, Christen R, Monteil H (1995) Taxonomy of *Corynebacterium diphtheriae* and related taxa, with recognition of *Corynebacterium ulcerans* sp. nov. nom. rev. FEMS Microbiol Lett 126(3):271–276
Romney MG, Roscoe DL, Bernard K, Lai S, Efstratiou A, Clarke AM (2006) Emergence of an invasive clone of nontoxigenic *Corynebacterium diphtheriae* in the urban poor population of Vancouver, Canada. J Clin Microbiol 44(5):1625–1629
Roush SW, Murphy TV (2007) Historical comparisons of morbidity and mortality for vaccine-preventable diseases in the United States. JAMA 298(18):2155–2163
Roux E, Yersin A (1888) Contribution à l'etude de la diphtérie. Annales de l'Insitut Pasteur 2:629–661
Schuhegger R, Lindermayer M, Kugler R, Heesemann J, Busch U, Sing A (2008) Detection of toxigenic *Corynebacterium diphtheriae* and *Corynebacterium ulcerans* strains by a novel real-time PCR. J Clin Microbiol 46(8):2822–2823
Shashikala P, Reddy PV, Prashanth K, Kanungo R, Devi S, Anitha P, Rajarajeshweri N, Cherian TM (2011) Persistence of nontoxigenic *Corynebacterium diphtheriae* biotype *gravis* strains in Pondicherry, Southern India. J Clin Microbiol 49(2):763–764
Sing A, Berger A, Schneider-Brachert W, Holzmann T, Reischl U (2011) Rapid detection and molecular differentiation of toxigenic *Corynebacterium diphtheriae* and *Corynebacterium ulcerans* strains by LightCycler PCR. J Clin Microbiol 49(7):2485–2489
Sykes JE, Mapes S, Lindsay LL, Samitz E, Byrne BA (2010) *Corynebacterium ulcerans* bronchopneumonia in a dog. J Vet Intern Med 24(4):973–976
Tiwari TS, Golaz A, Yu DT, Ehresmann KR, Jones TF, Hill HE, Cassiday PK, Pawloski LC, Moran JS, Popovic T, Wharton M (2008) Investigations of 2 cases of diphtheria-like illness due to toxigenic *Corynebacterium ulcerans*. Clin Infect Dis 46(3):395–401
Vitek CR (2006) Diphtheria. Curr Top Microbiol Immunol 304:71–94
von Behring EA, Kitosato S (1890) Über das Zustandekommen der Diphtherie-Immunität und der Tetanus-Immunität bei Thieren. Dtsch Med Wochenschr 16:113–114

von Hunolstein C, Rota MC, Alfarone G, Ricci ML, Salmaso S (2000) Diphtheria antibody levels in the Italian Population. Eur J Clin Microbiol Infect Dis 19(6):433–437

Wagner J, Ignatius R, Voss S, Höpfner V, Ehlers S, Funke G, Weber U, Hahn H (2001) Infection of the skin caused by *Corynebacterium ulcerans* and mimicking classical cutaneous diphtheria. Clin Infect Dis 33(9):1598–1600

Wagner KS, White JM, Crowcroft NS, De Martin S, Mann G, Efstratiou A (2010) Diphtheria in the United Kingdom, 1986–2008: the increasing role of *Corynebacterium ulcerans*. Epidemiol Infect 138(11):1519–1530

Wagner KS, White JM, Neal S, Crowcroft NS, Kupreviciene N, Paberza R, Lucenko I, Joks U, Akbas E, Alexandrou-Athanassoulis H, Detcheva A, Vuopio J, von Hunolstein C, Murphy PG, Andrews N, Efstratiou A (2011) Screening for *Corynebacterium diphtheriae* and *Corynebacterium ulcerans* in patients with upper respiratory tract infections 2007–2008: a multicentre European study. Clin Microbiol Infect 17(4):519–525

Zasada AA, Baczewska-Rej M, Wardak S (2010) An increase in non-toxigenic *Corynebacterium diphtheriae* infections in Poland—molecular epidemiology and antimicrobial susceptibility of strains isolated from past outbreaks and those currently circulating in Poland. Int J Infect Dis 14(10):e907–e912

Zuber PL, Gruner E, Altwegg M, von Graevenitz A (1992) Invasive infection with non-toxigenic *Corynebacterium diphtheriae* among drug users. Lancet 339(8805):1359

Chapter 2
Corynebacterium diphtheriae, Corynebacterium ulcerans and *Corynebacterium pseudotuberculosis*—General Aspects

Ana Luíza de Mattos Guaraldi, Raphael Hirata Júnior and Vasco Ariston de Carvalho Azevedo

Abstract *Corynebacterium diphtheriae*, *Corynebacterium ulcerans* and *Corynebacterium pseudotuberculosis* are potentially diphtheria toxin-producing microorganisms related to different infectious processes involving both human and animal hosts. This chapter aims to concise the current aspects concerning to the pathogenesis, epidemiology of diseases caused by those species and transmission amongst human and animal hosts. Aspects related to virulence factors, diagnosis and some molecular features observed after genome sequencing of some isolates were also approached.

Keywords *Corynebacterium diphtheriae* · *Corynebacterium ulcerans* · *Corynebacterium pseudotuberculosis* · Diphtheria toxin · Epidemiology · Pathogenicity · Virulence factors

2.1 Introduction

2.1.1 Human and Animal Infections Caused by Potentially Toxigenic Corynebacteria

Corynebacterium diphtheriae, Corynebacterium ulcerans and *Corynebacterium pseudotuberculosis* constitute a group of potentially toxigenic microorganisms related to different infectious processes involving both human and animal hosts.

A. L. de. M. Guaraldi (✉) · R. Hirata Júnior
Laboratório de Difteria e Corinebactérias de Importância Clínica-LDCIC,
Disciplina de Microbiologia e Imunologia, Universidade do Estado do Rio de Janeiro.
Av. 28 de Setembro, 87-Fundos, 3°andar, Vila Isabel, Rio de Janeiro,
RJ, CEP 20 551-030 Brazil
e-mail: guaraldi@uerj.br

V. A. de. C. Azevedo
Departmento de Biologia, Instituto de Ciências Biológicas,
Universidade Federal de Minas Gerais, Belo Horizonte, MG, Brazil

A. Burkovski (ed.), *Corynebacterium diphtheriae and Related Toxigenic Species*,
DOI 10.1007/978-94-007-7624-1_2,

C. diphtheriae is the main causative agent of diphtheria, a toxemic disease whose prevention depends on the implementation of effective immunization programs by using toxoid molecules (Hadfield et al. 2000; Vitek and Wharton 2008). Typically, classical respiratory diphtheria presents with a swollen 'bull neck' and an adherent 'pseudomembrane' in the respiratory tract mucosa. The pseudomembrane, considered as the main pathognomonic sign of diphtheria, may extend into the trachea, bronchi and bronchioles leading to severe limitations in airflow and bronchopneumonia. By multiplying on the infection site, toxigenic strains produce diphtheria toxin (DT) that is absorbed into circulation, acting in many tissues, with particular tropism for the myocardium, central nervous system, kidney and adrenal glands. Death may result from suffocation by tough local inflammatory effects of DT on the respiratory tract, 'pseudomembranes' dislodgement or from systemic effects of DT (mainly myocarditis and renal failure) (Hadfield 2000).

Diphtheria was one of the first infectious diseases to be conquered through mass immunization with toxoid. However, the concentration of protective antibodies decreases by 10% every year (Vitek and Wharton 1998). Over the last decades the increasing number of reported cases of atypical infections due to toxigenic and non-toxigenic *C. diphtheriae* strains in partially immunized persons also became a matter of concern (Cardénas et al. 1972; Zuber et al. 1992; Patey et al. 1995; Bitragunta et al. 2010). In many developing countries diphtheria continues to occur in children, adolescents and adults with high case-fatality rates due to inadequate nationwide coverage of immunization programs (Singh et al. 1999; Popovic et al. 2000; Mattos-Guaraldi et al. 2003; Sharma et al. 2007). Toxigenic *C. diphtheriae* may circulate in a community for 20 years after a reported case of diphtheria, even in countries where immunization programs are followed with great efficiency (De Zoysa et al. 1995; Golaz et al. 2000; Wagner et al. 2012). The introduction of toxigenic strains in a susceptible population may result in diphtheria outbreaks. All these aspects emphasize the need of vaccination strategies directed to persons of all ages and different ethnic groups and continuous surveillance of population's immunity and new diphtheria cases (Dittmann et al. 2000).

Nowadays there has been an increase in incidence of cases of infections due to non-toxigenic *C. diphtheriae*, including persistent sore throats and severe pharyngitis/tonsillitis. Invasive disease such as endocarditis, septic arthritis, splenic abscesses and osteomyelitis are not uncommon, including in cancer patients and patients making use of indwelling medical devices (Alexander 1984; Tiley et al. 1993; Lin et al. 1994; Poilane et al. 1995; Golaz et al. 2000; Mattos-Guaraldi et al. 2001; Dzupova et al. 2005; Hirata Jr. et al. 2008; Gomes et al. 2009; Farfour et al. 2012). The overall lack of information on the prevalence of colonization and disease by *C. diphtheriae* in the population may be partially due to a reduction in routine screening for this organism (Efstratiou and George 1999).

The transmission of *C. diphtheriae* is mainly from person-to-person by droplets of respiratory secretions. In endemic regions, especially in tropical areas, infections of skin lesions may serve as a reservoir for diphtheria bacilli in a variety of skin wounds, including those caused by insect bites and trauma (Werdmuller et al. 1996) as well as leishmaniotic and neoplastic ulcers (Vera et al. 2002; Mattos-Guaraldi et al. 2003). Skin infections may be more contagious than those of the respiratory tract (MacGregor 1990; Quick et al. 2000). Although classically considered as a strict human pathogen, *C. diphtheriae* is also capable to infect animals such as cats (De Zoysa et al. 2005; Hall et al. 2010), cows (Corboz et al. 1996) and horses (Leggett et al. 2010).

Human infections by *C. ulcerans*, including diphtheria, may be fatal and usually occur in adults with close animal contact (Wellinghausen et al. 2002; Lartigue et al. 2005). More recently, the majority (approximately 75%) of the cases of zoonotic diphtheria caused by *C. ulcerans* has occurred in adult patients who had been fully or partially vaccinated with diphtheric toxoid (Dias et al. 2011). During the last decades, the frequency and severity of human infections associated with *C. ulcerans* appear to be increasing in different countries (Hatanaka et al. 2003; Dewinter et al. 2005; Bonmarin et al. 2009; Hogg et al. 2009; Komiya et al. 2010; Kimura et al. 2011). In Europe, *C. ulcerans* is currently isolated in more frequency from diphtheria cases than *C. diphtheriae* (De Zoysa et al. 2005; Perkins et al. 2010; Taylor et al. 2010; Wagner et al. 2010).

The main reservoir of *C. ulcerans* seems to be cattle, in which it may induce mastitis. Cases of infection due to *C. ulcerans* in other animals species such as macaques (Bergin et al. 2000), squirrels (Olson et al. 1988), otters (Foster et al. 2002), camels (Tejedor et al. 2000), whales and lions (Seto et al. 2008), dogs (Katsukawa et al. 2009; Dias et al. 2010), cats (De Zoysa et al. 2005), pigs (Schuhegger et al. 2009) and goats (Morris et al. 2005) have been described.

C. ulcerans may be found colonizing the nasopharynx of asymptomatic rural workers. Some cases have no association with a farming community or the consumption of raw milk products or having contact with farm animals or their waste, which suggests other routes of infection (De Zoysa et al. 2005). Recently, there has been an increased concern over transmission of *C. ulcerans* between domestic animals and humans (Dewinter et al. 2005; De Zoysa et al. 2005; Lartigue et al. 2005; Aaron et al. 2006; Seto et al. 2008; Tiwari et al. 2008; Katsukawa et al. 2009). The circulation of this pathogen in apparently healthy dogs was observed in metropolitan areas of both industrialized and developing countries. Veterinary clinics should implement guidelines and be aware of carriage of *C. ulcerans* in the throat or nares of asymptomatic animals (Katsukawa et al. 2009; Dias et al. 2010).

C. ulcerans infected patients may exhibit skin lesions that completely mimic cutaneous diphtheria or present as a tracheal-bronchial tree covered by pseudomembranes (Wagner et al. 2001; Dewinter et al. 2005; Mattos-Guaraldi et al. 2008; Wagner et al. 2010). Independent of DT production, *C. ulcerans* was also found to produce clinical syndromes of the lower respiratory tract, such as pneumonia (Hommez et al. 1999; Hatanaka et al. 2003), pulmonary granulomatous nodules (Dessau et al. 1995), occasionally associated with signs of systemic inflammatory response syndrome and disseminated intravascular coagulation (Nureki et al. 2007). *C. ulcerans* infections may occur in children previously immunized against diphtheria (Leek et al. 1990). Older urban adults may be also at risk for toxic complications due to inadequate immune status (Gubler et al. 1990; Wellinghausen et al. 2002).

C. pseudotuberculosis is the etiological agent of caseous lymphadenitis (CLA) in small ruminant populations, such as sheep and goats that sometimes presents as pneumonia, hepatitis, pericarditis, mastitis, arthritis and subcutaneous abscesses. The pathogen is also associated with lymphadenitis in horses, ulcerative lymphangitis and pigeon fever in cattle, camels, swine, elks and buffaloes (Selim 2001; Foley et al. 2004, Perkins et al. 2004; Baird and Fontaine 2007; Sharpe et al. 2010; Kelly et al. 2012). Though the pathogen is distributed worldwide, it has the most serious economic impact in Oceania, Africa and South America (Estevao et al. 2007; Komala et al. 2008; Tarello and Theneyan 2008; Stefanska et al. 2008; Seyffert et al. 2010). The increase in number of infections could be the result of reporting bias, environmental factors facilitating infection or host factors such as greater herd susceptibility (Foley et al. 2004). Similar to diphtheria in humans, clonally expanding epidemic of animal infection caused by *C. pseudotuberculosis* may also occur (Kombarova et al. 2001; Saikya et al. 2010).

The potential of *C. pseudotuberculosis* to survive for several weeks in the environment likely contributes to its ability to spread within a herd or flock (Augustine and Renshaw 1986; Yeruham et al. 2004). Transmission among sheep or goats occurs mainly through contamination of superficial wounds, which can appear during common procedures, such as shearing, castration and ear tagging or through injuries of the animals' bodies generated by other traumatic events. Not infrequently, contaminated sheep cough bacteria onto skin cuts of other sheep, constituting another means of transmission (Paton et al. 1995; Williamson 2001).

Though transmission of *C. pseudotuberculosis* still needs to be determined with certainty, in cattle as well as in buffaloes the microorganism may be vectored to animals by flies such as *Musca domestica* and *Hippobosca equina*. In cases of edematous skin disease (OSD) in buffaloes the pathogen may be identified in up to 20% of flies in the vicinity of diseased animals (Yeruham et al. 1996; Braverman et al. 1999; Selim 2001; Spier et al. 2004). In OSD cases, *C. pseudotuberculosis* biovar *equi* exert their pathogenesis by secretion of DT in addition to phospholipase D (PLD) and their lipid contents of the cell walls (Selim 2001).

Very few studies have indicated the isolation of the causal agent of CLA and OSD from humans. Human infections caused by *C. pseudotuberculosis* are frequently similar to that observed in sheep and goats (CLA) and usually need the excision of infected lymph nodes accompanied of supplementary antimicrobial treatment, without observation of toxemic manifestations. The microorganism is usually acquired after close contact with infected animals, and no underlying diseases or predisposing conditions have been observed in infected patients (Hemond et al. 2009; Join-Lambert et al. 2006; Peel et al. 1997; Romero-Perez et al. 2004). Most of the reported cases have been related to occupational exposure, ingestion of raw goat meat and cow milk. About 25 cases of infection of humans with this microorganism have been reported in the literature (Mills et al. 1997; Peel et al. 1997; Liu et al. 2005). Peel and co-workers (1997) reviewed 22 cases, in which infected humans generally presented with lymphadenitis, abscesses and constitutional symptoms. Liu and co-workers (2005) reported a *C. pseudotuberculosis* infection in a patient's eye, due to an ocular implant. In most cases, the patients received antibiotic therapy and the affected lymph nodes were surgically removed. One case with toxemic symptoms from an injecting drug patient with endocarditis was recently reported. The patient had no history of animal contact and no possible source for *C. pseudotuberculosis* infection was identified (Wagner et al. 2011).

2.2 Microbiological and Diagnosis Aspects

Corynebacterium species constitute a group of catalase-positive, non-motile and non-spore forming, pleomorphic Gram-positive rods. The club or bar appearance of the cells is due to stored inorganic phosphate which forms metachromatic granules when stained by especial techniques (Funke and Bernard 2011). However, many *C. diphtheriae, C. ulcerans* and *C. pseudotuberculosis* strains may present confusing coccobacillary arrangements lacking phosphate granules. Fermentation of carbohydrates is diverse though the species are capable to ferment glucose and maltose (Table 2.1). Production of acid from sucrose may also be observed, including for some *C. diphtheriae* strains (Hirata Jr. et al. 2011). *C. diphtheriae* subspecies differing slightly in their colonial morphology and biochemical properties are currently recognized: *gravis*, *intermedius*, *mitis* and *belfanti*. Two biovars of *C. pseudotuberculosis*, based on nitrate-reducing ability, have been reported: nitrate-negative are referred as biovar *ovis* (usually isolated from goats and sheep) and nitrate-positive as biovar *equi* (usually isolated from horses and cattle) (Dorella et al. 2006). The strains isolated from cattle are variable for nitrate reduction (Tejedor-Junco et al. 2008). *C. pseudotuberculosis* and *C. ulcerans* are capable to produce urease that may distinguish these species from *C. diphtheriae*, while *C. ulcerans* and *C. diphtheriae* are capable to produce DNase (Cetinkaya et al. 2002; Pimenta et al. 2008b; Dias et al. 2011).

C. diphtheriae, C. ulcerans and *C. pseudotuberculosis* are cystinase positive in Tinsdale agar medium and negative for the pirazinamidase (PYZ) test. When suspect organisms are inoculated transversally with a hemolytic *Staphylococcus aureus* strain, reverse CAMP reactions in Blood Agar Medium are observed for *C. ulcerans* and *C. pseudotuberculosis* due to the ability to produce PLD. These phenotypic aspects are conventionally used in laboratory diagnosis of potentially toxigenic species (Funke and Bernard 2011).

Diphtheria is no longer diagnosed easily on clinical grounds. Mild cases of diphtheria resemble pharyngitis and pseudomembrane may be absent. In addition to cause invasive diseases non-toxigenic *C. diphtheriae* strains have the potential to undergo lysogenic conversion and to produce DT *in vivo*. Therefore, diagnosis of diphtheria is a laboratory emergency and deserves priority. An important test in the microbiological diagnosis of diphtheria is the detection of DT producing strains. Toxigenicity tests are not readily available in most diagnostic laboratories; it is strongly recommended that all isolates be referred promptly to reference laboratories, which are proficient in performing these tests. Several *in vitro* methods are available, including the conventional Elek test, modified Elek tests and genotypic

Table 2.1 Phenotypic properties of potentially toxigenic corynebacteria

Species	Conventional phenotypic tests											
	Hemolysis	DNase	Cystinase (H_2S)[b]	Pirazina-midase	Nitrate reduction	Urea hydrolysis	Glucose	Maltose	Sucrose	Glycogen	Starch	CAMP reaction
C. diphtheriae[a]												
Subsp. gravis	–	+	+	–	+	–	+	+	–/+	+	+	–
Subsp. mitis	+/–	+	+	–	+	–	+	+	–/+	–	–	–
Subsp. intermedius[c]	–	+	+	–	+	–	+	+	–/+	–	–	–
Subsp. belfanti	–	+	+	–	–	–	+	+	–/+	–	–	–
C. ulcerans[a]	+	+	+	–	–/+	+	+	+	–/+	+/–	+	Rev
C.pseudotuberculosis												
Biovar ovis	+	–	+	–	–	+	+	+	–/+	–/+	+	Rev
Biovar equi	+	–	+	–	+	+	+	+	–/+	–/+	+	Rev

[a] Species determined by API Coryne System
[b] Cystinase activity determined in Tinsdale agar medium or PISU test
[c] Lipophilic subspecies
Rev reverse CAMP test observed on Blood agar medium using hemolytic *S. aureus* strain

tests based on PCR for detection of *tox* gene. The gold standard assay is the *in vitro* Vero cell assay, which is based on the cytotoxicity of diphtheria toxin to cultured Vero cells (Farizo et al. 1993; Hauser et al. 1993; Pallen et al. 1994; Nakao and Popovic 1997; Efstratiou et al. 1998, 1999, 2000; Mothershed et al. 2002; Pimenta et al. 2008a; Konrad et al. 2010).

Rapid and accurate detection of potentially toxigenic corynebacteria is necessary for the determination of appropriate measures in controlling the dissemination of toxigenic clones in both veterinary and/or human populations. PCR assays offer many advantages over standard phenotypic tests: they are simple, easy to interprete, and becoming more widely available in laboratories in both industrialized and in developing countries. Recently, a multiplex PCR was proposed for laboratory diagnosis of *C. diphtheriae*, *C. ulcerans*, and *C. pseudotuberculosis*, with a preparation of primers capable to discriminate toxigenic isolates of the three species. The reaction mixture consists of primers (reverse and forward) for amplification of *rpoB* (for corynebacteria); *16SrDNA* (for *C. ulcerans* and *C. pseudotuberculosis*), *dtxR* (for *C. diphtheriae*), *pld* (specific for *C. pseudotuberculosis*), and *dipht4* (for *tox* gene) (Torres et al. 2013).

2.3 Antimicrobial Susceptibility

Antimicrobial treatment in diphtheria cases is capable to both eradicate the pathogen and limit the production of DT by toxigenic strains, alongside to reduce the risk of transmission of microorganisms by symptomatic carriers. Though important for the treatment of infections caused by *C. diphtheriae*, the antimicrobial treatment does not substitute the immunotherapy with diphtheria antitoxin in infections caused by toxigenic strains. *C. diphtheriae* strains are typically susceptible to beta-lactam antibiotics (especially Penicillin G and amoxicillin) and erythromycin. However, tolerance or resistance to penicillin and/or to erythromycin was reported in different countries (Coyle et al. 1979; Farizo et al. 1993; Maple 1994; Patey et al. 1995; von Hunolstein et al. 2002; Zasada et al. 2010; Pereira et al. 2008).

Resistance to erythromycin was reported to be carried by a plasmid and supposed to be associated to coryneform microorganisms colonizing the skin (Schiller et al. 1983; Serwold-Davis and Groman 1986). Multidrug-resistant (MDR) *C. diphtheriae* strains have been also described in recent global literature (Pereira et al. 2008; Zasada et al. 2010; Mina et al. 2011).

The recommendation for the treatment of diphtheria caused by *C. ulcerans* is the administration of antimicrobials penicillin G or erythromycin, in addition to the administration of diphtheria antitoxin. The resistance of *C. ulcerans* to antimicrobial treatment with penicillin in a dog has been reported (Lartigue et al. 2005), as well as the resistance or decreased sensitivity to clindamycin (Katsukawa et al. 2012). The resistance to erythromycin was observed in a *C. ulcerans* strain isolated from a fatal case of diphtheria, and the determination of antimicrobial resistance was highlighted (Tiwari et al. 2008).

C. pseudotuberculosis is generally susceptible to many antibiotics used for the treatment of corynebacterial infections, including penicillin G, ampicillin, erythromycin, gentamicin, and sulfamethoxazole-trimethoprim. However, in cases of lymphadenitis, antimicrobial treatment and surgical approaches to remove the infected lymph nodes are necessary, since antimicrobials do not reach inhibitory concentrations inside the caseous lesions (Dorella et al. 2006).

2.4 Virulence Factors

Diphtheria toxin The major known virulence factor of *C. diphtheriae* is the DT (Pappenheimer 1993; Wang and London 2009). The *tox* gene for DT production is present in corynebacteriophages (β tox$^+$, γ tox$^+$, ω tox$^+$), capable to integrate into the chromosomes of *C. diphtheriae*, *C. ulcerans* and *C. pseudotuberculosis*. The expression of DT is also dependent on chromosomal genes. In low intracellular iron concentrations, the regulatory *dtxR* gene is inhibited, resulting in an increase on DT production by *C. diphtheriae* strains.

DT is a polypeptide of 535 amino acids with a molecular weight of approximately 62,000 Da, composed of two fragments, A (DT-A—with the domain A (enzymatic active)) and B (DT-B—with domains B (binding to the cell receptor: HB-EGF (heparin binding epidermal growth factor) precursor along with DRAP27/CD9 membrane proteins), and T (translocation of the active fragment across the membrane). DT-A and DT-B are linked by disulfide bonds. DT-A is enzymatically active while DT-B, although not toxic, is essential for penetration of TDA in the cell cytoplasm (Funke et al. 1997; D'Silva and Lala 2000). DT has a lower minimum lethal dose of 50–100 ng/kg body weight (Pappenheimer 1977, 1993; Mekada et al. 1988). A single molecule of DT-A, when introduced directly into the cytoplasm, is sufficient to kill the eukaryotic cell (Yamaizumi et al. 1978). Besides the ability to inhibit the protein synthesis (Chang et al. 1989a), DT is also capable to induce internucleossomal cleavage of DNA (indicative of apoptosis-inducing activity) assigned to DT-A (Chang et al. 1989a, b; Nakamura and Wisnieski 1990), process that precedes cytolysis and does not seem to be due to the inhibition of protein synthesis.

The efficacy of diphtheric toxoid vaccine or antiserum against the zoonotic diphtheria caused by *C. ulcerans* still remains unknown. Differences in nucleotide sequences of the *tox* genes may occur among *C. diphtheriae* strains as well as among *C. ulcerans* strains (Mekada et al. 1988; Bitragunta et al. 2010). These facts may contribute towards situations in which individuals vaccinated with diphtheric toxoid or undergoing serum therapy may not present full protection against infections caused by toxigenic corynebacteria. Even if it is considered that the diphtheric toxoid may have a protective effect against diphtheria caused by *C. ulcerans* (i.e. through the presence of attenuated clinical symptoms), it should be remembered that the vaccination only prevents the action of DT and probably does not impede colonization by toxigenic corynebacteria (Neal and Efstratiou 2007; Begue 2010).

Since there are still many issues that require better assessment, not only in relation to clinical-laboratory diagnosis, but also in relation to treating and preventing diseases caused by potentially toxigenic corynebacteria, a better comprehension of virulence factors other than DT production is necessary.

Toxins other than DT PLD exotoxin is considered to be the major virulence factor expressed by *C. ulcerans* and *C. pseudotuberculosis* strains (Lipsky et al. 1982; Hodgson et al. 1992; Dorella et al. 2006). Endowed of a sphingomyelinase activity, PLD toxin is expressed from a chromosomal gene present in *C. ulcerans* and *C. pseudotuberculosis*. PLD is capable to induce the vascular permeability possibly contributing to the spread of the bacteria from the initial site of infection to secondary sites (Carne and Onon 1978; Lipsky et al. 1982; Coyle and Lipsky 1990; McNamara et al. 1995; Tachedjian et al. 1995; Peel et al. 1997; Tambourgi et al. 2002; Dorella et al. 2006). Cytotoxic effects and death of caprine macrophages due to action of PLD was observed during infection with *C. pseudotuberculosis* (Tashjian and Campbell 1983). Expression of PLD is regulated by multiple environmental factors, including cell-density and heat shock (McKean et al. 2007). The use of a PLD antitoxin may prevent the systemic dissemination of *C. pseudotuberculosis* but not the development of abscesses (Williamson 2001).

A putative *C. pseudotuberculosis* iron uptake gene cluster has a role in its virulence. The four genes in this putative operon were identified downstream from the *pld* gene. They were designated as Fe acquisition genes—*fagA*, *fagB*, *fagC* and *fagD*. Since *C. pseudotuberculosis* is an intracellular pathogen, this bacterium must be able to acquire iron from an environment in which this nutrient is scarce. Although there was no alteration in the utilization of iron by a *fagB(C)* mutant *in vitro*, this mutant had a decreased ability to survive and to cause abscesses in experimentally-infected goats (Billington et al. 2002).

In human infections caused by *C. ulcerans*, the occurrence of necrosis and mucosal ulceration as well as other clinical manifestations in the lower respiratory tract were attributed to the production of both DT and PLD (Dessau et al. 1995; Seto et al. 2008). The genome sequencing investigations of two Brazilian *C. ulcerans* isolates showed the presence of *pld* gene and absence of *tox* gene in both 809 strain (isolated from a case of human fatal pneumonia) and BR-AD22 strain (isolated from an asymptomatic dog) (Trost et al. 2011).

C. pseudotuberculosis and *C. ulcerans* strains also showed colonization behavior in the *Caenorhabditis elegans* and induction of death in *Galleria mellonella* models. Interestingly, a *C. ulcerans* PLD-deficient strain showed unaltered colonization behavior compared to the parental strain BR-AD22 in the *C. elegans*, although it caused a less severe response in the *Galleria* model. The strongest effects on the *Galleria* larvae obtained for *C. pseudotuberculosis* and *C. ulcerans* strains, which exhibited high degrees of melanisation, immobility and rapid death, were attributed to PLD production (Ott et al. 2012). Meanwhile, the PLD-positive 809 human isolate was considered more virulent than the non-arthritogenic PLD-positive BR-AD22 strain, since it was capable to induce arthritis and to cause mice death (Dias et al. 2011).

Adhesion to biotic and abiotic surfaces During the last decades the current reporting of diphtheria outbreaks attracted justifiable attention and stimulated the search for microbial factors responsible for invasiveness of *C. diphtheriae* and *C. ulcerans*. Changes in the circulating strains of *C. diphtheriae* could be responsible for the episodic diphtheria epidemic waves. However, the microbial factors that distinguish epidemic from non-epidemic *C. diphtheriae* strains remain under investigation (Popovic et al. 1996; Funke et al. 1999). Other factors besides antitoxin protection influence vulnerability to diphtheria, namely, the general immune status of the person infected, as well as the number and virulence of diphtheria bacilli involved. Moreover, immunized patients may have *C. diphtheriae* bacteremia and endocarditis in the absence of characteristic toxin-mediated lesions, confirming that invasive and toxigenic properties are independent of each other (Mattos-Guaraldi and Formiga 1998; Hogg et al. 1996; Patey et al. 1997; Funke et al. 1999). Multilocus sequence typing (MLST) to study genetic relationships reinforced the fact that invasive strains may be found in different clonal complexes of *C. diphtheriae* (Viguetti et al. 2012).

In the 1980s, opening studies started dealing with the adhesive properties of the *C. diphtheriae*. The multifactorial nature of the adhesiveness was considered once *C. diphtheriae* strains adhered to several biotic and abiotic substrates in varied intensities. Bacterial adhesive properties were independent of toxin production. Studies using human and animal erythrocytes showed *C. diphtheriae* capable to adhere differentially according to sucrose fermenting biotypes and subspecies (*gravis*, *mitis*, *intermedius* and *belfanti*) (Yanagawa and Honda 1976; Deacock et al. 1983; Kostyukova and Pereverzev 1985; Mattos-Guaraldi and Formiga 1986; Karas et al. 1991; Mattos-Guaraldi and Formiga 1991; Kostyukova and Karas 1991; Kostyukova et al. 1992; Mattos-Guaraldi and Formiga 1992; Mattos-Guaraldi and Formiga 1998a, b; Mattos-Guaraldi et al. 1999a, b; Hladka and Motyka 1998; Colombo et al. 2001). Further studies demonstrated that non-toxigenic (Bertuccini et al. 2004) and toxigenic *C. diphtheriae* strains have the ability to interact with human epithelial cells (Hirata Jr. et al. 2002, 2004). The internalization of *C. diphtheriae* by non-professional phagocyte cells was then documented. Invasive *C. diphtheriae* strains were able to survive within human epithelial cells (Hirata Jr. et al. 2002; Bertuccini et al. 2004) at different levels. *C. diphtheriae* strains were also capable to adhere to U937 macrophages inducing both necrosis and apoptosis. Both activities were independent to the expression of DT, since the non-toxigenic ATCC 27010 also induced both apoptosis and necrosis (Santos et al. 2010).

The surface lipid components of *C. pseudotuberculosis* may contribute to the pathogenesis, since it can induce hemorrhagic necrosis after intradermal injection in guinea pigs (Carne et al. 1956; Jolly 1965, 1966; Hard 1972, 1975). Additionally, mouse peritoneal macrophages were highly susceptible to the necrotizing/cytotoxic effect of *C. pseudotuberculosis* lipids, whereas rabbit cells were shown to be resistant (Hard 1975). However the molecules that permit the microbial interaction to animal and human cells are still unknown for *C. pseudotuberculosis* and *C. ulcerans*.

Few data are available dealing with carbohydrate structures on the cell surfaces of potentially toxigenic corynebacteria. *C. diphtheriae* strains were shown to contain N-acetylglucosamine (D-GlcNAc), N-acetylgalactosamine (D-GalNAc), galactose (D-Gal), mannose-like (D-Man-like) and sialic acid residues on their surfaces (Mattos-Guaraldi et al. 2000). In general, sialic acid molecules aid the microorganisms in immune system evasion by favoring the establishment and maintenance of the species in the mucous cutaneous surfaces of the host. The detection of sialic acid in bacteria is limited to a few examples of pathogenic bacteria, including *C. diphtheriae* (Gabius 1997; Schauer 1982). The incorporation of sialic acid in *C. diphtheriae* was shown to be mediated by *trans*-sialidase activity (Mattos-Guaraldi et al. 1998). A detailed review of these earlier studies has looked at potential adhesins, including hemagglutinins, hydrophobins, exposed sugar residues and enzymes expressing *trans*-sialidase activity (Mattos-Guaraldi et al. 2000).

Biofilm formation is a well-known pathogenic mechanism in indwelling medical device-related infections in hospitals (Christensen et al. 1982). Some non-diphtherial corynebacteria species have a natural tendency to form biofilms to available abiotic surfaces (Soriano et al. 2009). Different levels of adherence to glass and negatively charged plastic (polystyrene) surfaces were also observed among *C. diphtheriae* strains. Amorphous deposited substances or glycocalyx have been noted covering *C. diphtheriae* cells. The presence of an exposed sialic acid moiety was predominantly observed on surfaces of glass-adherent *C. diphtheriae* (Mattos-Guaraldi et al. 2000; Moreira et al. 2003; Gomes et al. 2009). Moreover, *C. diphtheriae* strains seemed to withstand oxidative environmental stress partly due to their biofilm-forming abilities (Silva de Souza et al. 2003). The expression of exopolysaccharides on the bacterial cells and the ability to adhere to glass surfaces were drastically decreased when *C. diphtheriae* strains were submitted to growth under iron-limiting conditions (Moreira et al. 2003; Silva de Souza et al. 2003). Recently, *C. diphtheriae* was found capable of adhering to catheters and inducing infections in susceptible hosts. In addition to bacterial growth and biofilm production on luminal surfaces, fibrin deposition seemed to be a characteristic of *C. diphtheriae* strains that may contribute to catheter malfunction (Gomes et al. 2009).

Studies have focused on lipoarabinomannan (LAM) as a virulence factor for several microorganisms, including *Mycobacterium tuberculosis*. Mannose-capped LAM (ManLAM) from *M. tuberculosis* may facilitate bacterial adherence to alveolar macrophages, particularly to mannose receptors and display immunomodulatory properties (Nigou et al. 2001; Nigou et al. 2002; Briken et al. 2004; Dao et al. 2004; Puissegur et al. 2007). *C. diphtheriae* also express a surface lipoarabinomannan-like lipoglycan (CdiLAM) capable to mediate bacterial adherence to human epithelial cells (Moreira et al. 2008). Others studies also showed the participation of non-fimbrial adhesins and a Spa-type pilus in the adherence process of *C. diphtheriae* to human epithelial cells. Pili adhesins (Ton-That and Schneewind 2003; Mandlik et al. 2007, 2008; Kang et al. 2009; Chang et al. 2011) are described in detail in a specific chapter in this book.

Non-fimbrial adhesins, named 67–72p, involved in the adhesion of *C. diphtheriae* to erythrocytes and HEp-2 cells were described (Colombo et al. 2001; Hirata Jr. et al. 2004). Later, the comparative analysis of MALDI-TOF MS experiments with *in silico* proteome deduced from the complete genome sequence of *C. diphtheriae* identified with significant scores 67–72p as the protein DIP0733 (Sabbadini et al. 2012). DIP0733 (67–72p) also contributes in bacterial invasion and apoptosis of epithelial cells during *C. diphtheriae* infection, emphasizing the potential for non-toxigenic strains to adhere, invade and survive within the intracellular environment.

Mapping and comprehensive analysis of the extracellular and cell surface proteome of *C. diphtheriae* initially described DIP1281 as an invasion-associated protein (Hansmeier et al. 2006). Later on, DIP1281 was found to be involved in the organization of the outer surface protein layer of *C. diphtheriae* (Ott et al. 2010).

C. diphtheriae is capable to adhere to fibrinogen and this property may be related to pseudomembrane formation. The pseudomembrane is composed of bacteria, necrotic epithelial and inflammatory cells embedded in a fibrin matrix, which adheres tightly to the underlying tissue (Hadfield et al. 2000). Both toxigenic and non-toxigenic strains were capable to aggregate in the presence of rabbit plasma and to bind to fluorescein labeled fibrinogen. The non-toxigenic ATCC 27010 type strain adhered with less intensity than the parental (toxigenic) strain ATCC 27012, suggesting that bacteriophage genes may also be related to the expression of the proteins capable to adhere to fibrinogen (Sabbadini et al. 2010). This fact may partially explain the higher ability of pseudomembrane formation by toxigenic strains. The adherence to fibrinogen may prone *C. diphtheriae* to be involved in endocarditis cases, since for many to bacterial pathogens the adherence to fibrinogen is related to the development of endocarditis cases, which vegetations composed of fibrin clots are usually present along with damage of cardiac valves.

2.5 Molecular Aspects

Cerdeño-Tárraga and co-workers (2003) have sequenced the genome of a UK clinical isolate (biotype *gravis* strain NCTC13129), representative of the clone responsible for the epidemics in the Eastern European region. It provided additional evidence that recent acquisition of pathogenicity factors goes beyond the toxin itself, and includes adhesins and fimbrial proteins (Cerdeno-Tarraga et al. 2003). Additionally, many virulence factors, in especial adhesins are expressed from pathogenicity islands (PAIs) of *C. diphtheriae* (currently 13 putative PAIs are recognized), and there are differences in the presence of putative PAIs among *C. diphtheriae* genomes (Soares et. al. 2012), justifying the differences in virulence properties of *C. diphtheriae* isolates. Recently, pangenomic analysis of *C. diphtheriae* provided insights into the genomic diversity of isolates from cases of classical diphtheria, endocarditis and pneumonia (Trost et al. 2012).

The molecular data deduced from the complete genome sequences of both human and dog *C. ulcerans* strains demonstrated that the less virulent dog isolate BR-AD22 strain has 2,338 protein-coding regions, whereas the genome of the human isolate *C. ulcerans* 809 comprises 2,182 predicted proteins (Trost et al. 2011).

Different from *C. ulcerans*, many *C. pseudotuberculosis* strains were submitted to genome sequencing (Trost et al. 2010; Cerdeira et al. 2011a, b; Silva et al. 2011; Lopes et al. 2012; Pethick et al. 2012a, b), revealing two gene clusters encoding proteins involved in the sortase-mediated polymerization of adhesive pili that may favor the adherence to host tissues facilitating additional ligand-receptor interactions. The genome annotation revealed the presence of proteases, sincluding serine proteases, neuraminidase H, nitric oxide reductase, an invasion-associated protein, and acyl-CoA carboxylase subunits involved in mycolic acid biosynthesis as potential virulence factors (Trost et al. 2010; Ruiz et al. 2011).

Therefore, the analysis of the genome sequences of different isolates of *C. diphtheriae, C. ulcerans* and *C. pseudotuberculosis* have been providing further insight for understanding their virulence and may be useful for the development of new diagnostic methods and vaccines, contributing to the control of the different veterinary and/or human diseases caused by these potentially toxigenic pathogens.

References

Aaron L, Heurtebise F, Bachelier MN, Guimard Y (2006) Pseudomembranous diphtheria caused by *Corynebacterium ulcerans*. La Rev de Med Int 27:333–335.

Alexander D (1984) Splenic abscess caused by *Corynebacterium diphtheriae*. Clin Pediatr 23:591–592

Augustine JL, Renshaw HW (1986) Survival of *Corynebacterium pseudotuberculosis* in axenic purulent exudate on common barnyard fomites. Am J Vet Res 47:713–715

Baird GJ, Fontaine MC (2007) *Corynebacterium pseudotuberculosis* and its role in ovine caseous lymphadenitis. J Comp Pathol 137:179–210

Begue P (2010) Consequences of opposition to vaccination in France and Europe. How to maintain effective vaccine coverage in 2010? Bull Acad Natl Med 194:719–732

Bergin I, Chien CC, Marini RP, Fox JG (2000) Isolation and characterization of *Corynebacterium ulcerans* from cephalic implants in macaques. Comp Med 50:530–535

Bertuccini L, Badassarri L, Hunolstein C von (2004) Internalization of non-toxigenic *Corynebacterium diphtheriae* by cultured human respiratory epithelial cells. Microb Pathog 37:111–118

Billington SJ, Esmay PA, Songer JG, Jost BH (2002) Identification and role in virulence of putative iron acquisition genes from *Corynebacterium pseudotuberculosis*. FEMS Microbiol Lett 208:41–45

Bitragunta S, Murhekar MV, Chakravarti A, Verma V, Namjoshi GS, Parekh SS, Sharma HJ, Kumar BK, Gupte MD (2010) Safety and immunogenicity of single dose of tetanus-diphtheria (Td) vaccine among non/partially immune children against diphtheria and/or tetanus, Hyderabad, India, 2007. Vaccine 28:5934–5938

Bonmarin I, Guiso N, Le Flèche-Matéos A, Patey O, Patrick AD, Levy-Bruhl D (2009) Diphtheria: a zoonotic disease in France? Vaccine 27:4196–4200

Braverman Y, Chizov-Ginzburg A, Saran A, Winkle M (1999) The role of houseflies (*Musca domestica*) in harbouring *Corynebacterium pseudotuberculosis* in dairy herds in Israel. Rev Sci Tech 18:681–690

Briken V, Porcelli SA, Besra GS, Kremer L (2004) Mycobacterial lipoarabinomannan and related lipoglycans: from biogenesis to modulation of the immune response. Mol Microbiol 53:391–403

Cardénas R, Prieto G, Vargas J, Martinez A (1972) *Corynebacterium diphtheriae* características de cepas aisladas recientemente em Maracaibo-Venezuela. Kasmera 4:185–198

Carne HR, Onon EO (1978) Action of *Corynebacterium ovis* exotoxin on endothelial cells of blood vessels. Nature 271:246–248

Carne HR, Kater JC, Wickham N (1956) A toxic lipid from the surface of *Corynebacterium ovis*. Nature 178:701–702

Cerdeira LT, Pinto AC, Schneider MP, Almeida SS de, Santos AR dos, Barbosa EG, Ali A, Barbosa MS, Carneiro AR, Ramos RT, Oliveira RS de, Barh D, Barve N, Zambare V, Belchior SE, Guimarães LC, Castro SS de, Dorella FA, Rocha FS, Abreu VA de, Tauch A, Trost E, Miyoshi A, Azevedo V, Silva A (2011a) Whole-genome sequence of *Corynebacterium pseudotuberculosis* PAT10 strain isolated from sheep in Patagonia, Argentina. J Bacteriol 193:6420–6421

Cerdeira LT, Schneider MP, Pinto AC, Almeida SS de, Santos AR dos, Barbosa EG, Ali A, Aburjaile FF, Abreu VA de, Guimarães LC, Soares Sde C, Dorella FA, Rocha FS, Bol E, Gomes deSPH, Lopes TS, Barbosa MS, Carneiro AR, Jucá Ramos RT, Coimbra NA, Lima AR, Barh D, Jain N, Tiwari S, Raja R, Zambare V, Ghosh P, Trost E, Tauch A, Miyoshi A, Azevedo V, Silva A (2011b) Complete genome sequence of *Corynebacterium pseudotuberculosis* strain CIP 52.97, isolated from a horse in Kenya. J Bacteriol 193:7025–7026

Cerdeño-Tárraga AM, Efstratiou A, Dover LG, Holden MT, Pallen M, Bentley SD, Besra GS, Churcher C, James KD, De Zoysa A, Chillingworth T, Cronin A, Dowd L, Feltwell T, Hamlin N, Holroyd S, Jagels K, Moule S, Quail MA, Rabbinowitsch E, Rutherford KM, Thomson NR, Unwin L, Whitehead S, Barrell BG, Parkhill J (2003) The complete genome sequence and analysis of *Corynebacterium diphtheriae* NCTC13129. Nucleic Acids Res 31:6516–6523

Cetinkaya B, Karahan M, Atil E, Kalin R, De Baere T, Vaneechoutte M (2002) Identification of *Corynebacterium pseudotuberculosis* isolates from sheep and goats by PCR. Vet Microbiol 88:75–83

Chang C, Mandlik A, Das A, Ton-That H (2011) Cell surface display of minor pilin adhesins in the form of a simple heterodimeric assembly in *Corynebacterium diphtheriae*. Mol Microbiol 79:1236–1247

Chang MP, Baldwin RL, Bruce C, Wisnieski BJ (1989a) Second cytotoxic pathway of diphtheria toxin suggested by nuclease activity. Science 246:1165–1167

Chang MP, Bramhall J, Graves S, Bonavida B, Wisnieski BJ (1989b) Internucleosomal DNA cleavage precedes diphtheria toxin-induced cytolysis: evidence that cell lysis is not a simple consequence of translation inhibition. J Biol Chem 264:15261–15267

Christensen GD, Simpson WA, Bismo AL, Beachey EH (1982) Adherence of slime-producing strains of *Staphylococcus epidermidis* to smooth surfaces. Infect Immun 37:318–326

Colombo AV, Hirata R Jr, De Souza CM, Monteiro-Leal LH, Previato JO, Formiga LC, Andrade AF, Mattos-Guaraldi AL (2001) *Corynebacterium diphtheriae* surface proteins as adhesins to human erythrocytes. FEMS Microbiol Lett 197:235–239

Corboz L, Thoma R, Braun U, Zbinden R (1996) Isolation of *Corynebacterium diphtheriae* subsp. *belfanti* from a cow with chronic active dermatitis. Schweiz Arch Tierheilkd 138:596–599

Coyle MB, Lipsky BA (1990) Coryneform bacteria in infectious diseases: clinical and labora-tory aspects. Clin Microbiol Rev 3:227–246

Coyle MB, Minshew BH, Bland JA, Hsu PC (1979) Erythromycin and clindamycin resistance in *Corynebacterium diphtheriae* from skin lesions. Antimicrob Agents Chem 16:525–527

Dao DN, Kremer L, Guérardel Y, Molano A, Jacobs WR Jr, Porcelli SA, Briken V (2004) *Mycobacterium tuberculosis* lipomannan induces apoptosis and interleukin-12 production in macrophages. Infect Immun 72:2067–2074

Deacock SJ, Steward A, Carne HR (1983) The role of adherence in determining the site of infection by *Corynebacterium diphtheriae*. J Hyg 90:415–424

Dessau RB, Brandt-Christensen M, Jensen OJ, Tonnesen P (1995) Pulmonary nodules due to *Corynebacterium ulcerans*. Eur Respir J 8:651–653

Dewinter LM, Bernard KA, Romney MG (2005) Human clinical isolates of *Corynebacterium diphtheriae* and *Corynebacterium ulcerans* collected in Canada from 1999–2003 but not fitting reporting criteria for cases of diphtheria. J Clin Microbiol 43:3447–3449

De Zoysa A, Efstratiou A, George RC, Jahkola M, Vuopio-Varkila J, Deshevoi S, Tseneva G, Rikushin Y (1995) Molecular epidemiology of *Corynebacterium diphtheriae* from

Northwestern Russia and surrounding countries studied by using ribotyping and pulsed-field gel electrophoresis. J Clin Microbiol 33:1080–1083

De Zoysa A, Hawkey PM, Engler K, George R, Mann G, Reilly W, Taylor D, Efstratiou A (2005) Characterization of toxigenic *Corynebacterium ulcerans* strains isolated from humans and domestic cats in the United Kingdom. J Clin Microbiol 43:4377–4381

Dias AA, Santos LS, Sabbadini PS, Santos CS, Silva Junior FC, Napoleão F, Nagao PE, Villas-Bôas MH, Hirata R Jr, Guaraldi AL (2011a) *Corynebacterium ulcerans* diphtheria: an emerging zoonosis in Brazil and worldwide. Rev Saude Publica 45:1176–1191

Dias AA, Silva FC Jr, Santos LS, Ribeiro-Carvalho MM, Sabbadini PS, Santos CS, Filardy AA, Myioshi A, Azevedo VA, Hirata R Jr, Villas-Bôas MH, Mattos-Guaraldi AL (2011b) Strain-dependent arthritogenic potential of the zoonotic pathogen *Corynebacterium ulcerans*. Vet Microbiol 153:323–331

Dias AASO, Silva FC Jr, Pereira GA, Souza MC, Camello TCF, Damasceno JALD, Pacheco LGC, Miyoshi A, Azevedo VA, Hirata R Jr, Bôas MHSV, Mattos-Guaraldi AL (2010) *Corynebacterium ulcerans* isolated from an asymptomatic dog kept in an animal shelter in the metropolitan area of Rio de Janeiro, Brazil. Vec Born Zoon Dis 10:743–748

Dittmann S, Wharton M, Vitek C, Ciotti M, Galazka A, Guichard S, Hardy I, Kartoglu U, Koyama S, Kreysler J, Martin B, Mercer D, Rønne T, Roure C, Steinglass R, Strebel P, Sutter R, Trostle M (2000) Successful control of epidemic diphtheria in the states of the Former Union of Soviet Socialist Republics: lessons learned. J Infect Dis 181 Suppl 1:10–22.

Dorella FA, Pacheco LG, Oliveira SC, Miyoshi A, Azevedo VA (2006) *Corynebacterium pseudotuberculosis*: microbiology, biochemical properties, pathogenesis and molecular studies of virulence. Vet Res 37:201–218

D'Silva PR, Lala AK (2000) Organization of diphtheria toxin in membranes. A hydrophobic photolabeling study. J Biol Chem 275:11771–11777

Dzupova O, Benes J, Kriz B, Horova B, Olexova A (2005) An unusual course of invasive infection due to nontoxinogenic strain of *Corynebacterium diphtheriae*. Klin Mikrobiol Infekc Lek 11:222–225

Efstratiou A, George RC (1999) Laboratory guidelines for the diagnosis of infections caused by *Corynebacterium diphtheriae* and *C. ulcerans*. World Health Organization. Commun Dis Public Health 2:250–257

Efstratiou A, Engler KH, Dawes CS, Sadic D (1998) Comparison of phenotypic and genotypic methods for detection of diphtheria toxin among isolates of pathogenic corynebacteria. J Clin Microbiol 36:3173–3177

Efstratiou A, Engler KH, Mazurova IK, Glushkevish T, Vuopio-Varkila J, Popovic T (2000) Current Approaches to the laboratory diagnosis of diphtheria. J Infect Dis 181:138–145.

Estevao BS, Gallardo A, Abalos A, Diaz Y, Alvarez L, Callejo R, Prieto M, Jodor N, Jensen O (2007) Diagnosis of caseous lymphadenitis in sheep from Patagonia. Rev Argent Microbiol 39:44–46

Farfour E, Badell E, Zasada A, Hotzel H, Tomaso H, Guillot S, Guiso N (2012) Characterization and comparison of invasive *Corynebacterium diphtheriae* isolates from France and Poland. J Clin Microbiol 50:173–175

Farizo KM, Strebel PM, Chen RT, Kimbler A, Cleary TJ, Cochi SL (1993) Fatal respiratory disease due to *Corynebacterium diphtheriae*: case report and review of guidelines for management, investigation and control. Clin Infect Dis 16:59–68

Foley JE, Spier SJ, Mihalyi J, Drazenovich N, Leutenegger CM (2004) Molecular epidemiologic features of *Corynebacterium pseudotuberculosis* isolated from horses. Am J Vet Res 65:1734–1737

Foster G, Patterson T, Howie F, Simpson V, Davison N, Efstratiou A, Lai S (2002) *Corynebacterium ulcerans* in free-ranging otters. Vet Rec 150:524

Funke G, Bernard KA (2011) Coryneform Gram-positive rods. In: Versalovic J, Carroll KC, Funke G, Jorgensen JH, Landry ML, Warnock DW (eds) Manual of Clinical Microbiology, 10th ed. American Society for Microbiology, Washington, DC, pp 413–442.

Funke G, Von Graevenitz A, Clarridge JE III, Bernard KA (1997) Clinical microbiology of coryneform bacteria. Clin Microbiol Rev 10:125–159

Funke G, Altwegg M, Frommelt L, Graevenitz A von (1999) Emergence of related non-toxigenic *Corynebacterium diphtheriae* biotype *mitis* strains in Western Europe. Emerg Infect Dis 5:477–480
Gabius HJ, Gabius S (1997) Glycoprotein structure and function. In: Gabius HJ, Gabius S (eds) Glycosciences. Chapman and Hall, Weinheim, pp 1–20
Golaz A, Lance-Parker S, Welty T, Schaefer L, Volmer L, La Fromboise C (2000) Epidemiology of diphtheria in South Dakota. S D J Med 53:281–285
Gomes DL, Martins CA, Faria LM, Santos LS, Santos CS, Sabbadini PS, Souza MC, Alves GB, Rosa AC, Nagao PE, Pereira GA, Hirata R Jr, Mattos-Guaraldi AL (2009) *Corynebacterium diphtheriae* as an emerging pathogen in nephrostomy catheter-related infection: evaluation of traits associated with bacterial virulence. J Med Microbiol 58:1419–1427
Gubler JG, Wüst J, Krech T, Hany A (1990) Classical pseudomembranous diphtheria caused by *Corynebacterium ulcerans*. Schweiz Med Wochenschr 120:1812–1816
Hadfield TL, McEvoy P, Polotsky Y, Tzinserling VA, Yakovlev AA (2000) The pathology of diphtheria. J Infect Dis 181:116–120.
Hall AJ, Cassiday PK, Bernard KA, Bolt F, Steigerwalt AG, Bixler D, Pawloski LC, Whitney AM, Iwaki M, Baldwin A, Dowson CG, Komiya T, Takahashi M, Hinrikson HP, Tondella ML (2010) Novel *Corynebacterium diphtheriae* in domestic cats. Emerg Infect Dis 16:688–689
Hansmeier N, Chao TC, Kalinowski J, Pühler A, Tauch A (2006) Mapping and comprehensive analysis of the extracellular and cell surface proteome of the human pathogen *Corynebacterium diphtheriae*. Proteomics 6:2465–2476
Hard GC (1972) Examination by electron microscopy of the interaction between peritoneal phagocytes and *Corynebacterium ovis*. J Med Microbiol 5:483–491
Hard GC (1975) Comparative toxic effect on the surface lipid of *Corynebacterium ovis* on peritoneal macrophages. Infect Immun 12:4139–4149
Hatanaka A, Tsunoda A, Okamoto M, Ooe K, Nakamura A, Miyakoshi M, Komiya T, Takahashi M (2003) *Corynebacterium ulcerans* diphtheria in Japan. Emerg Infect Dis 9:752–753
Hauser D, Popoff MR, Kiredjian M, Boquet P, Bimet F (1993) Polymerase chain reaction assay for diagnosis of potentially toxigenic *Corynebacterium diphtheriae* strains: correlation with ADP-ribosylation activity assay. J Clin Microbiol 31:2720–2723
Hemond V, Rosenstingl S, Auriault ML, Galanti MJ, Gatfosse M (2009) Axillary lymphadenitis due to *Corynebacterium pseudotuberculosis* in a 63-year-old patient. Med Mal Infect 39:136–139
Hirata R Jr, Napoleão F, Monteiro-Leal LH, Andrade AF, Nagao PE, Formiga LC, Fonseca LM, Mattos-Guaraldi AL (2002) Intracellular viability of toxigenic *Corynebacterium diphtheriae* strains in HEp-2 cells. FEMS Microbiol Lett 215:115–119
Hirata R Jr, Souza SM, Rocha-De-Souza CM, Andrade AF, Monteiro-Leal LH, Formiga LC, Mattos-Guaraldi AL (2004) Patterns of adherence to HEp-2 cells and actin polymerisation by toxigenic *Corynebacterium diphtheriae* strains. Microb Pathog 36:125–130
Hirata R Jr, Pereira GA, Filardy AA, Gomes DL, Damasco PV, Rosa AC, Nagao PE, Pimenta FP, Mattos-Guaraldi AL (2008) Potential pathogenic role of aggregative-adhering *Corynebacterium diphtheriae* of different clonal groups in endocarditis. Braz J Med Biol Res 41:986–991
Hirata R Jr, Pacheco LG, Soares SC, Santos LS, Moreira LO, Sabbadini PS, Santos CS, Miyoshi A, Azevedo VA, Mattos-Guaraldi AL (2011) Similarity of *rpoB* gene sequences of sucrose-fermenting and non-fermenting *Corynebacterium diphtheriae* strains. Antonie Van Leeuwenhoek 99:733–737
Hladka OA, Motyka OI (1998) The effect of antibiotics on the adhesion of *Corynebacterium diphtheriae* to buccal epithelial cells. Mikrobiol Z 60:60–64
Hodgson ALM, Krywult J, Corner LA, Rothel JS, Radford AJ (1992) Rational attenuation of *Corynebacterium pseudotuberculosis*: potential cheesy gland vaccine and live delivery vehicle. Infect Immun 60:2900–2905
Hogg GG, Strachan JE, Huayi L, Beaton SA, Robinson PM, Taylor K (1996) Non-toxigenic *Corynebacterium diphtheriae* biovar *gravis*: evidence for an invasive clone in a south-eastern Australian community. Med J Australia 164:72–75
Hogg RA, Wessels J, Hart J, Efstratiou A, De Zoysa A, Mann G, Allen T, Pritchard GC (2009) Possible zoonotic transmission of toxigenic *Corynebacterium ulcerans* from companion animals in a human case of fatal diphtheria. Vet Rec 165:691–692

Hommez J, Devriese LA, Vaneechoutte M, Riegel P, Butaye P, Haesebrouck F (1999) Identification of nonlipophilic corynebacteria isolated form dairy cows with mastitis. J Clin Microbiol 37:954–957

Hunolstein C von, Scopetti F, Efstratiou A, Engler K (2002) Penicillin tolerance amongst non-toxigenic *Corynebacterium diphtheriae* isolated from cases of pharyngitis. J Antimicrob Chemother 50:125–128

Join-Lambert OF, Ouache M, Canioni D, Beretti JL, Blanche S, Berche P, Kayal S (2006) *Corynebacterium pseudotuberculosis* necrotizing lymphadenitis in a twelve-year-old patient. Pediatr Infect J Dis 25:848–851

Jolly RD (1965) The pathogenesis of experimental *Corynebacterium ovis* infection in mice. N Z Vet J 13:141–147

Jolly RD (1966) Some observations on surface lipids of virulent and attenuated strains of *Corynebacterium ovis*. J Appl Bacteriol 29:189–196

Kang HJ, Paterson NG, Gaspar AH, Ton-That H, Baker EN (2009) The *Corynebacterium diphtheriae* shaft pilin SpaA is built of tandem Ig-like modules with stabilizing isopeptide and disulfide bonds. Proc Natl Acad Sci U S A 106:16967–16971

Karas SR, Kasimova DIa, Kostyukova NN (1991) The characteristics of *Corynebacterium diphtheriae* adhesions. Zh Mikrobiol Epidemiol Immunol 5:17–19

Katsukawa C, Kawahara R, Inoue K, Ishii A, Yamagishi H, Kida K, Nishino S, Nagahama S, Komiya T, Iwaki M, Takahashi M (2009) Toxigenic *Corynebacterium ulcerans* isolated from the domestic dog for the first time in Japan. Jpn J Infect Dis 62:171–172

Katsukawa C, Komiya T, Yamagishi H, Ishii A, Nishino S, Nagahama S, Iwaki M, Yamamoto A, Takahashi M (2012) Prevalence of *Corynebacterium ulcerans* in dogs in Osaka, Japan. J Med Microbiol 61:266–273

Kelly EJ, Rood KA, Skirpstunas R (2012) Abscesses in Captive Elk Associated with *Corynebacterium pseudotuberculosis*, Utah, USA. J Wildl Dis 48:803–805

Kimura Y, Watanabe Y, Suga N, Suzuki N, Maeda K, Suzuki K, Kitagawa W, Miura N, Morita H, Imai H (2011) Acute peritonitis due to *Corynebacterium ulcerans* in a patient receiving continuous ambulatory peritoneal dialysis: a case report and literature review. Clin Exp Nephrol 15:171–174

Komala TS, Ramlan M, Yeoh NN, Surayani AR, Sharifah Hamidah SM (2008) A survey of caseous lymphadenitis in small ruminant farms from two districts in Perak, Malaysia—Kinta and Hilir Perak. Trop Biomed 25:196–201

Kombarova S, Kim C, Melnikov V, Reeves M, Borisova O, Mazurova I, Popovic T (2001) Rapid identification of *Corynebacterium diphtheriae* clonal group associated with diphtheria epidemic, Russian Federation. Emerg Infect Dis 7:133–136

Komiya T, Seto Y, Zoysa A de, Iwaki M, Hatanaka A, Tsunoda A, Arakawa Y, Kozaki S, Takahashi M (2010) Two Japanese *Corynebacterium ulcerans* isolates from the same hospital: Ribotype, toxigenicity, and serum antitoxin titre. J Med Microbiol 59:1497–1504

Konrad R, Berger A, Huber I, Boschert V, Hormansdorfer S, Busch U, Hogardt M, Schubert S, Sing A (2010) Matrix-assisted laser desorption/ionization time-of-flight (MALDI-TOF) mass spectrometry as a tool for rapid diagnosis of potentially toxigenic *Corynebacterium* species in the laboratory management of diphtheria-associated bacteria. Euro Surveill 15:19699

Kostyukova NN, Karas SR (1991) The adhesive activity of diphtheritic strains in relation to the characteristics of the infectious process they induce. Zh Mikrobiol Epidemiol Immunobiol 11:24–27

Kostyukova NN, Pereverzev NA (1985) Adhesion in *Corynebacterium diphtheriae*. Zh Mikrobiol Epidemiol Immunobiol 11:30–33

Kostyukova NN, Karas SR, Kacimova DYa (1992) The study of agglutination of trypsin-treated sheep red cells by *Corynebacterium diphtheriae* strains. FEMS Microbiol Immunol 4:231–234

Lartigue MF, Monnet X, Lê Fleche A, Grimont PA, Benet JJ, Durrbach A, Fabre M, Nordmann P (2005) *Corynebacterium ulcerans* in an immunocompromised patient with diphtheria and her dog. J Clin Microbiol 43:999–1001

Leek MD, Sivaloganathan S, Devaraj SK, Zamiri I, Griffiths GD, Green MA (1990) Diphtheria with a difference- a rare *Corynebacterium* fatality with associated apoptotic cell death. Histopathology 16:187–189

Leggett BA, De Zoysa A, Abbott YE, Leonard N, Markey B, Efstratiou A (2010) Toxigenic *Corynebacterium diphtheriae* isolated from a wound in a horse. Vet Rec 166:656–657

Lin RVTP, Lim SCS, Yew FS, Tan SY, Tey BH (1994) *Corynebacterium diphtheriae* endocarditis in an adult with congenital heart disease: a case report. J Trop Med Hyg 97:189–191

Lipsky BA, Goldberger AC, Tompkins LS, Plorde JJ (1982) Infections caused by non-diphtheria corynebacteria. Rev Infect Dis 4:1220–1235

Liu DT, Chan WM, Fan DS, Lam DS (2005) An infected hydrogel buckle with *Corynebacterium pseudotuberculosis*. Br J Ophthalmol 89:245–246

Lopes T, Silva A, Thiago R, Carneiro A, Dorella FA, Rocha FS, Santos AR dos, Lima AR, Guimarães LC, Barbosa EG, Ribeiro D, Fiaux KK, Diniz CA, Abreu VA de, Almeida SS de, Hassan SS, Ali A, Bakhtiar SM, Aburjaile FF, Pinto AC, Soares S de C, Pereira Ude P, Schneider MP, Miyoshi A, Edman J, Spier S, Azevedo V (2012) Complete genome sequence of *Corynebacterium pseudotuberculosis* strain Cp267, isolated from a llama. J Bacteriol 194:3567–3568

MacGregor RRC (1990) *Corynebacterium diphtheriae*. In: Mandrell DL, Douglas RG, Bennet JE (eds) Enfermidades Infecciosas. Principios y Pratica II. Buenos Aires. Editorial Medica Panam, Argentina

Mandlik A, Swierczynski A, Das A, Ton-That H (2007) *Corynebacterium diphtheriae* employs specific minor pilins to target human pharyngeal epithelial cells. Mol Microbiol 64:111–124

Mandlik A, Swierczynski A, Das A, Ton-That H (2008) Pili in Gram-positive bacteria: assembly, involvement in colonization and biofilm development. Trends Microbiol 16:33–40

Maple PAC, Efstratiou A, Tseneva G, Rikushin Y, Deshevoi S, Jahkola M, Vuopio-Varkila J, George RC (1994) The *in vitro* susceptibilities of toxigenic strains of *Corynebacterium diphtheriae* isolated in north-western Russia and surrounding areas to ten antibiotics. J Antimicrob Chemother 34:1037–1040

Mattos-Guaraldi AL, Formiga LCD (1986) Agglutination of sheep erythrocytes by *Corynebacterium diphtheriae*. Braz J Med Biol Res 19:75–77

Mattos-Guaraldi AL, Formiga LCD (1991) Relationship of biotype and source to the hemagglutination and adhesive properties of *Corynebacterium diphtheriae*. Braz J Med Biol Res 24:399–406

Mattos-Guaraldi AL, Formiga LCD (1992) Hemagglutination patterns of *Corynebacterium diphtheriae*. Rev Microbiol São Paulo 23:8–12.

Mattos-Guaraldi AL, Formiga LCD (1998) Bacteriological properties of a sucrose fermenting *Corynebacterium diphtheriae* strain isolated from a case of endocarditis. Curr Microbiol 37:156–158

Mattos-Guaraldi AL, Formiga LCD, Andrade AFB (1998) *Trans*-sialidase activity for sialic acid incorporation on *Corynebacterium diphtheriae*. FEMS Microbiol Lett 168:167–172

Mattos-Guaraldi AL, Cappelli EA, Previato JO, Formiga LCD, Andrade AFB (1999a) Characterization of cell surface saccharides in two *Corynebacterium diphtheriae* strains. FEMS Microbiol Lett 170:159–166

Mattos-Guaraldi AL, Formiga LCD, Andrade AFB (1999b) Cell surface hydrophobicity of sucrose fermenting and non-fermenting *Corynebacterium diphtheriae* evaluated by different methods. Curr Microbiol 38:37–42

Mattos-Guaraldi AL, Formiga L, Pereira GA (2000) Cell surface components and adhesion in *Corynebacterium diphtheriae*. Microbes Infect 2:1507–1512

Mattos-Guaraldi AL, Formiga LCD, Camello TCF, Pereira GA, Hirata R Jr, Faria LMD, Halpern M (2001) *Corynebacterium diphtheriae* threats in cancer patients. Rev Argentina Microbiol 33:96–100

Mattos-Guaraldi AL, Moreira LO, Damasco PV, Hirata R Jr (2003) Diphtheria remains a threat to health in the developing world – An overview. Mem Inst Oswaldo Cruz 98:987–993

Mattos-Guaraldi AL, Sampaio JLM, Santos CS, Pimenta FP, Pereira GA, Pacheco LGC, Miyoshi A, Azevedo V, Moreira LO, Gutierrez FL, Costa JLF, Costa-Filho R, Damasco PV, Camello TCF, Hirata R Jr (2008) First detection of *Corynebacterium ulcerans* producing diphtheria-like toxin in human with pulmonary infection in Rio De Janeiro metropolitan area, Brazil. Mem Inst Oswaldo Cruz 103:396–400

McKean SC, Davies JK, Moore RJ (2007) Expression of phospholipase D, the major virulence factor of *Corynebacterium pseudotuberculosis*, is regulated by multiple environmental factors and plays a role in macrophage death. Microbiology 153:2203–2211

McNamara PJ, Cuevas WA, Songer JG (1995) Toxic phospholipases D of *Corynebacterium pseudotuberculosis, Corynebacterium ulcerans* and *Arcanobacterium haemolyticum*: cloning and sequence homology. Gene 156:113–118

Mekada E, Okada Y, Uchida T (1988) Identification of diphtheria toxin receptor and a nonproteinous diphtheria toxin-binding molecule in Vero cell membrane. J Cell Biol 107:511–519

Mills AE, Mitchell RD, Lim EK (1997) *Corynebacterium pseudotuberculosis* is a cause of human necrotising granulomatous lymphadenitis. Pathology 29:231–233

Mina NV, Burdz T, Wiebe D, Rai JS, Rahim T, Shing F, Hoang L, Bernard K (2011) Canada's first case of a multidrug-resistant *Corynebacterium diphtheriae* strain, isolated from a skin abscess. J Clin Microbiol 49:4003–4005

Moreira LO, Andrade AF, Vale MD, Souza SM, Hirata R Jr, Asad LM, Asad NR, Monteiro-Leal LH, Previato JO, Mattos-Guaraldi AL (2003) Effects of iron limitation on adherence and cell surface carbohydrates of *Corynebacterium diphtheriae* strains. Appl Environ Microbiol 69:5907–5913

Moreira LO, Mattos-Guaraldi AL, Andrade AF (2008) Novel lipoarabinomannan-like lipoglycan (CdiLAM) contributes to the adherence of *Corynebacterium diphtheriae* to epithelial cells. Arch Microbiol 190:521–530

Morris WE, Uzal FA, Cipolla AL (2005) Pyogranulomatous meningoencephalitis in a goat due to *Corynebacterium ulcerans*. Vet Rec 56:317–318

Mothershed EA, Cassiday PK, Pierson K, Mayer LW, Popovic T (2002) Development of a real-time fluorescence PCR assay for rapid detection of the diphtheria toxin gene. J Clin Microbiol 40:4713–4719

Nakamura LT, Wisnieski BJ (1990) Characterization of the deoxyribonuclease activity of diphtheria toxin. J Biol Chem 265:5237–5241

Nakao H, Popovic T (1997) Development of a direct PCR assay for detection of the diphtheria toxin gene. J Clin Microbiol 35:1651–1655

Neal S, Efstratiou A (2007) DIPNET—establishment of a dedicated surveillance network for diphtheria in Europe. Euro Surveill 12:9–12

Nigou J, Zelle-Rieser C, Gilleron M, Thurnher M, Puzo G (2001) Mannosylated lipoarabinomannans inhibit IL-12 production by human dendritic cells: evidence for a negative signal delivered through the mannose receptor. J Immunol 166:7477–7485

Nigou J, Gilleron M, Rojas M, García LF, Thurnher M, Puzo G (2002) Mycobacterial lipoarabinomannans: modulators of dendritic cell function and the apoptotic response. Microbes Infect 4:945–953

Nureki S, Miyazaki E, Matsuno O, Takenaka R, Ando M, Kumamoto T, Nakano T, Ohkusu K, Ezaki T (2007) *Corynebacterium ulcerans* infection of the lung mimicking the histology of Churg-Strauss syndrome. Chest 131:1237–1239

Olson ME, Goemans I, Bolingbroke D, Lundberg S (1988) Gangrenous dermatitis caused by *Corynebacterium ulcerans* in Richardson ground squirrels. J Am Vet Med Assoc 193:367–368

Ott L, Höller M, Gerlach RG, Hensel M, Rheinlaender J, Schäffer TE, Burkovski A (2010) *Corynebacterium diphtheriae* DIP1281 invasion-associated protein is involved in cell surface organization, adhesion and internalization in epithelial cells. BMC Microbiol 10:1–9

Ott L, McKenzie A, Baltazar MT, Britting S, Bischof A, Burkovski A, Hoskisson PA (2012) Evaluation of invertebrate infection models for pathogenic corynebacteria. FEMS Immunol Med Microbiol 65:413–421

Pallen MJ, Hay AJ, Puckey LH, Efstratiou A (1994) Polymerase chain reaction for screening clinical isolates of corynebacteria for the production of diphtheria toxin. J Clin Pathol 47:353–356

Pappenheimer AM Jr (1977) Diphtheria toxin. Ann Rev Biochem 46:69–94

Pappenheimer AM Jr (1993) The story of a toxic protein, 1888–1992. Protein Sci 2:292–298

Patey O, Bimet F, Emond JP, Estrangin E, Riegel PH, Halioua B, Dellion S, Kiredjian M (1995) Antibiotic susceptibilities of 38 non-toxigenic strains of *Corynebacterium diphtheriae*. J Antimicrob Chemother 36:1108–1110

Patey O, Bimet F, Riegel P, Halioua B, Emond JP, Estrangin E, Dellion S, Alonso JM, Kiredjian M, Dublanchet A, Lafaix C (1997) Coryne study group clinical and molecular study of *Corynebacterium diphtheriae* systemic infections in France. J Clin Microbiol 35:441–445

Paton MW, Sutherland SS, Rose IR, Hart RA, Mercy AR, Ellis TM (1995) The spread of *Corynebacterium pseudotuberculosis* infection to unvaccinated and vaccinated sheep. Aust Vet J 72:266–269

Peel MM, Palmer GG, Stacpoole AM, Kerr TG (1997) Human lymphadenitis due to *Corynebacterium pseudotuberculosis*: report of ten cases from Australia and review. Clin Infect Dis 24:185–191

Pereira GA, Pimenta FP, Santos FR, Damasco PV, Hirata R Jr, Mattos-Guaraldi AL (2008) Antimicrobial resistance among Brazilian *Corynebacterium diphtheriae* strains. Mem Inst Oswaldo Cruz 103:507–510

Perkins SL, Magdesian KG, Thomas WP, Spier SJ (2004) Pericarditis and pleuritis caused by *Corynebacterium pseudotuberculosis* in a horse. J Am Vet Med Assoc 224:1133–1138

Perkins SL, Cordery R, Nixon G, Abrahams A, Andrews J, White J, Efstratiou A, Anaraki S (2010) Investigations and control measures following a non-travel-associated case of toxigenic *Cornyebacterium diphtheriae*, London, United Kingdom, December 2009-January 2010. Euro Surveill 15:19544

Pethick FE, Lainson AF, Yaga R, Flockhart A, Smith DG, Donachie W, Cerdeira LT, Silva A, Bol E, Lopes TS, Barbosa MS, Pinto AC, Dos Santos AR, Soares SC, Almeida SS, Guimaraes LC, Aburjaile FF, Abreu VA, Ribeiro D, Fiaux KK, Diniz CA, Barbosa EG, Pereira UP, Hassan SS, Ali A, Bakhtiar SM, Dorella FA, Carneiro AR, Ramos RT, Rocha FS, Schneider MP, Miyoshi A, Azevedo V, Fontaine MC (2012a) Complete genome sequences of *Corynebacterium pseudotuberculosis* strains 3/99–5 and 42/02-A, isolated from sheep in Scotland and Australia, respectively. J Bacteriol 194:4736–4737

Pethick FE, Lainson AF, Yaga R, Flockhart A, Smith DG, Donachie W, Cerdeira LT, Silva A, Bol E, Lopes TS, Barbosa MS, Pinto AC, Dos Santos AR, Soares SC, Almeida SS, Guimaraes LC, Aburjaile FF, Abreu VA, Ribeiro D, Fiaux KK, Diniz CA, Barbosa EG, Pereira UP, Hassan SS, Ali A, Bakhtiar SM, Dorella FA, Carneiro AR, Ramos RT, Rocha FS, Schneider MP, Miyoshi A, Azevedo V, Fontaine MC (2012b) Complete genome sequence of *Corynebacterium pseudotuberculosis* strain 1/06-A, isolated from a horse in North America. J Bacteriol 194:4476

Pimenta FP, Hirata R Jr, Rosa AC, Milagres LG, Mattos-Guaraldi AL (2008a) A multiplex PCR assay for simultaneous detection of *Corynebacterium diphtheriae* and differentiation between non-toxigenic and toxigenic isolates. J Med Microbiol 57:1438–1439

Pimenta FP, Souza MC, Pereira GA, Hirata R Jr, Camello TC, Mattos-Guaraldi AL (2008b) DNase test as a novel approach for the routine screening of *Corynebacterium diphtheriae*. Lett Appl Microbiol 46:307–311

Poilane I, Fawaz F, Nathanson M, Cruaud P, Martin T, Collignon A, Gaudelus J (1995) *Corynebacterium diphtheriae* osteomyelitis in an immunocompetent child: a case report. Eur J Pediatr 154:381–383

Popovic T, Kombarova SY, Reeves MW, Nakao H, Mazurova IK, Wharton M, Wachsmuth K, Wenger JD (1996) Molecular epidemiology of diphtheria in Russia, 1985–1994. J Infect Dis 174:1064–1072

Popovic T, Mazurova IK, Efstratiou A, Vuopio-Varkila J, Reeves MW, Zoysa A de, Glushkevich T, Grimont P (2000) Molecular epidemiology of diphtheria. J Infect Dis 181:168–177.

Puissegur MP, Lay G, Gilleron M, Botella L, Nigou J, Marrakchi H, Mari B, Duteyrat JL, Guerardel Y, Kremer L, Barbry P, Puzo G, Altare F (2007) Mycobacterial lipomannan induces granuloma macrophage fusion via a TLR2-dependent, ADAM9- and beta1 integrin-mediated pathway. J Immunol 178:3161–3169

Quick ML, Sutter RW, Kobaidze K, Malakmadze N, Strebel PM, Nakashidze R, Murvanidze S (2000) Epidemic diphtheria in the Republic of Georgia, 1993–1996: risk factors for fatal out-come among hospitalized patients. J Infect Dis 181:130–137.

Romero-Perez JC, Suner-Machado M, Batista-Diaz N (2004) *Corynebacterium pseudotuberculosis* lymphadenitis in a young patient. Rev Clin Esp 204:388–389

Ruiz JC, D'Afonseca V, Silva A, Ali A, Pinto AC, Santos AR, Rocha AA, Lopes DO, Dorella FA, Pacheco LG, Costa MP, Turk MZ, Seyffert N, Moraes PM, Soares SC, Almeida SS, Castro TL, Abreu VA, Trost E, Baumbach J, Tauch A, Schneider MP, McCulloch J, Cerdeira LT, Ramos RT, Zerlotini A, Dominitini A, Resende DM, Coser EM, Oliveira LM, Pedrosa AL, Vieira CU, Guimarães CT, Bartholomeu DC, Oliveira DM, Santos FR, Rabelo ÉM, Lobo FP, Franco GR, Costa AF, Castro IM, Dias SR, Ferro JA, Ortega JM, Paiva LV, Goulart LR, Almeida JF, Ferro MI, Carneiro NP, Falcão PR, Grynberg P, Teixeira SM, Brommonschenkel S, Oliveira SC, Meyer R, Moore RJ, Miyoshi A, Oliveira GC, Azevedo V (2011) Evidence for reductive genome evolution and lateral acquisition of virulence functions in two *Corynebacterium pseudotuberculosis* strains. PLoS One 6:18551

Sabbadini PS, Genovez MR, Silva CF, Adelino TL, Santos CS, Pereira GA, Nagao PE, Dias AA, Mattos-Guaraldi AL, Hirata R Jr (2010) Fibrinogen binds to nontoxigenic and toxigenic *Corynebacterium diphtheriae* strains. Mem Inst Oswaldo Cruz 105:706–711

Sabbadini PS, Assis MC, Trost E, Gomes DL, Moreira LO, Santos CS, Pereira GA, Nagao PE, Azevedo VA, Hirata R Jr, Santos AL, Tauch A, Mattos-Guaraldi AL (2012) *Corynebacterium diphtheriae* 67–72p hemagglutinin, characterized as the protein DIP0733, contributes to invasion and induction of apoptosis in HEp-2 cells. Microb Pathog 52:165–176

Saikya L, Nath R, Saikya NJ, Choudhury G, Sarkar M (2010) A Diphtheria outbreak in Assam, India. South Asian J Trop Med Public Health 41:647–652

Santos CS, Santos LS dos, Souza MC, Santos Dourado F, Dias AASO, Sabbadini PS, Pereira GA, Cabral MC, Hirata R Jr, Mattos-Guaraldi AL (2010) Non-opsonic phagocytosis of homologous non-toxigenic and toxigenic *Corynebacterium diphtheriae* strains by human U-937 macrophages. Microbiol Immunol 54:1–10

Schauer R (1982) Chemistry, metabolism, and biological functions of sialic acids. Adv Carbohydr Chem Biochem 40:131–234

Schiller J, Groman N, Coyle M (1983) Plasmids in *Corynebacterium diphtheriae* mediating erythromycin resistance. Antimicrob Agents Chemother 18:814–821

Schuhegger R, Schoerner C, Dlugaiczyk J, Lichtenfeld I, Trouillier A, Zeller-Peronnet V, Busch U, Berger A, Kugler R, Hormansdorfer S, Sing A (2009) Pigs as source for toxigenic *Corynebacterium ulcerans*. Emerg Infect Dis 15:1314–1315

Selim SA (2001) Oedematous skin disease of buffalo in Egypt. J Vet Med B Infect Dis Vet Public Health 48:241–258

Serwold-Davis TM, Groman NB (1986) Mapping and cloning of *Corynebacterium diphtheriae* plasmid pNG2 and characterization of its relatedness to plasmids from skin coryneforms. Antimicrob Agents Chemother 30:69–72

Seto Y, Komiya T, Iwaki M, Kohda T, Mukamoto M, Tanahashi M, Kozaki S (2008) Properties of corynephage attachment site and molecular epidemiology of *Corynebacterium ulcerans* isolated from humans and animals in Japan. Jpn J Infect Dis 61:116–122

Seyffert N, Guimaraes AS, Pacheco LG, Portela RW, Bastos BL, Dorella FA, Heinemann MB, Lage AP, Gouveia AM, Meyer R, Miyoshi A, Azevedo V (2010) High seroprevalence of caseous lymphadenitis in Brazilian goat herds revealed by *Corynebacterium pseudotuberculosis* secreted proteins-based ELISA. Res Vet Sci 88:50–55

Sharma NC, Banavaliker JN, Ranjan R, Kumar R (2007) Bacteriological and epidemiological characteristics of diphtheria cases in and around Delhi -a retrospective study. Indian J Med Res 126:545–552

Sharpe AE, Brady CP, Johnson AJ, Byrne W, Kenny K, Costello E (2010) Concurrent outbreak of tuberculosis and caseous lymphadenitis in a goat herd. Vet Rec 166:591–592

Silva A, Schneider MP, Cerdeira L, Barbosa MS, Ramos RT, Carneiro AR, Santos R, Lima M, D'Afonseca V, Almeida SS, Santos AR, Soares SC, Pinto AC, Ali A, Dorella FA, Rocha F, Abreu VA de, Trost E, Tauch A, Shpigel N, Miyoshi A, Azevedo V (2011) Complete genome sequence of *Corynebacterium pseudotuberculosis* I19, a strain isolated from a cow in Israel with bovine mastitis. J Bacteriol 193:323–324

Silva de Souza SM, Hirata R Jr, Moreira LO, Gomes ML, Braga De Andrade AF, Bernardo-Filho M, Mattos-Guaraldi AL (2003) Influence of stannous chloride on the adhesive properties of *Corynebacterium diphtheriae* strains. Int J Mol Med 12:657–661

Singh J, Harit AK, Jain DC, Panda RC, Tewari KN, Bhatia R (1999) Diphtheria is declining but continues to kill many children: analysis of data from a sentinel center in Delhi. Epidemiol Infect 123:209–215

Soares SC, Abreu VA, Ramos RT, Cerdeira L, Silva A, Baumbach J, Trost E, Tauch A, Hirata R Jr, Mattos-Guaraldi AL, Miyoshi A, Azevedo V (2012) PIPS: pathogenicity island prediction software. PLoS One 7:30848

Soriano F, Huelves L, Naves P, Rodríguez-Cerrato V, del Prado G, Ruiz V, Ponte C (2009) *In vitro* activity of ciprofloxacin, moxifloxacin, vancomycin and erythromycin against planktonic and biofilm forms of *Corynebacterium urealyticum*. J Antimicrob Chemother 63:353–356

Spier SJ, Leutenegger CM, Carroll SP, Loye JE, Pusterla JB, Carpenter TE, Mihalyi JE, Madigan JE (2004) Use of a real-time polymerase chain reaction-based fluorogenic 5' nuclease assay to evaluate insect vectors of *Corynebacterium pseudotuberculosis* infections in horses. Am J Vet Res 65:829–834

Stefanska I, Rzewuska M, Binek M (2008) Evaluation of three methods for DNA fingerprinting of *Corynebacterium pseudotuberculosis* strains isolated from goats in Poland. Pol J Microbiol 57:105–112

Tachedjian M, Krywult J, Moore RJ, Hodgson AL (1995) Caseous lymphadenitis vaccine development: site-specific inactivation of the *Corynebacterium pseudotuberculosis* phospholipase D gene. Vaccine 13:1785–1792

Tambourgi DV, Da Silva MS, Billington SJ, Gonçalves De Andrade RM., Magnoli FC, Songer JG, Van Den Berg CW (2002) Mechanism of induction of complement susceptibility of erythrocytes by spider and bacterial sphingomyelinases. Immunology 107:93–101

Tarello W, Theneyan M (2008) *Corynebacterium pseudotuberculosis* and *Corynebacterium renale* isolated from two Arabian oryx (*Oryx leucoryx*). Vet Rec 162:862–863

Tashjian JJ, Campbell SG (1983) Interaction between caprine macrophages and *Corynebacterium pseudotuberculosis*: an electron microscopy study. Am J Vet Res 44:690–693

Taylor J, Saveedra-Campos M, Harwood D, Pritchard G, Raphaely N, Kapadia S, Efstratiou A, White J, Balasegaram S (2010) Toxigenic *Corynebacterium ulcerans* infection in a veterinary student in London, United Kingdon, may 2010. Euro Surveill 15:1–3

Tejedor MT, Martin J, Lupiola P, Gutierrez C (2000) Caseous lymphadenitis caused by *Corynebacterium ulcerans* in the dromedary camel. Can Vet J 41:126–127

Tejedor-Junco MT, Lupiola P, Schulz U, Gutierrez C (2008) Isolation of nitrate-reductase positive *Corynebacterium pseudotuberculosis* from dromedary camels. Trop Anim Health Prod 41:711–714

Tiley S, Kociuba KR, Heron LG, Munro R (1993) Infective endocarditis due to non toxigenic *Corynebacterium diphtheriae*: report of seven cases and review. Clin Infect Dis 16:271–275

Tiwari TS, Golaz A, Yu DT, Ehresmann KR, Jones TF, Hill HE, Cassiday PK, Pawloski LC, Moran JS, Popovic T, Wharton M (2008) Investigations of 2 cases of diphtheria-like illness due to toxigenic *Corynebacterium ulcerans*. Clin Infect Dis 46:395–401

Ton-That H, Schneewind O (2003) Assembly of pili on the surface of *Corynebacterium diphtheriae*. Mol Microbiol 50:1429–1438

Torres LFC, Ribeiro D, Hirata R Jr, Pacheco LGC, Souza MC, Santos LS, Santos CS dos, Salah M, Costa MM, Ribeiro MG, Selim SA, Azevedo VAC, Mattos-Guaraldi AL (2013) Multiplex PCR for identification and toxigenicity of *Corynebacterium* group of zoonotic potential and an overview of human and animal infections. Mem Inst Oswaldo Cruz. (Accepted for publication)

Trost E, L O, Schneider J, Schröder J, Jaenicke S, Goesmann A, Husemann P, Stoye J, Dorella FA, Rocha FS, Soares Sde C, D'Afonseca V, Miyoshi A, Ruiz J, Silva A, Azevedo V, Burkovski A, Guiso N, Join-Lambert OF, Kayal S, Tauch A (2010) The complete genome sequence of *Corynebacterium pseudotuberculosis* FRC41 isolated from a 12-year-old girl with necrotizing lymphadenitis reveals insights into gene-regulatory networks contributing to virulence. BMC Genomics 11:728

Trost E, Al-Dilaimi A, Papavasiliou P, Schneider J, Viehoever P, Burkovski A, Soares SC, Almeida SS, Dorella FA, Miyoshi A, Azevedo V, Schneider MP, Silva A, Santos CS, Santos LS, Sabbadini P, Dias AA, Hirata R Jr, Mattos-Guaraldi AL, Tauch A (2011) Comparative analysis of two complete *Corynebacterium ulcerans* genomes and detection of candidate virulence factors. BMC Genomics 12:383

Trost E, Blom J, Soares Sde C, Huang IH, Al-Dilaimi A, Schröder J, Jaenicke S, Dorella FA, Rocha FS, Miyoshi A, Azevedo V, Schneider MP, Silva A, Camello TC, Sabbadini PS, Santos CS, Santos LS, Hirata R Jr, Mattos-Guaraldi AL, Efstratiou A, Schmitt MP, Ton-That H, Tauch A (2012) Pangenomic study of *Corynebacterium diphtheriae* that provides insights into the genomic diversity of pathogenic isolates from cases of classical diphtheria, endocarditis, and pneumonia. J Bacteriol 194:3199–3215

Vera LA, Macêdo VO, Magalhães AV, Ciuffo IA, Santos CG, Santos JB (2002) Úlceras leishmanióticas cutâneas com presença de *Corynebacterium diphtheriae*. Rev Soc Bras Med Trop 35:311–313

Viguetti SZ, Pacheco LG, Santos LS, Soares SC, Bolt F, Baldwin A, Dowson CG, Rosso ML, Guiso N, Miyoshi A, Hirata R Jr, Mattos-Guaraldi AL, Azevedo V (2012) Multilocus sequence types of invasive *Corynebacterium diphtheriae* isolated in the Rio de Janeiro urban area, Brazil. Epidemiol Infect 140:617–220

Vitek C, Wharton M (2008) Diphtheria toxoid. In: Plotkin SA, Orenstein WA, Offit PA (eds) Vaccines 5th ed. Saunders, Philadelphia, pp 139–156

Vitek CR, Wharton M (1998) Diphtheria in the Former Soviet Union: Reemergence of a pandemic disease. Emerg Infect Dis 4:539–550

Wagner J, Ignatius R, Voss S, Höpfner V, Ehlers S, Funke G, Weber U, Hahn H (2001) Infection of the skin caused by *Corynebacterium ulcerans* and mimicking classical cutaneous diphtheria. Clin Infect Dis 33:1598–1600

Wagner KS, White JM, Lucenko I, Mercer D, Crowcroft NS, Neal S, Efstratiou A (2012) Members of the Diphtheria Surveillance Network Diphtheria in the Postepidemic Period, Europe, 2000–2009. Emerg Infect Dis 18:218–235

Wagner KS, White JM, Neal S, Crowcroft NS, De Martin S, Mann G, Efstratiou A (2010) Diphtheria in the United Kingdom, 1986–2008: the increasing role of *Corynebacterium ulcerans*. Epidemiol Infect 138:1519–1530

Wagner KS, White JM, Neal S, Crowcroft NS, Kupreviciene N, Paberza R, Lucenko I, Joks U, Akbas E, Alexandrou-Athanassoulis H, Detcheva A, Vuopio J, Hunolstein C von, Murphy PG, Andrews N, Efstratiou A (2011) Screening for *Corynebacterium diphtheriae* and *Corynebacterium ulcerans* in patients with upper respiratory tract infections 2007–2008: a multicentre European study. Clin Microbiol Infect 17:519–525

Wang J, London E (2009) The membrane topography of the diphtheria toxin T domain linked to the A chain reveals a transient transmembrane hairpin and potential translocation mechanisms. BioChemistry 48:10446–10456

Wellinghausen N, Sing A, Kern WV, Perner S, Marre R, Rentschler J (2002) A fatal case of necrotizing sinusitis due to toxigenic *Corynebacterium ulcerans*. Int J Med Microbiol 292:59–63

Werdmuller BF, Brakman M, Vreede RW (1996) A tropical ulcer; cutaneous diphtheria. Ned Tijdschr Geneeskd 140:2414–2416

Williamson LH (2001) Caseous lymphadenitis in small ruminants. Vet. Clin. North Am. Food Anim Pract 17:359–371

Yamaizumi M, Mekada E, Uchida T, Okada Y (1978) One molecule of diphtheria toxin fragment A introduced into a cell can kill the cell. Cell 15:245–250

Yanagawa R, Honda E (1976) Presence of pili in species of human and animal parasites and pathogens of the genus *corynebacterium*. Infect Immun 13:1293–1295

Yeruham I, Braverman Y, Shpigel NY, Chizov-Ginzburg A, Saran A, Winkler M (1996) Mastitis in dairy cattle caused by *Corynebacterium pseudotuberculosis* and the feasibility of transmission by houseflies. I Vet Q 18:87–89

Yeruham I, Friedman S, Perl S, Elad D, Berkovich Y, Kalgard Y (2004) A herd level analysis of a *Corynebacterium pseudotuberculosis* outbreak in a dairy cattle herd. Vet Dermatol 15:315–320

Zasada AA, Baczewska-Rej M, Wardak S (2010) An increase in non-toxigenic *Corynebacterium diphtheriae* infections in Poland—molecular epidemiology and antimicrobial susceptibility of strains isolated from past outbreaks and those currently circulating in Poland. Int J Infect Dis 14:907–912

Zuber PLF, Gruner E, Altwegg M, Graevenitz A von (1992) Invasive infection with non-toxigenic *Corynebacterium diphtheriae* among drug users. Lancet 339:1359

Chapter 3
Comparative Genomics and Pathogenicity Islands of *Corynebacterium diphtheriae*, *Corynebacterium ulcerans*, and *Corynebacterium pseudotuberculosis*

Eva Trost and Andreas Tauch

Abstract The systematic application of next-generation DNA sequencing technologies has provided detailed insights into the genomics of corynebacteria. The genomes of 13 *Corynebacterium diphtheriae* strains isolated from cases of classical diphtheria, endocarditis and pneumonia were completely sequenced and annotated, providing first insights into the pan-genome of this species. Comparative gene content analyses revealed an enormous collection of variable pilus gene clusters relevant for adhesion properties of *C. diphtheriae*. Variation in the distributed genome is apparently a common strategy of *C. diphtheriae* to establish differences in host-pathogen interactions. Molecular data deduced from the complete genome sequences of two *Corynebacterium ulcerans* strains provided considerable knowledge of candidate virulence factors, including a novel type of ribosome-binding protein with striking structural similarity to Shiga-like toxins. Likewise, functional data deduced from the complete genome sequences of six *Corynebacterium pseudotuberculosis* isolates from various sources greatly extended the knowledge of virulence factors and indicated that this species is equipped with a distinct gene set promoting its survival under unfavorable environmental conditions encountered in the mammalian host.

Keywords Genome sequence · Core genome · Pan-genome · Pathogenicity island · Virulence factor

---

A. Tauch (✉) · E. Trost
Institut für Genomforschung und Systembiologie, Centrum für Biotechnologie, Universität Bielefeld, Universitätsstraße 27, 33615, Bielefeld, Germany
e-mail: tauch@cebitec.uni-bielefeld.de

E. Trost
e-mail: etrost@cebitec.uni-bielefeld.de

A. Burkovski (ed.), *Corynebacterium diphtheriae and Related Toxigenic Species*, DOI 10.1007/978-94-007-7624-1_3,

3.1 Introduction

A new generation of DNA sequencing approaches, collectively called next-generation DNA sequencing technologies, has provided unprecedented opportunities for high-throughput functional genome research (Mardis 2008; Shendure and Ji 2008). Since first introduced to the market in 2005, these technologies have been used for standard sequencing applications, such as whole-genome sequencing and genome resequencing, and for novel applications previously unexplored by the 'classical' Sanger sequencing strategy (Morozova and Marra 2008). Despite the many advances in chemistries and the robust performance of modern Sanger sequencers, the application of this relatively expensive method to large genome sequencing projects has remained beyond the means of the typical grant-funded investigator. An inherent limitation of Sanger sequencing is the requirement of *in vivo* amplification of DNA fragments that are to be sequenced, which is usually achieved by cloning into bacterial hosts (Morozova and Marra 2008; Shendure and Ji 2008). The Roche/454 technology, the first next-generation DNA sequencing technology released to the market, circumvents the cloning requirement by taking advantage of a highly efficient *in vitro* DNA amplification method known as emulsion PCR (Rothberg and Leamon 2008; Droege and Hill 2008). Moreover, the use of the picotiter plate system in the Roche/454 approach allows hundreds of thousands of pyrosequencing reactions to be carried out in parallel, massively increasing the sequencing throughput (Droege and Hill 2008).

Recent scientific discoveries in the field of corynebacterial genomics resulted from the systematic application of next-generation DNA sequencing technologies; i.e. the Roche/454 Genome Sequencer FLX System and the Life Technologies SOLiD System (Tauch et al. 2008a; Cerdeira et al. 2011a). Not surprisingly, the first next-generation sequencing studies have focused on the genomes of corynebacterial pathogens because of their importance in human disease, including *Corynebacterium urealyticum*, *Corynebacterium kroppenstedtii*, *Corynebacterium aurimucosum* and *Corynebacterium resistens* (Soriano and Tauch 2008; Tauch et al. 2008a, b; Trost et al. 2010a; Schröder et al. 2012). The Genomes OnLine Database GOLD (Pagani et al. 2012) lists additional corynebacterial species, whose genomes have been sequenced to high-quality draft status in the course of the human microbiome project (Lewis et al. 2012).

This chapter describes the current status of genome sequencing projects of the three closely related pathogenic species *Corynebacterium diphtheriae*, *Corynebacterium ulcerans* and *Corynebacterium pseudotuberculosis*, the so-called 'diphtheria' group, and summarizes the major findings of comparative genomic studies, thereby focussing on the detection of pathogenicity islands and virulence factors. To date, complete genome sequences of 21 strains have been published and sequencing of more clinical isolates is currently ongoing (Pagani et al. 2012). General features of the completely sequenced genomes of *C. diphtheriae*, *C. ulcerans* and *C. pseudotuberculosis* are listed in Table 3.1. The deduced genomic data considerably improve our understanding of the architecture and evolution of corynebacterial genomes, species-specific traits and potential factors contributing to pathogenicity in humans and animals.

Table 3.1 Overview of completely sequenced corynebacterial isolates and general features of the genome sequences

Strain	Genome size (bp)	G+C content (%)	No. of genes	No. of tRNAs	No. of rRNA operons (16S-23S-5S)	No. of unique genes	Types of CRISPRs (number of repeats)	GenBank Accession No.	Reference
Corynebacterium diphtheriae									
NCTC 13129	2,488,635	53.5	2,368	54	5	124	I (7); II (26)	BX248353	Cerdeño-Tarrága et al. 2003
C7(β)^{tox}+	2,499,189	53.5	2,350	56	5	126	I (6)	CP003210	Trost et al. 2012
PW8	2,530,683	53.7	2,361	53	5	101	III (15)	CP003216	Trost et al. 2012
CDC-E8392	2,433,326	54.6	2,270	54	5	52	III (12)	CP003211	Trost et al. 2012
31A	2,535,346	53.6	2,402	51	5	104	I (28)	CP003206	Trost et al. 2012
241	2,426,551	53.4	2,260	50	5	6	I (15); II (4)	CP003207	Trost et al. 2012
VA01	2,395,441	53.4	2,196	50	5	27	I (7)	CP003217	Trost et al. 2012
HC01	2,427,149	53.4	2,260	53	5	7	I (15); II (4)	CP003212	Trost et al. 2012
HC02	2,468,612	53.7	2,244	53	5	69	I (5)	CP003213	Trost et al. 2012
HC03	2,478,364	53.5	2,268	50	5	35	III (42)	CP003214	Trost et al. 2012
HC04	2,484,332	53.5	2,280	50	5	13	III (15)	CP003215	Trost et al. 2012
INCA 402	2,449,071	53.6	2,235	50	5	44	III (17)	CP003208	Trost et al. 2012
BH8	2,485,519	53.6	2,375	53	5	85	I (1)	CP003209	Trost et al. 2012
Corynebacterium ulcerans									
809	2,502,095	53.3	2,182	52	4	90	IV (28); V (12); VI (67)	CP002790	Trost et al. 2011
BR-AD22	2,606,374	53.4	2,338	52	4	132	IV (38); V (10); VI (32)	CP002791	Trost et al. 2011
Corynebacterium pseudotuberculosis									
FRC41	2,337,914	52.2	2,110	49	4	49	IV (1)	CP002097	Trost et al. 2010b
I19	2,337,730	52.2	2,124	49	4	1	IV (1)	CP002251	Silva et al. 2011
1002	2,335,112	52.2	2,111	48	4	1	IV (1)	CP001809	Ruiz et al. 2011
C231	2,328,208	52.2	2,103	48	4	4	IV (1)	CP001829	Ruiz et al. 2011
PAT10	2,335,323	52.2	2,079	48	4	4	IV (1)	CP002924	Cerdeira et al. 2011b
CIP 52.97	2,320,595	52.1	2,057	47	4	86	–	CP003061	Cerdeira et al. 2011c

3.2 The Pan-Genome of *C. diphtheriae* and Deduced Pathogenicity Islands

3.2.1 The Reference Genome of C. diphtheriae *NCTC 13129*

The first genome of the 'diphtheria' group to be sequenced was that of *C. diphtheriae* NCTC 13129, which was initially isolated from a pharyngeal membrane of a patient with clinical diphtheria (Cerdeño-Tarrága et al. 2003). This toxigenic strain is a representative of the clone responsible for an outbreak of diphtheria in the states of the former Soviet Union in the 1990s (Dittmann et al. 2000). The whole-genome shotgun method with Sanger technology has been applied to determine the genome sequence of *C. diphtheriae* NCTC 13129 (Cerdeño-Tarrága et al. 2003). The complete genome sequence derived from two genomic shotgun libraries and terminal sequences from a large-insert bacterial artificial chromosome (BAC) library that was used for generating a scaffold. The genome of *C. diphtheriae* NCTC 13129 has a size of 2,488,635 bp with a G+C content of 53.5 % and contains 2,320 predicted coding regions, of which 45 were annotated as pseudogenes (Cerdeño-Tarrága et al. 2003). Very recently, a comprehensive re-annotation of this genome sequence has been performed as a new approach to make the *C. diphtheriae* NCTC 13129 reference genome more descriptive and current with relevant features regarding the organism's lifestyle (Salzberg 2007; D'Afonseca et al. 2012). This *in silico* strategy is facilitated by the massive amounts of publicly available data linked to sequenced genomes of other species of the genus *Corynebacterium* (Pagani et al. 2012). With respect to structural genomics of *C. diphtheriae* NCTC 13129, 23 protein-coding regions were deleted and 71 new genes were added to the initial genome annotation (D'Afonseca et al. 2012). Nevertheless, all gene regions previously assigned as pseudogenes were validated and ten new pseudogenes were created. In relation to functional genomics, about 57 % of the initial genome annotation was updated to become functionally more informative, as the product descriptions of 973 predicted proteins were updated. Among them, 370 gene products previously annotated as 'hypothetical proteins' now have more informative descriptions (D'Afonseca et al. 2012). The re-annotation resulted in the discovery of new genes in the *C. diphtheriae* NCTC 13129 genome sequence, correction of coding strands and the significant improvement of functional description of protein-coding regions, including classical virulence genes (D'Afonseca et al. 2012). The re-annotated archives of *C. diphtheriae* NCTC 13129 are available at: http://lgcm.icb.ufmg.br/pub/C_diphtheriae_reannotation.embl.

Genomic islands in the genome of *C. diphtheriae* NCTC 13129 were detected by examining local anomalies in the nucleotide composition of the DNA, such as G+C content, GC skew and/or dinucleotide frequency deviations that can be indicative of the recent acquisition of DNA regions by horizontal gene transfer (Cerdeño-Tarrága et al. 2003). The most prominent genomic island of *C. diphtheriae* NCTC 13129 comprises the complete genome of a tox^{+} corynephage encoding diphtheria toxin. This prophage has a size of 36,566 bp with a G+C content of 52.2 % and encodes

43 predicted proteins. Sequence similarities on the amino acid level were detected to proteins of phage BFK20 from *Brevibacterium flavum* (Bukovska et al. 2006). The diphtheria toxin gene *tox* is located at the right end of the prophage genome, adjacent to the attachment site and within a DNA region of low G+C content. This specific location of *tox* is indicative of a bacterial gene that was acquired from a previous host and is dispensable for the life cycle of the phage, but may affect the phenotype or fitness of the lysogenic bacterium (Brüssow et al. 2004). In addition to the tox^+ corynephage (PICD 1), twelve genomic regions (PICD 2–13) with local anomalies in the nucleotide composition were detected in *C. diphtheriae* NCTC 13129 (Table 3.2). Several genes potentially involved in pathogenicity of *C. diphtheriae* NCTC 13129 are located on the detected genomic islands. These putative pathogenicity islands encode, for instance, a siderophore biosynthesis and export system, a putative lantibiotic biosynthesis system, and three types of sortase-related adhesive pili (Table 3.2). It is therefore likely that *C. diphtheriae* NCTC 13129 has recently acquired by horizontal transfer specialized genes that may be involved in the pathogenic lifestyle by encoding variable pilus structures for the adherence of the bacterium to host cell surfaces (Cerdeño-Tarrága et al. 2003).

3.2.2 The Pan-Genome of the Species **C. diphtheriae**

Very recently, the knowledge of the gene content of *C. diphtheriae* isolates was considerably extended, as the genomes of twelve clinical strains initially recovered from cases of classical diphtheria, endocarditis, and pneumonia were completely sequenced and annotated (Trost et al. 2012). The selected collection of *C. diphtheriae* strains (Table 3.1) includes the prominent ancestor of many toxoid vaccine producers *C. diphtheriae* PW8 (Park and Williams 1896) and the laboratory strain *C. diphtheriae* C7(β)$^{tox+}$ (Freeman 1951; Barksdale and Pappenheimer 1954). Including the genome sequence of the reference strain *C. diphtheriae* NCTC 13129, a comparative analysis of these genomes allowed the first characterization of the pan-genome of the species *C. diphtheriae* (Trost et al. 2012). The microbial pan-genome is defined as the total gene repertoire in a bacterial species and comprises the 'core genome', which is shared by all individuals, the 'dispensable genome' containing genes present only in a subset of individuals, and the 'unique genome', which is unique to an individual (Medini et al. 2005; Tettelin et al. 2008).

The twelve *C. diphtheriae* genomes were sequenced by pyrosequencing with the Roche/454 Genome Sequencer FLX System and sequencing depths ranging from 29× to 55× (Trost et al. 2012). All genomic sequences were assembled to circular chromosomes with 2.395 Mb to 2.535 Mb in size (Table 3.1). The average G+C content of each genome is in the range of 53 %, which is consistent with the G+C content of the reference genome of *C. diphtheriae* NCTC 13129 (Cerdeño-Tarrága et al. 2003). The annotation of the twelve *C. diphtheriae* genomes and re-annotation of the *C. diphtheriae* NCTC 13129 genome sequence (D'Afonseca et al. 2012) revealed a median number of 2,294 protein-coding genes for each strain, with the lowest number of 2,196 genes annotated in the genome of *C. diphtheriae* VA01 and the highest number of 2,402 genes in *C. diphtheriae* 31A (Table 3.1). A comparative gene content analysis showed that the mean number of genes shared by two strains comprises 1,903 ± 54 orthologous genes, while the mean number of genes not shared by a distinct pair of strains comprises 644 ± 134 genes, indicating the large variability of the gene repertoire in the sequenced *C. diphtheriae* isolates (Trost et al. 2012).

The number of core genes of *C. diphtheriae* was determined with the software EDGAR using bidirectional best BLASTP hits for genome comparisons (Blom et al. 2009). Based on a series of calculations using all *C. diphtheriae* genomes individually as a reference, the core genome of the sequenced *C. diphtheriae* strains comprises 1,632 genes that can therefore be regarded as highly conserved in this species (Trost et al. 2012). To deduce the development of the core genome in dependence on the number of sequenced *C. diphtheriae* strains, the median number of core genes in each genome was calculated based on the permutation of all possible genome comparisons. According to this approach, the number of core genes present in *C. diphtheriae* will comprise about 1,611 protein-coding genes when adding further genome sequences to the current data set. This value revealed a genetic backbone of the *C. diphtheriae* genome, which includes approximately 70 % of the gene repertoire of the sequenced strains, with about 30 % of the gene content being variable to some extent and therefore belonging to the dispensable portion of the *C. diphtheriae* genome. The full complement of protein-coding regions that are part of the dispensable genome of *C. diphtheriae* was determined as 2,361 distributed genes (Trost et al. 2012).

The bioinformatic characterization of the unique genome (Table 3.1) revealed the average number of 61 ± 43 strain-specific genes per sequenced *C. diphtheriae* isolate (Trost et al. 2012). To deduce the development of the number of unique genes in dependence on the number of sequenced *C. diphtheriae* genomes, the median number of strain-specific coding regions was determined using the permutation of all possible genome comparisons. The respective calculation indicated that the median number of unique genes estimated to occur in additionally sequenced *C. diphtheriae* genomes comprises about 65 genes. Accordingly, the sum total of protein-coding regions representing the pan-genome of *C. diphtheriae* currently comprises 4,786 genes, which is about three times the size of the deduced core genome (Trost et al. 2012). This calculation was corroborated by applying Heaps' law: $n = \kappa \times N^{\gamma}$, with N being the number of sequenced genomes (Tettelin et al. 2008). Hence, the number of protein-coding regions added to the pan-genome of *C. diphtheriae* will increase by 69 genes per newly sequenced genome, indicating an open pan-genome for the species *C. diphtheriae* (Trost et al. 2012). In general, a microbial pan-genome can be classified as 'closed' or 'open' (Tettelin et al. 2008). A pan-genome is considered to be closed, if the number of new genes added per newly sequenced genome converges to zero. Therefore, a closed microbial pan-genome indicates a static gene content of a bacterial species that is no longer expendable by genome sequencing. On the other hand, a pan-genome is considered open when each newly sequenced strain can be expected to reveal some genes unique within the species, regardless of the number of already analyzed genomes. An open pan-genome is therefore associated with a dynamic gene content of a bacterial species (Halachev et al. 2011).

3.2.3 Genetic Variability of CRISPR/cas Regions in C. diphtheriae

Due to the genetic diversity of *C. diphtheriae* isolates, a number of typing methods have been established for inter-strain differentiation, such as amplified fragment length polymorphism analysis, multilocus enzyme electrophoresis, pulsed-field gel electrophoresis, ribotyping, randomly amplified polymorphic DNA analysis (Mokrousov 2009), and multi-locus sequence typing (Jolley et al. 2004). These methods allow the identification of clonal groups of closely related *C. diphtheriae* strains with different sensitivities. A newer method to determine the phylogenetic relationship of *C. diphtheriae* strains is the so-called spoligotyping (spacer oligonucleotide typing), which is based on the presence of arrays of clustered regularly interspaced short palindromic repeats (CRISPRs) in the genome sequence (Mokrousov et al. 2005). These arrays are composed of direct repeats that are separated by non-repetitive, similar-sized spacer sequences (Deveau et al. 2010). CRISPRs and associated *cas* genes represent a widespread genetic system across bacteria that causes RNA interference against foreign nucleic acids, for instance resistance to bacteriophages (Deveau et al. 2010; Marraffini and Sontheimer 2010). The CRISPR/*cas* system of *C. diphtheriae* therefore participates in a constant evolutionary battle between the bacterium and corynephages through the addition or deletion of spacer sequences in the bacterial genome and mutations or deletion in phage genomes. Targets for spoligotyping are the spacer regions between the direct repeats, as variations in the number or nucleotide sequence of spacers provide patterns for the differentiation between clonal groups of *C. diphtheriae* isolates (Mokrousov et al. 2005). In a macroarray-based approach of spoligotyping, 154 clinical *C. diphtheriae* strains were subdivided into 34 spoligotypes (Mokrousov et al. 2005).

Three types of CRISPR/*cas* systems were detected in the genomes of the sequenced *C. diphtheriae* strains (Fig. 3.1). A detailed classification of the CRISPR/*cas* regions is listed in Table 3.1. CRISPR/*cas* type I was detected in the genomes of eight strains and is composed of three *cas* genes (*cas1* to *cas3*). The number of associated spacer sequences ranges from one to 28. CRISPR/*cas* type II is additionally present in three *C. diphtheriae* genomes and contains eight *cas* genes (*cas4* to *cas11*). The number of repeats in these arrays ranged from four to 26. CRISPR/*cas* type III is present in five genomes, with varying numbers of repeats ranging from 12 to 42. The type III CRISPR/*cas* region is flanked by eight *cas* genes (*cas12* to *cas19*). A nucleotide sequence comparison of the identified spacer sequences revealed that only 48 out of the 219 spacers are shared by two or three *C. diphtheriae* strains, supporting the view that CRISPR/*cas* regions provide an attractive target for the solid discrimination between different *C. diphtheriae* isolates (Mokrousov et al. 2005; Trost et al. 2012).

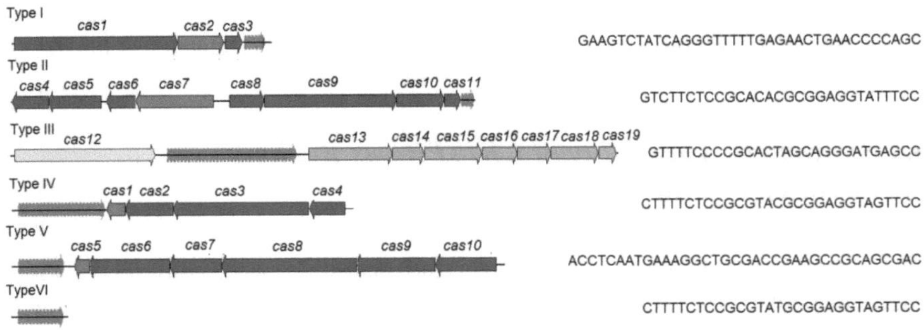

Fig. 3.1 Schematic representation of CRISPR/*cas* regions detected in the genomes of *C. diphtheriae*, *C. ulcerans*, and *C. pseudotuberculosis*. The CRISPR/*cas* regions show different numbers and arrangements of *cas* genes (*labeled arrows*). The position of the CRISPR locus is also variable. The nucleotide sequences of the conserved repeats are shown. CRISPR types I–III were detected in the sequenced *C. diphtheriae* genomes. CRISPR types IV–VI were found in the genomes of *C. ulcerans* isolates. CRISPR type VI of *C. ulcerans* is lacking associated *cas* genes. *C. pseudotuberculosis* isolates contain only CRISPR type IV. See also Table 3.1 for a detailed classification of CRISPR/*cas* regions

3.2.4 *Genetic Variability of tox*$^+$ *Corynephages in* C. diphtheriae

The pan-genome project also provided more detailed information about the genetic variability of corynephages harboring the diphtheria toxin gene *tox* that was identified in *C. diphtheriae* NCTC 13129 (Cerdeño-Tarrága et al. 2003) and in *C. diphtheriae* strains C7(β)$^{tox+}$, CDC-E8392, PW8 and 31A (Trost et al. 2012). In the case of *C. diphtheriae* PW8, two non-tandem copies of the corynephage ω^{tox+} were detected in the complete genome sequence, as suggested previously from restriction endonuclease maps of phage DNA (Rappuoli et al. 1983). The 36-kb genome sequences of both ω^{tox+} corynephages are almost identical, as they show only five nucleotide mismatches. Both copies of the prophage are separated by a 2-kb genomic region encoding a putative membrane protein that is flanked by two copies of a tRNA^{Arg} gene representing the known attachment site of corynephages in *C. diphtheriae* (Ratti et al. 1997). Nucleotide sequence comparisons of the *tox*$^+$ corynephages revealed that the ω^{tox+} phage of *C. diphtheriae* PW8 is similar to the β^{tox+} phage present in *C. diphtheriae* C7(β)$^{tox+}$, which is consistent with an early report demonstrating that both corynephages differ in only three genomic regions (Rappuoli et al. 1983). A highly different *tox*$^+$ prophage was detected in the genome sequence of *C. diphtheriae* 31A (Trost et al. 2012). Significant nucleotide sequence similarity to β-like corynephages was observed only at the right-hand end of the prophage genome, which harbors the *tox* gene region. Other regions of the prophage genome revealed homology at the amino acid level to proteins of the prophage ΦCULC22IV, which is present in the *tox*$^-$ strain *C. ulcerans* BR-AD22 (Trost et al. 2011). It has been proposed previously that the diphtheria toxin gene *tox* was acquired by corynephage β due to the terminal location of this gene in the prophage and the significantly decreased G + C content of this region in the phage genome (Cerdeño-Tárraga et al. 2003; Brüssow et al. 2004). The detection of identical *tox* genes in different prophages now indicates that the acquisition of the diphtheria toxin gene *tox* occurred independently in two different corynephages or that gene shuffling is frequently found in this group of phages (Trost et al. 2012).

3.2.5 Pathogenicity Islands and Pilus Gene Clusters of C. diphtheriae

The plasticity of the *C. diphtheriae* genome was analyzed by two comparative approaches designed to detect differences in the repertoire of pathogenicity islands that were initially assigned in the reference genome of *C. diphtheriae* NCTC 13129 (Cerdeño-Tárraga et al. 2003). The distribution of pathogenicity islands PICD 3 and PICD 8 was investigated by a PCR-based approach in eleven *C. diphtheriae* strains (Soares et al. 2011). The pathogenicity island PICD 8 was detected in only one strain, *C. diphtheriae* HC01 (Table 3.1), whereas PICD 3 was more widely distributed and present in six *C. diphtheriae* strains. This data indicated that the pathogenicity islands of *C. diphtheriae* strains can be differentiated by their variable genomic stability, thereby contributing to genome evolution and the lifestyle of this bacterium (Soares et al. 2011). Another study analyzed the global genome organization of *C. diphtheriae* C7(−) and *C. diphtheriae* PW8 by comparative genomic hybridization including probes representing the 13 pathogenicity islands of *C. diphtheriae* NCTC 13129 (Iwaki et al. 2010). Remarkably, eleven of the 13 pathogenicity islands were considered to be absent in the genome of *C. diphtheriae* C7(−), although this strain retained clear signs of pathogenicity, including adhesion to Detroit 562 cells and the formation of abscesses in animal skin. In contrast, the genome of *C. diphtheriae* PW8 was considered to lack only three pathogenicity islands, but exhibited more reduced signs of pathogenicity in a model system. These results already suggested a great genomic heterogeneity of the species *C. diphtheriae*, not only in genome organization, but also in pathogenicity (Iwaki et al. 2010).

This data were recently confirmed on a much broader scale during the pan-genome project of *C. diphtheriae* (Trost et al. 2012). Genomic islands and candidate pathogenicity islands of the sequenced *C. diphtheriae* strains were identified with the new pathogenicity island prediction software PIPS, which performs a combined analysis of DNA sequences based on typical features of genomic islands (Soares et al. 2012). In total, 57 genomic islands were identified in the sequenced *C. diphtheriae* genomes, including 24 islands (PICD 1–24) in the reference genome of *C. diphtheriae* NCTC 13129 (Table 3.2). Additionally detected genomic islands in *C. diphtheriae* NCTC 13129 carry, for instance, the *irp6ABC* operon encoding a siderophore-dependent iron uptake system (PICD 15) (Qian et al. 2002) and the siderophore biosynthesis and transport gene cluster *ciuABCDEFG* (PICD 19) (Kunkle

Table 3.2 Overview of predicted pathogenicity islands of *C. diphtheriae* NCTC 13129

Reference	Name	Begin CDS	End CDS	Begin position	End position	Length (bp)	Prominent function of island gene(s)
NCTC 13129	PICD 1	DIP0179	DIP0222	152426	190661	38236	Diphtheria toxin encoding corynephage
NCTC 13129	PICD 2	DIP0223	DIP0247	190816	210235	19420	Adhesive pilus
NCTC 13129	PICD 3	DIP0282	DIP0290	249276	256671	7396	Iron transport system
NCTC 13129	PICD 4	DIP0334	DIP0359	305695	326445	20751	Secreted proteins, including polysaccharide degradation enzyme
NCTC 13129	PICD 5	DIP0438	DIP0445	400793	409559	8767	Metal transport system and secreted proteins
NCTC 13129	PICD 6	DIP0750	DIP0766	726536	742372	15837	Lantibiotic biosynthesis proteins
NCTC 13129	PICD 7	DIP0794	DIP0823	776261	796859	20599	Phage-related proteins
NCTC 13129	PICD 8	DIP1645	DIP1664	1680222	1700446	20225	Secreted proteins, including extracellular matrix-binding protein
NCTC 13129	PICD 9	DIP1817	DIP1843	1866720	1883310	16591	Phage-related proteins
NCTC 13129	PICD 10	DIP2010	DIP2015	2065290	2070753	5464	Adhesive pilus
NCTC 13129	PICD 11	DIP2064	DIP2093	2114764	2144615	29852	Fimbrial-associated protein and surface-anchored protein
NCTC 13129	PICD 12	DIP2143	DIP2170	2210329	2244077	33749	Siderophore biosynthesis and transport proteins
NCTC 13129	PICD 13	DIP2208	DIP2234	2297805	2322967	25163	CRISPR locus
NCTC 13129	PICD 14	DIP0028	DIP0051	27025	43284	16260	CRISPR locus
NCTC 13129	PICD 15	DIP0071	DIP0115	63353	94809	31457	Iron transport system
NCTC 13129	PICD 16	DIP0267	DIP0275	230464	239860	9397	Antibiotic resistance protein
NCTC 13129	PICD 17	DIP0320	DIP0326	285652	291817	6166	Transport system with unknown function
NCTC 13129	PICD 18	DIP0448	DIP0466	418336	436609	18274	Two-component system and transport system with unknown function
NCTC 13129	PICD 19	DIP0582	DIP0607	550390	573656	23267	Siderophore biosynthese and transport proteins
NCTC 13129	PICD 20	DIP1891	DIP1901	1941600	1956347	14748	Transport system with unknown function
NCTC 13129	PICD 21	DIP1944	DIP1971	1998093	2018900	20808	Diverse functions and proteins with unknown function
NCTC 13129	PICD 22	DIP2021	DIP2049	2076036	2096446	20411	Secreted proteins, including secretory lipases
NCTC 13129	PICD 23	DIP2123	DIP2135	2176625	2198998	22374	Transport system with unknown function
NCTC 13129	PICD 24	DIP2302	DIP2345	2398584	2451928	53345	Two-component system and transport system with unknown function

and Schmitt 2005). Therefore, the extended search for genomic islands in the sequenced *C. diphtheriae* strains revealed additional gene clusters with characteristics of horizontal gene transfer, which are probably involved in iron acquisition. Comparative *in silico* analysis of the predicted genomic islands revealed that some are strain-specific, whereas others are partially or completely conserved in more than one strain (Trost et al. 2012). Only eight genomic islands can be regarded as highly conserved in all *C. diphtheriae* genomes, demonstrating the great genomic plasticity of *C. diphtheriae* (Trost et al. 2012). Many genomic islands encode typical phage products and the respective genomic regions of the *C. diphtheriae* genomes can be regarded as remnants of prophages. Some genomic islands encode proteins involved in specific metabolic pathways and were assigned as metabolic islands of the *C. diphtheriae* genome, whereas others encode proteins involved in antibiotic resistance or heavy metal ion resistance, such as cadmium, copper, mercury, and arsenic resistance.

The plasticity of the *C. diphtheriae* genome is also obvious when visualizing the gene content of the sequenced isolates with the BRIG software (Alikhan et al. 2011) and using the genome of *C. diphtheriae* NCTC 13129 as a reference (Fig. 3.2). Variations in the gene repertoire of the *C. diphtheriae* isolates seem to cluster in genomic regions assigned as pathogenicity islands, indicating that horizontal gene transfer is a major force in shaping the gene content and physiological traits of *C. diphtheriae* strains.

The search for pathogenicity islands in the genomes of the sequenced *C. diphtheriae* strains led to the detection of several islands harboring gene clusters for adhesive pili (Trost et al. 2012), which play important roles in bacterial colonization and pathogenesis (Ton-That and Schneewind 2003). Pilus assembly has been studied extensively in *C. diphtheriae* and occurs by a two-step mechanism, whereby pilin subunits are first polymerized and then covalently anchored to cell wall peptidoglycan. A pilin-specific sortase catalyzes the polymerization of the pilus, consisting of the shaft protein, the tip pilin and the base pilin (Rogers et al. 2011). Based on amino acid sequence homology searches using the pilin motif and cell wall sorting signal as queries, at least two pilus gene clusters were identified in each of the sequenced *C. diphtheriae* isolates, with *C. diphtheriae* HC04 haboring four pilus gene clusters (Fig. 3.3). Six different types of pilus gene clusters were detected according to the arrangement of genes encoding pilus subunits (*spa*) or pilin-specific sortases (*srt*). It is noteworthy to mention that the genome of *C. diphtheriae* PW8 contains a highly degenerated SpaD gene cluster with multiple intact and disrupted genes encoding SpaD, SpaE, SpaF pilins and sortases SrtB and SrtE, in addition to a SpaA gene cluster with a disrupted *spaC* gene (Fig. 3.3). Mobile DNA elements were also detected in the SpaD locus of *C. diphtheriae* PW8, suggesting horizontal gene transfer for gene duplication. Phylogenetic trees reconstructed with the neighbor-joining algorithm revealed that the protein components of the pilus, i.e. shaft protein, tip pilin and base pilin, and the cognate pilin-specific sortases display a great diversity in their amino acid sequences (Trost et al. 2012). Therefore, most *spa* and *srt* genes present on the predicted pathogenicity islands of the sequenced *C. diphtheriae* strains were

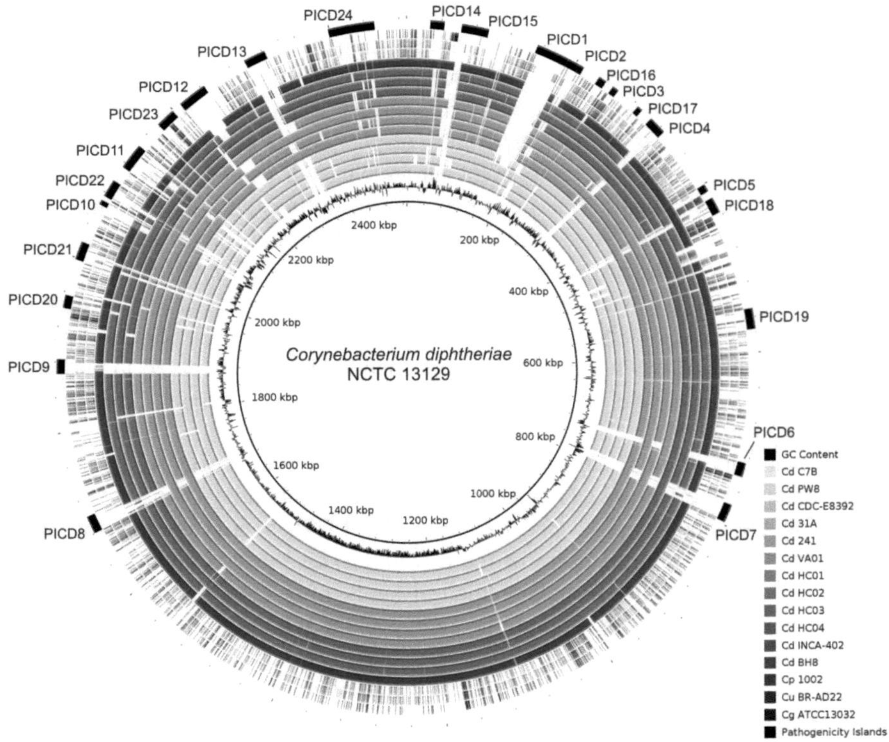

Fig. 3.2 Circular genome comparison between *C. diphtheriae* strains using *C. diphtheriae* NCTC 13129 as a reference. The genome comparison was generated with the BLAST Ring Image Generator BRIG. It shows the positions of candidate pathogenicity islands in the genome of *C. diphtheriae* NCTC 13129 and the presence/absence of these islands in other *C. diphtheriae* strains or species of the genus *Corynebacterium*. Abbreviations: GC Content, G+C profile of a genome region; Cd, *C. diphtheriae*; Cp, *C. pseudotuberculosis*; Cu, *C. ulcerans*; Cg, *C. glutamicum*; PICD, putative pathogenicity island of *C. diphtheriae*

assigned as unique genes during the pan-genome analysis. This result strongly implies that important variations exist on the cell surface of *C. diphtheriae* strains, which might be relevant for the initial step of an infection (Trost et al. 2012). Previous studies demonstrated different degrees of attachment of *C. diphtheriae* to HEp-2 cell monolayers (Hirata et al. 2004), differences in adhesion of *C. diphtheriae* C7(−) and *C. diphtheriae* PW8 to Detroit 562 cells (Iwaki et al. 2010) and strain-specific differences of *C. diphtheriae* in adhesion, invasion and intracellular survival (Ott et al. 2010). Moreover, mutations in the base pilin SpaB and the tip pilin SpaC of the SpaA-type pilus reduced the adhesive activity of *C. diphtheriae* (Mandlik et al. 2007).

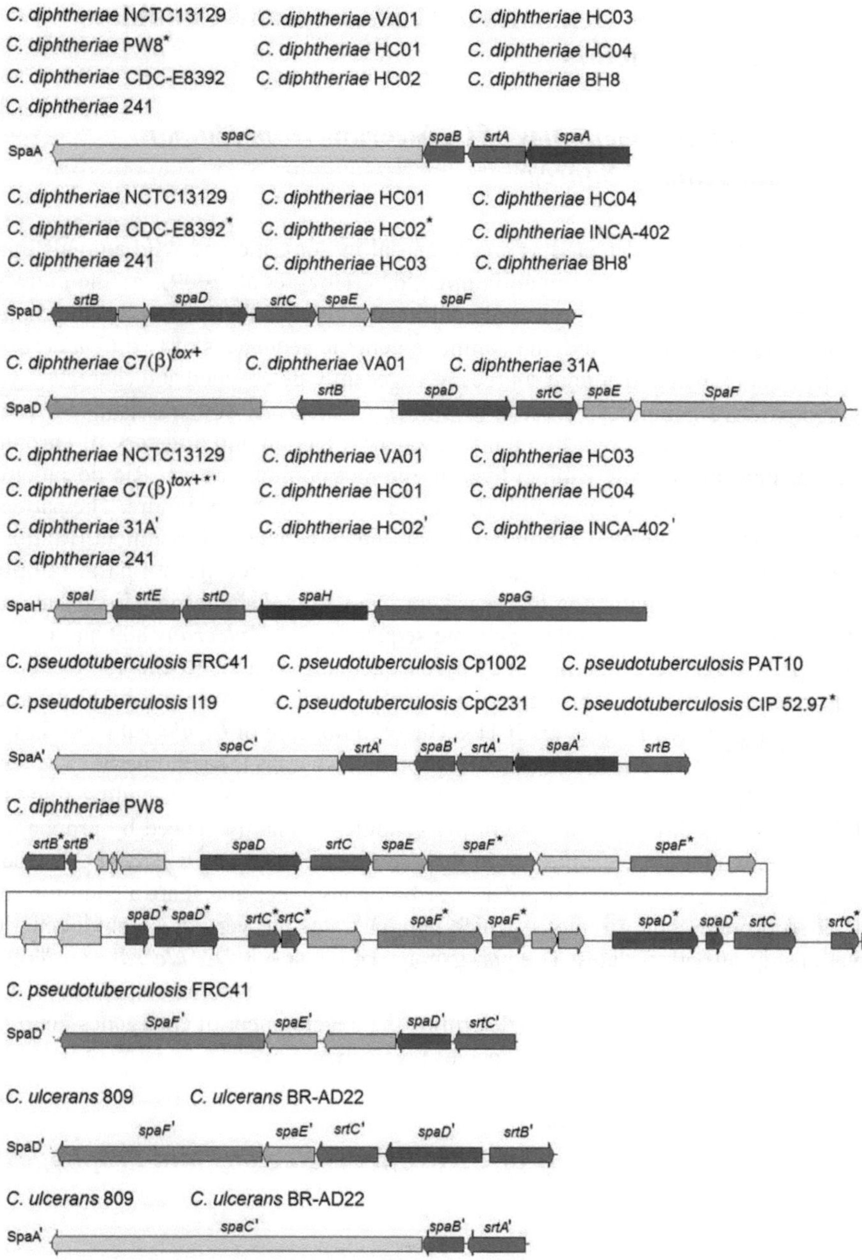

Fig. 3.3 Schematic representation of pilus gene clusters in the genomes of *C. diphtheriae*, *C. ulcerans* and *C. pseudotuberculosis*. The detected pilus gene clusters revealed different arrangements of genes encoding subunits of adhesive pili (*spa*) or pilin-specific sortases (*srt*). Assigned strains are listed above the gene clusters. Each *C. diphtheriae* genome contains at least two pilus gene clusters. *C. diphtheriae* PW8 contains a degenerated gene cluster with multiple intact and disrupted genes. Symbols: *asterisk* (*), clusters of the labeled strains contain a fragmented gene; *prime* ('), denotes variants of the respective pilus gene cluster

3.3 Comparative Genomics of *C. ulcerans* and Candidate Virulence Factors

3.3.1 Reference Genomes of C. ulcerans *from Human and Animal Sources*

C. ulcerans has been detected as a commensal in domestic and wild animals that may serve as reservoirs for zoonotic infections (Hogg et al. 2009). As the knowledge of the bacterium's lifestyle and additional virulence factors besides the diphtheria toxin was very limited, the complete genome sequences of two *C. ulcerans* strains from human and animal sources were recently determined and characterized by comparative genomics (Trost et al. 2011). *C. ulcerans* 809 was isolated from an elderly woman with rapidly fatal pulmonary infection and a history of chronic bilateral limb ulcers. The woman lived in the metropolitan area of Rio de Janeiro and was hospitalized in coma, with shock and acute respiratory failure. The patient died 23 days after hospitalization despite an intensive medical treatment (Mattos-Guaraldi et al. 2008). *C. ulcerans* BR-AD22 was recovered from a nasal sample of a young asymptomatic dog that was kept in an animal shelter in Rio de Janeiro (Dias et al. 2010). The complete genome sequences of *C. ulcerans* 809 and *C. ulcerans* BR-AD22 were determined by pyrosequencing with the Roche/454 Genome Sequencer FLX System. This approach revealed sequence coverages of 42.8× and 22.9×, respectively (Trost et al. 2011). The chromosome of *C. ulcerans* 809 has a size of 2,502,095 bp and encodes 2,182 proteins, whereas the genome of *C. ulcerans* BR-AD22 is 104,279 bp larger and comprises 2,338 protein-coding regions (Table 3.1). The difference in size of the genomes is mainly caused by prophage-like elements that are present only in the genome of *C. ulcerans* BR-AD22. Both genomes show a highly similar order of orthologous genes and share a common set of 2,076 protein-coding regions, which demonstrates the very close phylogenetic relationship of both isolates. The pan-genome of the species *C. ulcerans* currently comprises 2,445 protein-coding genes. Obviously, more genome sequences of *C. ulcerans* isolates are necessary to determine the development of core genes, unique genes and the pan-genome of this species precisely.

*3.3.2 Genetic Variability of CRISPR/*cas *Regions and Prophages in* C. ulcerans

A screening of the genome sequences of *C. ulcerans* 809 and *C. ulcerans* BR-AD22 with the CRISPRFinder revealed the presence of three CRISPR/*cas* regions, herein named CRISPR types IV–VI (Fig. 3.1). CRISPR type IV is present in both *C. ulcerans* genomes and flanked by four *cas* genes. The direct repeats of this locus are 29 bp in length and separated by spacers with variable nucleotide sequences that are completely different in both *C. ulcerans* strains. Similar features were observed for the second CRISPR/*cas* region in the *C. ulcerans* genomes (Fig. 3.1). CRISPR type V is characterized by six *cas* genes and repeats of 36 bp. The spacer sequences present in *C. ulcerans* 809 are also different to those present in the corresponding CRISPR/*cas* region of the *C. ulcerans* BR-AD22 genome. CRISPR type VI of *C. ulcerans* is lacking associated *cas* genes in the direct proximity (Fig. 3.1). The spacer sequences of this CRISPR type have a length of 29 bp and show the largest variation between both strains, with 67 spacers present in the genome of *C. ulcerans* 809 and 32 spacers in *C. ulcerans* BR-AD22 (Table 3.1). The detection of CRISPR/*cas* regions in the genome of *C. ulcerans* strains and the sequence variations of the CRISPR loci suggests the use of these markers for a precise typing of clonal groups of *C. ulcerans* isolates from human and animal sources (Trost et al. 2011).

In accordance with the tox^{-} phenotype of *C. ulcerans* 809 and *C. ulcerans* BR-AD22, both genomes were devoid of nucleotide sequences of a tox^{+} corynephage encoding diphtheria toxin. However, the genomes of *C. ulcerans* 809 and *C. ulcerans* BR-AD22 harbor the highly similar prophages ΦCULC809I and ΦCULC22I with sizes of about 42 kb (Trost et al. 2011). Both prophages were detected at the same genomic position and apparently integrated at slightly different sites into a coding region for a hypothetical protein that might represent the integration site of these phages in the *C. ulcerans* chromosome. Minor differences between the prophages were detected in the number of genes, as ΦCULC809I comprises 45 genes, whereas 42 genes were assigned to the ΦCULC22I genome. According to global amino acid sequence alignments, both prophages share 36 genes that code for gene products with at least 98 % amino acid sequence identity, indicating the very close relationship of both prophages from different *C. ulcerans* isolates (Trost et al. 2011). The genome sequence of *C. ulcerans* BR-AD22 contains the additional prophages ΦCULC22II, ΦCULC22III and ΦCULC22IV, of which ΦCULC22III is incomplete and probably a defective remnant of a formerly active corynephage (Trost et al. 2011). Most strain-specific genes of the animal isolate *C. ulcerans* BR-AD22 were assigned to the additional prophage-like regions ΦCULC22II, ΦCULC22III and ΦCULC22IV. Therefore, only 92 protein-coding regions of this strain were regarded as unique genes, of which 13 genes were annotated with putative physiological functions (Trost et al. 2011).

3.3.3 Pathogenicity Islands and Virulence Factors of C. ulcerans

The search for unique genes by reciprocal best BLASTP matches revealed 90 strain-specific genes for the human isolate *C. ulcerans* 809, of which ten were annotated with putative physiological functions (Trost et al. 2011). This set of gene regions includes the *vsp2* gene coding for a secreted serine protease and the *rbp* gene encoding a putative ribosome-binding protein. Both gene products represent candidate virulence factors of *C. ulcerans* 809. The *rbp* gene is located between a gene coding for a putative phage integrase and a transposase gene and is moreover specified by the low G+C content of 45.1 %, suggesting the lateral transfer of *rbp* to *C. ulcerans* 809. The respective tyrosine recombinase detected in *C. ulcerans* 809 shares 92 % amino acid sequence identity with the integrase of the β-type corynephage present in the reference genome of *C. diphtheriae* NCTC 13129 and is also encoded directly adjacent to a $tRNA^{Arg}$ gene. This gene annotation supports the assumption that a lysogenic β-type corynephage had been integrated downstream of the *rbp* gene in the *C. ulcerans* genome in former times (Trost et al. 2011). The protein product of the *rbp* gene showed weak similarity to the A chains of the Shiga-like toxins SLT-1 and SLT-2 from *Escherichia coli*, but contains all highly conserved amino acid residues relevant for the catalytic N-glycosidase activity (O'Loughlin et al. 2001; LaPointe et al. 2005). In contrast, the amino acid sequence of the Rbp protein lacks the typical ER-targeting sequence at the C-terminal end, which is necessary for the retranslocation of the catalytic domain of SLT-1 from the endoplasmatic reticulum (ER) into the cytosol of the host cell (O'Loughlin et al. 2001). As *C. ulcerans* can probably persist as a facultative intracellular pathogen in mammalian host cells, a retranslocation of the Rbp protein into the cytosol is nonessential for activity. The secretion of the putative toxin into the cytosol of host cells is supported instead by a typical signal sequence at the amino-terminal end of the protein (Trost et al. 2011). The enzymatic activity of the ribosome-binding protein Rbp probably leads to inhibition of protein biosynthesis by depurination of a single adenosine residue in the 28S rRNA of the eukaryotic ribosome (O'Loughlin et al. 2001). A genome screening for further virulence factors revealed the presence of endoglycosidase E (see below), neuraminidase H and adhesive pili of the SpaA' and SpaD' type that are encoded in both *C. ulcerans* genomes (Fig. 3.3). The *C. ulcerans* genome is apparently equipped with genes for a broad spectrum of virulence factors, including a novel ribosome-binding protein that is encoded only in the human isolate *C. ulcerans* 809.

Putative pathogenicity islands of *C. ulcerans* were detected the in larger genome of the animal isolate *C. ulcerans* BR-AD22 (Table 3.3). The *C. ulcerans* BR-AD22 genome contains 14 putative pathogenicity islands, including a phospholipase D gene region, an operon encoding urease and genes for iron uptake systems, which are generally associated with virulence. Most genes assigned to the pathogenicity islands of *C. ulcerans* have diverse roles in cellular metabolism or even hitherto unknown functions. However, all candidate virulence factors in the detected pathogenicity islands have characteristics that are indicative of horizontal gene transfer.

3.4 Towards the Pan-Genome of *C. pseudotuberculosis*

3.4.1 The Reference Genome of C. pseudotuberculosis *1002*

C. pseudotuberculosis is an important animal pathogen and the causative agent of a disease that is commonly called caseous lymphadenitis (Dorella et al. 2006a). This

Table 3.3 Overview of predicted pathogenicity islands of *C. ulcerans* BR-AD-22

Reference	Name	Begin CDS	End CDS	Begin position	End position	Length (bp)	Prominent function of island gene(s)
BR-AD22	PICU1	CULC22_00019	CULC22_00042	19833	44743	24911	CRISPR locus, phospholipase D and iron acquisition genes
BR-AD22	PICU2	CULC22_00051	CULC22_00112	54498	118292	63795	CRISPR locus, ABC transport systems, two-component systems and transcriptional regulators
BR-AD22	PICU3	CULC22_00166	CULC22_00182	175545	192468	16924	Diverse functions and proteins with unknown function
BR-AD22	PICU4	CULC22_00224	CULC22_00236	248273	264854	16582	Diverse functions and proteins with unknown function
BR-AD22	PICU5	CULC22_00667	CULC22_00683	725593	742611	17019	Putrescine synthesis and ABC transport protein systems
BR-AD22	PICU6	CULC22_01155	CULC22_01200	1276769	1325503	48735	Phage-related proteins
BR-AD22	PICU7	CULC22_01654	CULC22_01724	1835091	1891202	56112	Diverse functions and proteins with unknown function
BR-AD22	PICU8	CULC22_01773	CULC22_01788	1944778	1967893	23116	Secreted proteins and proteins with unknown function
BR-AD22	PICU9	CULC22_01794	CULC22_01816	1972423	1989721	17299	Diverse functions and proteins with unknown function
BR-AD22	PICU10	CULC22_01921	CULC22_01985	2108976	2164364	55389	Diverse functions and proteins with unknown function
BR-AD22	PICU11	CULC22_02033	CULC22_02044	2214650	2227049	12400	Chaperone and proteins with unknown function
BR-AD22	PICU12	CULC22_02071	CULC22_02085	2254055	2265204	11150	Cytochrome C biosynthesis and proteins with unknown function
BR-AD22	PICU13	CULC22_02134	CULC22_02168	2333661	2374767	41107	Iron and oligopeptide transport system, urease operon and diverse functions
BR-AD22	PICU14	CULC22_02307	CULC22_02325	2550850	2574374	23525	Diverse functions and proteins with unknown function

disease is found in the major sheep and goat production areas worldwide and causes significant economic losses. The strain selected for the first genome sequencing project was *C. pseudotuberculosis* 1002, which was isolated from goat caseous granulomas in Bahia state (Brazil). This strain has been licensed as a live attenuated vaccine strain in Brazil (Dorella et al. 2006a). The genome of *C. pseudotuberculosis* 1002 was sequenced using both 'classical' Sanger and pyrosequencing technologies (Ruiz et al. 2011). The genome sequencing project initially started with only 215 genomic survey sequences (GSSs) obtained from random samples of a BAC library of *C. pseudotuberculosis* 1002 (Dorella et al. 2006b). This representative library contained about 18,000 clones with inserts ranging in size from 25 to 120 kb and provided a 390-fold coverage of the *C. pseudotuberculosis* genome. Many GSSs (80.4 %) revealed significant similarity to the genome sequence of *C. diphtheriae* NCTC 13129 confirming the very close phylogenetic relationship of both species (Dorella et al. 2006b). Pyrosequencing was carried out with the Roche/454 Genome Sequencer FLX System and a sequencing depth finally resulting in 31 × coverage of the *C. pseudotuberculosis* 1002 genome (Ruiz et al. 2011). The chromosome of *C. pseudotuberculosis* 1002 has a size of 2,335,112 bp with a G+C content of 52.19 % and contains 2,111 predicted protein-coding regions, of which 53 were annotated as pseudogenes (Ruiz et al. 2011).

Meanwhile, five additional genome sequences of *C. pseudotuberculosis* isolates have been determined and published (Table 3.1), including *C. pseudotuberculosis* C231 from a sheep in Australia (Ruiz et al. 2011), *C. pseudotuberculosis* I19 from a cow with mastitis in Israel (Silva et al. 2011), *C. pseudotuberculosis* PAT10 from a sheep with lung abscess in Argentina (Cerdeira et al. 2011b), *C. pseudotuberculosis* CIP 52.97 from a horse with ulcerative lymphangitis in Kenya (Cerdeira et al. 2011c) and *C. pseudotuberculosis* FRC41 from a young French girl with necrotizing lymphadenitis (Trost et al. 2010b), which was the first genome sequence publicly available for this species. The complete genome sequences of *C. pseudotuberculosis* C231 and *C. pseudotuberculosis* FRC41 were both determined with the Roche/454 Genome Sequencer FLX System, whereas the Life Technologies SOLiD System was used for the remaining three genome projects. To address the problem of short reads in the case of the latter next-generation sequencing technology, a new hybrid *de novo* assembly strategy was developed combining De Bruijn graph and Overlap-Layout-Consensus methods (Cerdeira et al. 2011a). This *in silico* approach was used in a case study to assemble the complete genome sequence of *C. pseudotuberculosis* I19 from short reads (Cerdeira et al. 2011a). Briefly, contigs were assembled *de novo* from the short reads and were oriented using the complete genome sequence of *C. pseudotuberculosis* FRC41 as a reference for anchoring. Remaining gaps in the genome sequence of *C. pseudotuberculosis* I19 were closed using an iterative anchoring of additional short reads adjacent to sequence gaps (Cerdeira et al. 2011a). This new assembly strategy is feasible as the sequenced *C. pseudotuberculosis* genomes show a highly similar architecture and a highly conserved order of orthologous coding regions (Ruiz et al. 2011).

3.4.2 Comparative Genomics and the Pan-Genome of C. pseudotuberculosis

The number of core genes of *C. pseudotuberculosis* was calculated with the software EDGAR using bidirectional best BLASTP hits for genome comparisons (Blom et al. 2009). Based on a set of calculations using all *C. pseudotuberculosis* genomes individually as references, the core genome of the hitherto sequenced *C. pseudotuberculosis* isolates comprises 1,810 genes that can be regarded as highly conserved in this species. The bioinformatic characterization of the unique genome of *C. pseudotuberculosis* revealed a very low number of strain-specific genes in the four *C. pseudotuberculosis* biovar *ovis* isolates 1002, C231, I19 and PAT10, whereas 86 unique genes were detected in the genome of the *C. pseudotuberculosis* biovar *equi* isolate CIP 52.7 and 49 in *C. pseudotuberculosis* FRC41 from a human clinical source (Table 3.1). This preliminary data indicate that the genomes of *C. pseudotuberculosis* biovar *ovis* isolates have very similar gene contents. Accordingly, the sum total of protein-coding regions representing the pan-genome of *C. pseudotuberculosis* currently comprises only 2,630 genes, which is just about 1.5 times the size of the predicted core genome.

The close similarity between the sequenced *C. pseudotuberculosis* strains is also evident when comparing the structure of the CRISPR/*cas* regions. All *C. pseudotuberculosis* isolates share a CRISPR type IV with only one repeat sequence, with the exception of *C. pseudotuberculosis* CIP 52.97 that completely lacks a CRISPR/*cas* region (Table 3.1). It is therefore unlikely that spoligotyping is a suitable approach to analyze the genetic diversity of *C. pseudotuberculosis* isolates.

3.4.3 Pathogenicity Islands and Virulence Factors of C. pseudotuberculosis

Pathogenicity islands of *C. pseudotuberculosis* were detected and annotated in the reference genome of *C. pseudotuberculosis* 1002 by the means of the recently developed software PIPS (Ruiz et al. 2011; Soares et al. 2012). The *C. pseudotuberculosis* 1002 genome includes eleven putative pathogenicity islands (Table 3.4), which contain several classical virulence factors, including genes for pilus subunits, adhesion factors, iron uptake systems and secreted toxins. All of the candidate virulence factors in the islands have characteristics that indicate horizontal transfer. Comparative *in silico* analysis of the predicted pathogenicity islands with the BRIG software (Alikhan et al. 2011) revealed that most of the respective genes belong to the distributed genome of *C. pseudotuberculosis* and are only present in genome sequences of biovar *ovis* isolates, whereas others are partially or completely conserved in almost all strains (Fig. 3.4). This data indicates that prominent differences exist in the genetic repertoires of isolates belonging to the *C. pseudotuberculosis* biovars *ovis* or *equi*.

Table 3.4 Overview of predicted pathogenicity islands of *C. pseudotuberculosis* 1002

Reference	Name	Begin CDS	End CDS	Begin position	End position	Length (bp)	Prominent function of island gene(s)
1002	PICP1	Cp1002_0022	Cp1002_0031	19903	29136	9234	Phospholipase D and iron acquisition genes
1002	PICP2	Cp1002_0040	Cp1002_0067	38609	68350	29742	Iron and choline transport system and transcriptional regulators
1002	PICP8	Cp1002_0159	Cp1002_0167	163700	176472	12773	Purine nucleoside phosphorylase and deoxyribonucleoside regulator
1002	PICP3	Cp1002_0174	Cp1002_0185	186099	196852	10754	Iron transport system
1002	PICP9	Cp1002_0553	Cp1002_0573	575444	604590	29147	Diverse functions and proteins with unknown function
1002	PICP4	Cp1002_0980	Cp1002_0992	1057400	1076800	19401	Iron transport system
1002	PICP5	Cp1002_1445	Cp1002_1472	1588030	1617021	28992	Iron transport system and transcriptional regulators
1002	PICP6	Cp1002_1553	Cp1002_1565	1701830	1713424	11595	Transport system with unknown function
1002	PICP10	Cp1002_1617	Cp1002_1633	1766713	1788178	21466	Diverse functions
1002	PICP7	Cp1002_1903	Cp1002_1932	2089913	2123600	33688	Iron and oligopeptide transport system, urease operon and diverse functions
1002	PICP11	Cp1002_2069	Cp1002_2080	2290665	2301762	11098	Diverse functions and proteins with unknown function

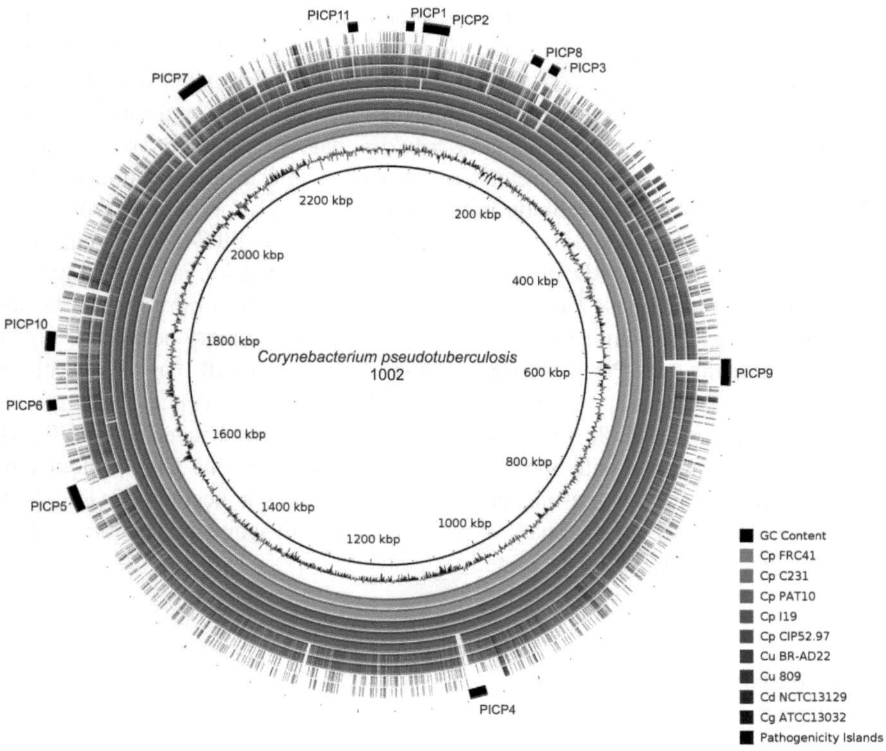

Fig. 3.4 Circular genome comparison between *C. pseudotuberculosis* strains using *C. pseudotuberculosis* 1002 as a reference. The circular genome comparison shows the positions of putative pathogenicity islands in the genome of the reference strain *C. pseudotuberculosis* 1002 (biovar *ovis*) and the presence/absence of these islands in other *C. pseudotuberculosis* biovar *ovis* strains (C231, PAT10 and I19), a *C. pseudotuberculosis* biovar *equi* strain (CIP 52.97), an isolate from a human clinical source (FRC41), and in other corynebacterial species. Abbreviations: *GC Content* G+C profile of a genome region; *Cp C. pseudotuberculosis*; *Cu C. ulcerans*; *Cd C. diphtheriae*; *Cg C. glutamicum*; *PICP* putative pathogenicity island of *C. pseudotuberculosis*

Despite the importance of *C. pseudotuberculosis* for animal health, there is little information about the pathogenesis and the facultative intracellular lifestyle of this bacterium. Only few virulence factors were identified previously in *C. pseudotuberculosis* (Dorella et al. 2006a), of which the most prominent one is phospholipase D (Pld), a sphingomyelin-degrading exotoxin (McKean et al. 2007). The annotation of the complete *C. pseudotuberculosis* FRC41 genome sequence provided additional knowledge of candidate virulence factors in this species (Trost et al. 2010b). In addition to the virulence factor phospholipase D, the endoglycosidase EndoE (misleadingly described as corynebacterial protease CP40 in previous studies) is encoded in the genome of this human isolate. The *ndoE* gene product of *C. pseudotuberculosis* FRC41 revealed sequence similarity to the α-domain of the secreted endoglycosidase EndoE from *Enterococcus faecalis* (Collin and Fischetti 2004.).

EndoE from *E. faecalis* is a two-domain protein that is characterized by two distinct activities involved in the degradation of N-linked glycans from ribonuclease B and the hydrolysis of the conserved glycans on IgG. The latter activity of the enzyme was assigned exclusively to the β-domain of EndoE, suggesting that the homologous protein from *C. pseudotuberculosis* has only endoglycosidase activity. In this way, *C. pseudotuberculosis* is probably able to interact directly with the mammalian host by glycolytic modulation of host glycoproteins (Trost et al. 2010b). The genome annotation of *C. pseudotuberculosis* FRC41 revealed serine proteases, neuraminidase H, nitric oxide reductase, an invasion-associated protein and acyl-CoA carboxylase subunits involved in mycolic acid biosynthesis as additional candidate virulence factors. Moreover, a gene-regulatory network analysis suggested that the cAMP-sensing transcriptional regulator GlxR plays a key role in controlling the expression of several genes contributing to virulence of *C. pseudotuberculosis* (Trost et al. 2010b). The human isolate *C. pseudotuberculosis* FRC41 is furthermore equipped with SpaA' and SpaD' gene clusters encoding protein subunits involved in the sortase-mediated polymerization of adhesive pili (Fig. 3.3). The pilus gene cluster of the SpaA'-type is present in all sequenced genomes of *C. pseudotuberculosis* (Fig. 3.3).

3.5 Future Perspectives

The development of ultra-fast next-generation sequencing technologies has opened a new era of microbial genomics, enabling the rapid and detailed characterization of bacterial genomes and associated bacterial lifestyles. This progress in microbial genomics is obviously helping to shape our understanding of bacterial evolution. In particular, comparative genomics, or on a broader scale pan-genomics, affords the opportunity to detect species-specific features of a genome or strain-specific traits such as virulence factors contributing to the pathogenicity of bacteria. The systematic application of next-generation sequencing technologies also provides the possibility to generate bacterial DNA sequence data of extraordinary resolution, making it possible to identify single nucleotide changes within entire genomes and to map genome-wide single-nucleotide polymorphisms (SNPs). This type of studies provides data of very fine-scale resolution and enables the detection of the evolutionary history of multiple isolates within a clonal bacterial lineage. Distinguishing clonal groups within a pathogenic bacterial species was initially performed by phenotypic and subsequently by genotypic typing techniques and has been the cornerstone of infectious disease epidemiology, allowing the identification and tracking of clones responsible for infection and disease. Sequence-based typing approaches, such as multilocus sequence typing, have relied on the variation within a few selected marker genes. Although this technique is highly informative, it has a limited resolution when applied to closely related isolates. Approaches based on next-generation sequencing are also suitable for identifying subtle evolutionary events or for distinguishing clonal strains within a recent epidemic when applied on bacterial collections of known origin. Studies of the phylogeny of a bacterial species or of a clonal lineage within a species are highly dependent on the quantity and diversity of sampled isolates. However, the recent pan-genomic study of *C. diphtheriae* demonstrated that it has become possible to fully sequence significant numbers of isolates in a strain collection in reasonable time, thereby revealing new information on the plasticity of the *C. diphtheriae* genome. In principle, the size and the composition of the investigated strain collection is crucial for subsequent biological interpretations. This is particularly relevant for bacterial pathogens that reside in multiple niches and is therefore considered in the ongoing genome sequencing of *C. pseudotuberculosis* isolates from various geographical regions and sources, including sheep, goats, cows, horses, and humans. This strategy prevents bias in the data sets and provides a more complete picture of the true diversity of the bacterial species. A pan-genomic approach is also feasible for the characterization of *C. ulcerans*, which has been recognized in a broad spectrum of animal hosts. Whole-genome sequencing also facilitates the identification of gene losses and gene gains that can play a significant role in the evolution or pathogenicity of a bacterial species, as indicated by the detection of the candidate virulence factor *rbp* in *C. ulcerans* 809. Therefore, next-generation sequencing technologies provide a means of rapidly detecting associations between phenotype and genotype. The next few years will see an increase in the biological interpretation of such data using either high-throughput *in vitro* assays or the selected testing by targeted genetic experiments. This should further improve our understanding of the various forces that are important in the evolution of bacterial pathogens and enable the development of appropriate interventions. The next few years also promise an enhanced understanding of how and why epidemic clones emerge or disappear, and ultimately the better management and treatment of infectious diseases.

Acknowledgement The authors thank Siomar de Castro Soares for bioinformatic support, running the PIPS pipeline and visualizing comparative genomic data with the BRIG software.

References

Alikhan NF, Petty NK, Ben Zakour NL, Beatson SA (2011) BLAST Ring Image Generator (BRIG): simple prokaryote genome comparisons. BMC Genomics 12:402

Barksdale WL, Pappenheimer AM Jr (1954) Phage-host relationships in nontoxigenic and toxigenic diphtheria bacilli. J Bacteriol 67:220–232

Blom J, Albaum SP, Doppmeier D, Pühler A, Vorhölter FJ, Zakrzewski M, Goesmann A (2009) EDGAR: a software framework for the comparative analysis of prokaryotic genomes. BMC Bioinformatics 10:154

Brüssow H, Canchaya C, Hardt WD (2004) Phages and the evolution of bacterial pathogens: from genomic rearrangements to lysogenic conversion. Microbiol Mol Biol Rev 68:560–602

Bukovska G, Klucar L, Vlcek C, Adamovic J, Turna J, Timko J (2006) Complete nucleotide sequence and genome analysis of bacteriophage BFK20—a lytic phage of the industrial producer *Brevibacterium flavum*. Virology 348:57–71

Cerdeira LT, Carneiro AR, Ramos RT, Almeida SS de, D'Afonseca V, Schneider MP, Baumbach J, Tauch A, McCulloch JA, Azevedo VA, Silva A (2011a) Rapid hybrid *de novo* assembly of a microbial genome using only short reads: *Corynebacterium pseudotuberculosis* I19 as a case study. J Microbiol Methods 86:218–223

Cerdeira LT, Pinto AC, Schneider MP, Almeida SS de, Santos AR dos, Barbosa EG, Ali A, Barbosa MS, Carneiro AR, Ramos RT, Oliveira RS de, Barh D, Barve N, Zambare V, Belchior SE, Guimarães LC, Castro SS de, Dorella FA, Rocha FS, Abreu VA de, Tauch A, Trost E, Miyoshi A, Azevedo V, Silva A (2011b) Whole-genome sequence of *Corynebacterium pseudotuberculosis* PAT10 strain isolated from sheep in Patagonia, Argentina. J Bacteriol 193:6420–6421

Cerdeira LT, Schneider MP, Pinto AC, Almeida SS de, Santos AR dos, Barbosa EG, Ali A, Aburjaile FF, Abreu VA de, Guimarães LC, Castro SS de, Dorella FA, Rocha FS, Bol E, Gomes deSPH, Lopes TS, Barbosa MS, Carneiro AR, Jucá Ramos RT, Coimbra NA, Lima AR, Barh D, Jain N, Tiwari S, Raja R, Zambare V, Ghosh P, Trost E, Tauch A, Miyoshi A, Azevedo V, Silva A (2011c) Complete genome sequence of *Corynebacterium pseudotuberculosis* strain CIP 52.97, isolated from a horse in Kenya. J Bacteriol 193:7025–7026

Cerdeño-Tarrága AM, Efstratiou A, Dover LG, Holden MT, Pallen M, Bentley SD, Besra GS, Churcher C, James KD, De Zoysa A, Chillingworth T, Cronin A, Dowd L, Feltwell T, Hamlin N, Holroyd S, Jagels K, Moule S, Quail MA, Rabbinowitsch E, Rutherford KM, Thomson NR, Unwin L, Whitehead S, Barrell BG, Parkhill J (2003) The complete genome sequence and analysis of *Corynebacterium diphtheriae* NCTC13129. Nucleic Acids Res 31:6516–6523

Collin M, Fischetti VA (2004) A novel secreted endoglycosidase from *Enterococcus faecalis* with activity on human immunoglobulin G and ribonuclease B. J Biol Chem 279:22558–22570

D'Afonseca V, Soares SC, Ali A, Santos AR, Pinto AC, Magalhães AA, Faria CJ, Barbosa E, Guimarães LC, Eslabão M, Almeida SS, Abreu VA, Zerlotini A, Carneiro AR, Cerdeira LT, Ramos RT, Hirata R Jr, Mattos-Guaraldi AL, Trost E, Tauch A, Silva A, Schneider MP, Miyoshi A, Azevedo V (2012) Reannotation of the *Corynebacterium diphtheriae* NCTC13129 genome as a new approach to studying gene targets connected to virulence and pathogenicity in diphtheria. Open Access Bioinformatics 4:1–13

Deveau H, Garneau JE, Moineau S (2010) CRISPR/Cas system and its role in phage-bacteria interactions. Annu Rev Microbiol 64:475–93

Dias AA, Silva FC Jr, Pereira GA, Souza MC, Camello TC, Damasceno JA, Pacheco LG, Miyoshi A, Azevedo VA, Hirata R Jr, Bôas MH, Mattos-Guaraldi AL (2010) *Corynebacterium ulcerans* isolated from an asymptomatic dog kept in an animal shelter in the metropolitan area of Rio de Janeiro, Brazil. Vector Borne Zoonotic Dis 10:743–748

Dittmann S, Wharton M, Vitek C, Ciotti M, Galazka A, Guichard S, Hardy I, Kartoglu U, Koyama S, Kreysler J, Martin B, Mercer D, Ronne T, Roure C, Steinglass R, Strebel P, Sutter R, Trostle M. 2000. Successful control of epidemic diphtheria in the states of the Former Union of Soviet Socialist Republics: lessons learned. J Infect Dis. 181:10–22

Dorella FA, Pacheco LG, Oliveira SC, Miyoshi A, Azevedo V (2006a) *Corynebacterium pseudotuberculosis*: microbiology, biochemical properties, pathogenesis and molecular studies of virulence. Vet. Res 37:201–218

Dorella FA, Fachin MS, Billault A, Dias Neto E, Soravito C, Oliveira SC, Meyer R, Miyoshi A, Azevedo V (2006b) Construction and partial characterization of a *Corynebacterium pseudotuberculosis* bacterial artificial chromosome library through genomic survey sequencing. Genet Mol Res 5:653–663

Droege M, Hill B (2008) The Genome Sequencer FLX System—longer reads, more applications, straight forward bioinformatics and more complete datasets. J. Biotechnol 136:3–10

Freeman VJ (1951) Studies on the virulence of bacteriophage-infected strains of *Corynebacterium diphtheriae*. J Bacteriol 61:675–688

Halachev MR, Loman NJ, Pallen MJ (2011) Calculating orthologs in bacteria and archaea: a divide and conquer approach. PLoS One 6:e28388

Hirata R Jr, Souza SM, Rocha-de-Souza CM, Andrade AF, Monteiro-Leal LH, Formiga LC, Mattos-Guaraldi AL (2004) Patterns of adherence to HEp-2 cells and actin polymerisation by toxigenic *Corynebacterium diphtheriae* strains. Microb Pathog 36:125–130

Hogg RA, Wessels J, Hart J, Efstratiou A, De Zoysa A, Mann G, Allen T, Pritchard GC (2009) Possible zoonotic transmission of toxigenic *Corynebacterium ulcerans* from companion animals in a human case of fatal diphtheria. Vet Rec 165:691–692

Iwaki M, Komiya T, Yamamoto A, Ishiwa A, Nagata N, Arakawa Y, Takahashi M (2010) Genome organization and pathogenicity of *Corynebacterium diphtheriae* C7(–) and PW8 strains. Infect Immun 78:3791–3800

Jolley KA, Chan MS, Maiden MC (2004) mlstdbNet – distributed multi-locus sequence typing (MLST) databases. BMC Bioinformatics 5:86

Kunkle CA, Schmitt MP (2005) Analysis of a DtxR-regulated iron transport and siderophore biosynthesis gene cluster in *Corynebacterium diphtheriae*. J Bacteriol 187:422–433

LaPointe P, Wei X, Gariepy J (2005) A role for the protease-sensitive loop region of Shiga-like toxin 1 in the retrotranslocation of its A1 domain from the endoplasmic reticulum lumen. J Biol Chem 280:23310–23318

Lewis CM Jr, Obregón-Tito A, Tito RY, Foster MW, Spicer PG (2012) The Human Microbiome Project: lessons from human genomics. Trends Microbiol 20:1–4

Mandlik A, Swierczynski A, Das A, Ton-That H (2007) *Corynebacterium diphtheriae* employs specific minor pilins to target human pharyngeal epithelial cells. Mol Microbiol 64:111–124

Mardis ER (2008) Next-generation DNA sequencing methods. Annu Rev Genom Hum Genet 9:387–402

Marraffini LA, Sontheimer EJ (2010) CRISPR interference: RNA-directed adaptive immunity in bacteria and archaea. Nat Rev Genet 11:181–190

Mattos-Guaraldi AL, Sampaio JL, Santos CS, Pimenta FP, Pereira GA, Pacheco LG, Miyoshi A, Azevedo V, Moreira LO, Gutierrez FL, Costa JL, Costa-Filho R, Damasco PV, Camello TC, Hirata R Jr (2008) First detection of *Corynebacterium ulcerans* producing a diphtheria-like toxin in a case of human with pulmonary infection in the Rio de Janeiro metropolitan area, Brazil. Mem Inst Oswaldo Cruz 103:396–400

McKean SC, Davies JK, Moore RJ (2007) Expression of phospholipase D, the major virulence factor of *Corynebacterium pseudotuberculosis*, is regulated by multiple environmental factors and plays a role in macrophage death. Microbiology 153:2203–2211

Medini D, Donati C, Tettelin H, Masignani V, Rappuoli R (2005) The microbial pan-genome. Curr Opin Genet Dev 15:589–594

Mokrousov I (2009) *Corynebacterium diphtheriae*: genome diversity, population structure and genotyping perspectives. Infect Genet Evol 9:1–15

Mokrousov I, Narvskaya O, Limeschenko E, Vyazovaya A (2005) Efficient discrimination within a *Corynebacterium diphtheriae* epidemic clonal group by a novel macroarray-based method. J Clin Microbiol 43:1662–1668

Morozova O, Marra MA (2008) Applications of next-generation sequencing technologies in functional genomics. Genomics 92:255–264

O'Loughlin EV, Robins-Browne RM (2001) Effect of Shiga toxin and Shiga-like toxins on eukaryotic cells. Microbes Infect 3:493–507

Ott L, Höller M, Rheinlaender J, Schäffer TE, Hensel M, Burkovski A (2010) Strain-specific differences in pili formation and the interaction of *Corynebacterium diphtheriae* with host cells. BMC Microbiol 10:257

Pagani I, Liolios K, Jansson J, Chen IM, Smirnova T, Nosrat B, Markowitz VM, Kyrpides NC. 2012. The Genomes OnLine Database (GOLD) v.4: status of genomic and metagenomic projects and their associated metadata. Nucleic Acids Res. 40:D571–579.

Park WH, Williams AW (1896) The production of diphtheria toxin. J Exp Med 1:164–185

Qian Y, Lee JH, Holmes RK (2002) Identification of a DtxR-regulated operon that is essential for siderophore-dependent iron uptake in *Corynebacterium diphtheriae*. J Bacteriol 184:4846–4856

Rappuoli R, Michel JL, Murphy JR (1983) Restriction endonuclease map of corynebacteriophage γ_c^{tox+} isolated from the Park-Williams no. 8 strain of *Corynebacterium diphtheriae*. J Virol 45:524–530

Ratti G, Covacci A, Rappuoli R (1997) A $tRNA_2^{Arg}$ gene of *Corynebacterium diphtheriae* is the chromosomal integration site for toxinogenic bacteriophages. Mol Microbiol 25:1179–1181

Rogers EA, Das A, Ton-That H (2011) Adhesion by pathogenic corynebacteria. Adv Exp Med Biol 715:91–103

Rothberg JM, Leamon JH (2008) The development and impact of 454 sequencing. Nat Biotechnol 26:1117–1124

Ruiz JC, D'Afonseca V, Silva A, Ali A, Pinto AC, Santos AR, Rocha AA, Lopes DO, Dorella FA, Pacheco LG, Costa MP, Turk MZ, Seyffert N, Moraes PM, Soares SC, Almeida SS, Castro TL, Abreu VA, Trost E, Baumbach J, Tauch A, Schneider MP, McCulloch J, Cerdeira LT, Ramos RT, Zerlotini A, Dominitini A, Resende DM, Coser EM, Oliveira LM, Pedrosa AL, Vieira CU, Guimarães CT, Bartholomeu DC, Oliveira DM, Santos FR, Rabelo ÉM, Lobo FP, Franco GR, Costa AF, Castro IM, Dias SR, Ferro JA, Ortega JM, Paiva LV, Goulart LR, Almeida JF, Ferro MI, Carneiro NP, Falcão PR, Grynberg P, Teixeira SM, Brommonschenkel S, Oliveira SC, Meyer R, Moore RJ, Miyoshi A, Oliveira GC, Azevedo V (2011) Evidence for reductive genome evolution and lateral acquisition of virulence functions in two *Corynebacterium pseudotuberculosis* strains. PLoS One 6:e18551

Salzberg SL (2007) Genome re-annotation: a wiki solution? Genome Biol 8:102

Schröder J, Maus I, Meyer K, Wördemann S, Blom J, Jaenicke S, Schneider J, Trost E, Tauch A (2012) Complete genome sequence, lifestyle, and multi-drug resistance of the human pathogen *Corynebacterium resistens* DSM 45100 isolated from blood samples of a leukemia patient. BMC Genomics 13:141

Shendure J, Ji H (2008) Next-generation DNA sequencing. Nat Biotechnol 26:1135–1145

Silva A, Schneider MP, Cerdeira L, Barbosa MS, Ramos RT, Carneiro AR, Santos R, Lima M, D'Afonseca V, Almeida SS, Santos AR, Soares SC, Pinto AC, Ali A, Dorella FA, Rocha F, Abreu VA de, Trost E, Tauch A, Shpigel N, Miyoshi A, Azevedo V (2011) Complete genome sequence of *Corynebacterium pseudotuberculosis* I19, a strain isolated from a cow in Israel with bovine mastitis. J Bacteriol 193:323–324

Soares SC, Dorella FA, Pacheco LG, R. Hirata R Jr, Mattos-Guaraldi AL, Azevedo V, Miyoshi A (2011) Plasticity of *Corynebacterium diphtheriae* pathogenicity islands revealed by PCR. Genet Mol Res 10:1290–1294

Soares SC, Abreu VA, Ramos RT, Cerdeira L, Silva A, Baumbach J, Trost E, Tauch A, Hirata R Jr, Mattos-Guaraldi AL, Miyoshi A, Azevedo V (2012) PIPS: Pathogenicity Island Prediction Software. PLoS One 7:e30848

Soriano F, Tauch A (2008) Microbiological and clinical features of *Corynebacterium urealyticum*: urinary tract stones and genomics as the Rosetta Stone. Clin Microbiol Infect 14:632–643

Tauch A, Trost E, Tilker A, Ludewig U, Schneiker S, Goesmann A, Arnold W, Bekel T, Brinkrolf K, Brune I, Götker S, Kalinowski J, Kamp PB, Lobo FP, Viehoever P, Weisshaar B, Soriano F, Dröge M, Pühler A (2008a) The lifestyle of *Corynebacterium urealyticum* derived from its complete genome sequence established by pyrosequencing. J Biotechnol 136:11–21

Tauch A, Schneider J, Szczepanowski R, Tilker A, Viehoever P, Gartemann KH, Arnold W, Blom J, Brinkrolf K, Brune I, Götker S, Weisshaar B, Goesmann A, Dröge M, Pühler A (2008b) Ultrafast pyrosequencing of *Corynebacterium kroppenstedtii* DSM44385 revealed insights into the physiology of a lipophilic corynebacterium that lacks mycolic acids. J Biotechnol 136:22–30

Tettelin H, Riley D, Cattuto C, Medini D (2008) Comparative genomics: the bacterial pan-genome. Curr Opin Microbiol 11:472–477

Ton-That H, Schneewind O (2003) Assembly of pili on the surface of *Corynebacterium diphtheriae*. Mol Microbiol 50:1429–1438

Trost E, Götker S, Schneider J, Schneiker-Bekel S, Szczepanowski R, Tilker A, Viehoever P, Arnold W, Bekel T, Blom J, Gartemann KH, Linke B, Goesmann A, Pühler A, Shukla SK, Tauch A (2010a) Complete genome sequence and lifestyle of black-pigmented *Corynebacterium aurimucosum* ATCC 700975 (formerly *C. nigricans* CN-1) isolated from a vaginal swab of a women with spontaneous abortion. BMC Genomics 11:91

Trost E, L O, Schneider J, Schröder J, Jaenicke S, Goesmann A, Husemann P, Stoye J, Dorella FA, Rocha FS, Castro SS de, D'Afonseca V, Miyoshi A, Ruiz J, Silva A, Azevedo V, Burkovski A, Guiso N, Join-Lambert OF, Kayal S, Tauch A (2010b) The complete genome sequence of *Corynebacterium pseudotuberculosis* FRC41 isolated from a 12-year-old girl with necrotizing lymphadenitis reveals insights into gene-regulatory networks contributing to virulence. BMC Genomics 11:728

Trost E, Al-Dilaimi A, Papavasiliou P, Schneider J, Viehoever P, Burkovski A, Soares SC, Almeida SS, Dorella FA, Miyoshi A, Azevedo V, Schneider MP, Silva A, Santos CS, Santos LS, Sabbadini P, Dias AA, Hirata R Jr, Mattos-Guaraldi AL, Tauch A (2011) Comparative analysis of two complete *Corynebacterium ulcerans* genomes and detection of candidate virulence factors. BMC Genomics 12:383

Trost E, Blom J, Castro SS de, Huang IH, Al-Dilaimi A, Schröder J, Jaenicke S, Dorella FA, Rocha FS, Miyoshi A, Azevedo V, Schneider MP, Silva A, Camello TC, Sabbadini PS, Santos CS, Santos LS, Hirata R Jr, Mattos-Guaraldi AL, Efstratiou A, Schmitt MP, Ton-That H, Tauch A (2012) Pangenomic study of *Corynebacterium diphtheriae* that provides insights into the genomic diversity of pathogenic isolates from cases of classical diphtheria, endocarditis, and pneumonia. J Bacteriol 194:3199–3215

Chapter 4
Corynephages: Infections of the Infectors

Vartul Sangal and Paul A. Hoskisson

Abstract *Corynebacterium diphtheriae* is the etiological agent of diphtheria; an acute toxin-mediated infection of the upper respiratory tract and skin. The toxin gene is encoded on a temperate bacteriophage, which during lysogeny is capable of toxin production. In this chapter, we will summarise current knowledge regarding corynephage, focusing on the lifecycle and biology of the toxin-carrying corynephages that are known to convert pathogenic corynebacteria, and discuss the insights recent genome corynebacterial sequencing has given us.

Keywords Bacteriophage · Corynephage · *Corynebacterium* · Diphtheria · Phage conversion · Pathogenicity · Toxigenic · Toxin

4.1 Introduction

Diphtheria is an acute, toxin-mediated disease of the upper respiratory tract, skin and occasionally other mucous membranes caused by the *Corynebacterium diphtheriae* and its relatives. Whilst there is a requirement for colonization of the host by *C. diphtheriae* (or its close relative *Corynebacterium ulcerans*), it is the infection of the bacterium by a group of closely related bacteriophages, which carry a toxin gene that causes the primary clinical symptoms in humans. It is this intimate link between the host (the bacterium in this case), the pathogen (the bacteriophage) and the human host that makes the biology of corynephages so fundamental to global human health.

In 1951, Freeman reported that following exposure of non-toxigenic *C. diphtheriae* to bacteriophage, the isolation of diphtheria toxin-producing strains was possible (Freeman 1951). Subsequently it was shown that there was a direct link

---

P. A. Hoskisson (✉) · V. Sangal
Strathclyde Institute of Pharmacy and Biomedical Science, University of Strathclyde,
161 Cathedral Street, Glasgow G4 0RE UK
e-mail: Paul.hoskisson@strath.ac.uk

A. Burkovski (ed.), *Corynebacterium diphtheriae and Related Toxigenic Species*,
DOI 10.1007/978-94-007-7624-1_4,

between toxin production and lysogeny (Freeman and Morse 1952). Initial hypotheses suggested that the bacteriophage selected for toxin-producing mutants, however it was later shown that it was a bacteriophage-specific phenotype and that the β phage carried the genetic material for toxin production (Groman 1953; Bardsdale and Pappenheimer 1954; Groman 1955). Despite its importance, limited efforts were made to study pathobiology of corynebacterial phages over last 60 years. In this chapter we will review the current knowledge of corynephages, focusing on toxigenic phage, the role they play in corynebacterial biology, discuss the role of genomics in understanding corynephage dynamics from evolutionary point and also some of the clues to bacteriophage resistance from the multitude of corynebacterial genome sequences that are increasingly available.

4.2 Non-toxigenic Corynephages

Bacteriophage have been studied extensively for many years and were a driving force in the molecular biology revolution, in terms of their use as model systems and tools to study a wide range of biological processes. However the role of bacteriophage in the evolution of bacteria cannot be under-estimated. The extensive sequencing of bacterial and archaeal genomes that has been undertaken in the last decade has lead to a greater understanding of the role of bacteriophages in genome evolution and has lead to a resurgence in interest in bacteriophage biology (Brüssow et al. 2004).

The acquisition of beneficial DNA sequences through horizontal gene transfer (conjugation, transposition, transformation, transduction and lysogenisation—the last two in this list being bacteriophage-mediated) offers a significant selective advantage under a wide range of conditions, extending the genetic diversity of bacterial strains. These mechanisms often lead to loss as well as gain of genes (Lawrence and Ochman 1997; Jain et al. 2002), and as suggested by Brüssow et al. (2004), it is the transient selective advantage of gene loss/gain in a dynamic environment that is the real driving force of bacterial evolution.

Despite the industrial and medical importance of corynebacteria, relatively little is known about their bacteriophages. Several bacteriophages have been isolated for *Corynebacterium glutamicum* strains (some strains previously designated *Brevibacterium flavum*), an industrial producer of amino acids, due to their negative impact on industrial production (Sonnen et al. 1990) or for the development vectors for site-specific insertion and genetic manipulation (Oram et al. 2002). The corynephages against *C. glutamicum* have generally been isolated directly from production fermenters, from areas close to production facilities (Kato et al. 1984; Trautwetter et al. 1987; Sonnen et al. 1990) or from soil (Trautwetter and Blanco 1988), which, as one might expect is a good hunting ground for corynephages. The corynephages that have been isolated against these strains generally have a very narrow host range due to the close genetic relationship between the industrial bacterial strains (Trautwetter and Blanco 1988; Sonnen et al. 1990). Little is understood of the wider ecology of corynephage infecting *C. diphtheriae* and *C. ulcerans*, such as their abundance in their natural environmental niches, their host-range and the level of genome diversity amongst corynephages.

Virulent corynephage for *C. glutamicum* have largely been isolated from production facilities where they were found to significantly reduce productivity through slow growth and lysis of cultures (Halgasova et al. 2005). There have been a number of corynephage isolated, largely belonging to the B group of bacteriophages based on the Bradley classification corresponding to the *siphoviridae* (Bradley 1967; Ackermann 2009). Koptides et al. (1992) isolated the virulent phage BFK20 from *C. (Brevibacterium) flavum*, which was found to be a double-stranded DNA bacteriophage with a non-contractile tail and a 50 nm polyhedral head. There has been considerable further work on the analysis of BFK20 in terms of its genome sequence (EMBL Acc. No.: AJ278322; (Bukovska et al. 2006)), its replication (Halgasova et al. 2005) and host cell binding and entry (Gerova et al. 2011). Trautwetter et al. (1987) isolated CG33 from *C. glutamicum* that was found to be a group B bacteriophage of comparable morphology to BFK20 (Koptides et al. 1992). Similarly, Trautwetter and Blanco (1988) isolated 20 novel corynephage for *Corynebacterium* and reported that 19 of these were virulent. All appeared to be Group B (*siphoviridae*), with non-contractile tails of around 100 nm and polyhedral heads in the range of 41–57 nm.

Temperate corynephages are so far the best studied examples, primarily due to the ability to induce them through ultra-violet light treatment (Kato et al. 1984; Patek et al. 1985; Moreau et al. 1995; Frunzke et al. 2008). The temperate corynephage BK1 was isolated from *C. (Brevibacterium) ketoglutamicum* through the isolation of cloudy plaques (Trautwetter and Blanco 1988), and was further demonstrated to be inducible using mitomycin C. Patek et al. (1985) isolated three new corynephage from *C. glutamicum* CBII, each being non-contractile tailed DNA phages, yet were little characterized thereafter. Isolation of the temperature corynephage ϕGA1 from a *C. (Brevibacterium) flavum* ATCC14067 lysogen is particularly interesting given the relationship this corynephage has with the virulent corynephage, Cog, from *C. glutamicum* LP-6 (Sonnen et al. 1990). The temperate corynephage ϕGA1 was found not to cause plaques on a wide range of strains or to cause lysis following induction. The virulent phage Cog was only able to plaque on *C. glutamicum* LP-6. Cog was also found to adsorb to a range of strains, but was not able to form plaques. Interestingly, Southern hybridizations between these two corynephages indicated high levels of homology between the two, suggesting a close relationship at the genetic level. It was hypothesized that introduction of DNA from one phage, without establishing an infection, is a driver of bacteriophage evolution through recombination of homologous regions between corynephages and as such may generate novel recombinant bacteriophages (Sonnen et al. 1990). This homologous recombination between related bacteriophages, which share partial homology can result in the exchange of small functional units or 'gene modules' resulting in the emergence of new bacteriophages and forms the basis of why we see the so-called mosaic genomes of bacteriophage (Hatfull and Hendrix 2011). Given this close relationship, in terms of morphology and genetics, between the temperate and lytic corynephages and the *Corynebacterium* strains used for industrial amino acid production, it is not surprising that the corynephages isolated to date have all appeared to be related, and with limited studies also suggesting a close relationship at the genetic level. This is an area that requires further investigation given the industrial and medical importance of corynebacteria, however it is clear that we are currently underestimating the diversity of these phages and it is an area that should receive more attention.

4.3 Toxigenic Corynephages

The conversion of non-toxigenic pathogenic corynebacteria (*C. diphtheriae* and *C. ulcerans*) to toxigenic strains through lysogeny of a bacteriophage is a fascinating story of host-pathogen interaction. The colonization of human hosts by non-toxigenic *C. diphtheriae* that are subsequently converted to toxigenic strains, causing clinical diphtheria symptoms are known in the literature (Simmons et al. 1980). However, this finding proved elusive in a natural setting for many years. The early work of Freeman (1951, 1952) showed that there was a direct link between lysogeny of a bacteriophage, the formation diphtheria toxin and that the β phage carried the gene or genes required for toxin production (Groman 1953; Bardsdale and Pappenheimer 1954; Groman 1955). The majority of our knowledge regarding the conversion of strains to toxigeny has come from study of β phage, yet ω and γ phage, are also known to have converting activity, with up to nine temperate *C. diphtheriae* corynephage known (Holmes and Barksdale 1970). Holmes and Barksdale (1970) subsequently showed that six of their nine temperate corynephages are toxin-positive (α, β, δ, L, P, π) and were able convert, while a seventh, although non-converting carried the toxin gene (γ). Interestingly data investigating the relationships and evolution of these groups of converting and non-converting strains is scarce, with restriction mapping being the primary discriminatory feature. Until recently only one published, fully sequenced toxigenic corynephage genome was available (Cerdeno-Tarraga et al. 2003), however recent sequencing projects have contributed significantly to our knowledge of toxin carrying corynephage (Sangal et al. 2012a, b; Trost et al. 2012).

The plaques of α, β, δ, L, P, π, K, γ and ρ are generally uniform, turbid and reach around 2 mm in diameter, forming densely turbid peripheries upon prolonged incubation (Holmes and Barksdale 1970). The same authors performed one-step growth curves on the phages and found that β, γ and L phage were indistinguishable with a mean burst size of 37 phage per cell over a latent period of 65 min, followed by a rise period of 18 min. The β phage is the best studied of the converting corynephages and is a typical lambdoid bacteriophage, with a polyhedral head of around 55 nm in diameter and with a long (270 nm) tail (Holmes and Barksdale 1970). The genome is linear double-stranded DNA and was estimated to be 34.7 kbp by restriction endonuclease mapping and could be circularized through cohesive ends (*cos*) (Buck et al. 1978). Initial estimates suggested that there was enough DNA to code for 30–40 proteins (Groman 1984), yet it was not until the genome of *C. diphtheriae gravis* was sequenced that the full complement of a toxin-carrying phage genome was known to be 42 genes (Cerdeno-Tarraga et al. 2003). The toxin-producing corynephages appears to be conventional in their lytic cycles and of course can form lysogenic infections resulting in the formation of toxin producing strains (Groman 1984). Interestingly, as expected most phage genes are repressed during lysogeny, yet it is of note that toxin is produced by *C. diphtheriae* during this period and expression is related to the presence of the global iron regulator DtxR and iron availability in the host (Kunkle and Schmitt 2003, 2005).

4.4 Identifying the Diversity of Corynebacterial Prophages

Phages affect bacterial genome architecture and are major drivers in bacterial evolution. The majority of corynephages identified to date have been through the analysis of genome sequences as prophages (Frunzke et al. 2008). Using bioinformatics tools such as Phage_finder (http://phage-finder.sourceforge.net/: (Fouts 2006)) or PHAST (http://phast.wishartlab.com) (Zhou et al. 2011)) it is possible to identify prophages within completely sequenced bacterial genomes. The range and number of prophages within fully sequenced *C. diphtheriae* and its potentially toxigenic relatives genomes varies widely. It is however interesting to note the diversity of prophages carried by different strains (Table 4.1). Some strains of *C. diphtheriae* carry multiple prophages, with *C. diphtheriae* C7 beta carrying four, one with homology to BFK20, two with homology to mycophages and one similar to an *Escherichia coli* bacteriophage (http://phast.wishartlab.com/Download.html). *C. diphtheriae* HCO2 carries three corynephages, one showing homology to an *Escherichia coli* phage, one to a phage from *Geobacillus* and another similar to an *Enterococcus* phage (Table 4.1). Various corynebacterial strains detailed in Table 4.1 contain single or multiple phages, with homology to bacteriophages from a diverse range of organisms such as *Burkholderia, Pseudomonas, E. coli, Brucella, Rhodococcus, Streptomyces, Mycobacterium, Ralstonia* and *Listeria*. Interestingly, the toxigenic bacteriophages in *C. diphtheriae* strains C7, CDCE8392, PW8 and NCTC13129 all show homology to the *C. glutamicum* phage BFK20 (see above). This homology is especially conserved in the late genes, encoding proteins that function in the assembly of the bacteriophage capsid and tail (Bukovska et al. 2006). Interestingly significant homology to BFK20 also is apparent in a prophage contained in the genome of *C. lipophiloflavum* (http://phast.wishartlab.com/Download.html), suggesting the ubiquity of this phage type amongst corynebacteria.

Table 4.1 Distribution of prophages in sequenced coynebacterial genomes using the pre-computed database from PHAST (Zhou et al. 2011). Data was correct as of 1st May 2012

Species and strain	PHAST Accession number	Number of predicted prophage
Corynebacterium jeikeium ATCC 43734	ACYW00000000.1	2
Corynebacterium pseudotuberculosis 1002	CP001809.2	0
Corynebacterium ulcerans 809	CP002790.1	1
Corynebacterium diphtheriae 31A	CP003206.1	1
Corynebacterium diphtheriae 241	CP003207.1	1
Corynebacterium diphtheriae INCA 402	CP003208.1	1
Corynebacterium diphtheriae BH8	CP003209.1	1
Corynebacterium diphtheriae C7 (beta)	CP003210.1	4
Corynebacterium diphtheriae CDCE 8392	CP003211.1	2
Corynebacterium diphtheriae HC01	CP003212.1	1
Corynebacterium diphtheriae HC02	CP003213.1	3
Corynebacterium diphtheriae HC03	CP003214.1	1
Corynebacterium diphtheriae HC04	CP003215.1	0
Corynebacterium diphtheriae PW8	CP003216.1	2
Corynebacterium diphtheriae VA01	CP003217.1	0
Corynebacterium diphtheriae NCTC 13129	NC_002935.2	1
Corynebacterium glutamicum ATCC 13032	NC_006958.1	1
Corynebacterium ulcerans BR-AD22	NC_015683.1	3
Corynebacterium lipophiloflavum DSM 44291	ACHJ00000000.1	3

4.5 Comparative Genomics of Toxigenic Corynephage

Prophages containing *tox* genes were recently identified in multiple strains of *C. diphtheriae* C7, CDCE8392, PW8 and NCTC13129 (Trost et al. 2012) in addition Sangal et al. (2012a, b) also identified homologous *tox* gene-containing corynephages in two strains, one a *C. diphtheriae mitis* strain (NCTC03529) and a *C. diphtheriae intermedius* strain (NCTC05011) indicating the prominence of this corynephage lineage within the species (Fig. 4.1). One point of note is the presence of two copies of corynephage ω in *C. diphtheriae* PW8, a strain which is used widely

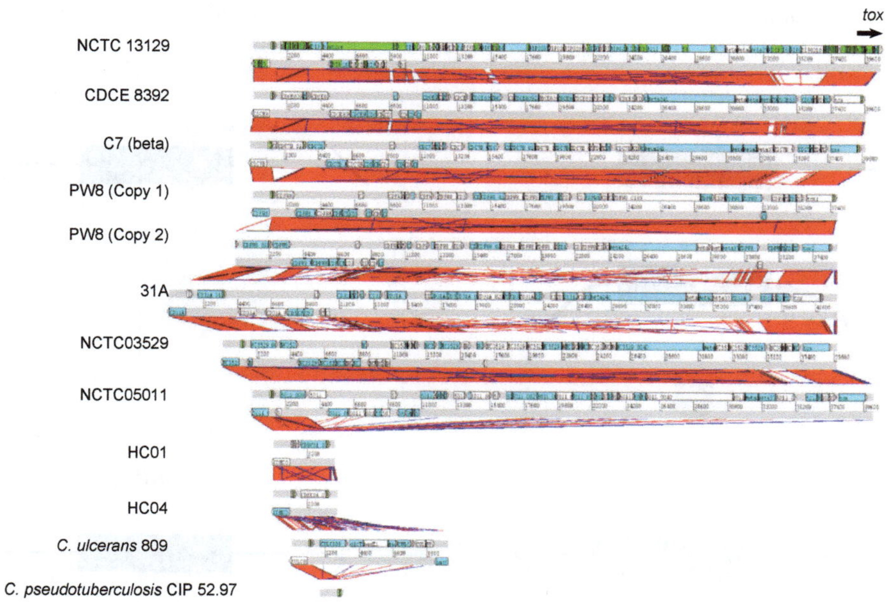

Fig. 4.1 Linear comparison of corynephage sequences from *C. diphtheriae* strains NCTC13129, CDCE8392, C7(beta), PW8, 31A, NCTC03529 and NCTC05011 (GenBank Accession Nos.: BX248353.1, CP003211.1, CP003210.1, CP003216.1, CP003206.1, AJGI00000000.1 and AJVH00000000.1, respectively) using Artemis Comparison Tool (ACT). These sequences were extracted from genomes after a nucleotide blast search of sequences from DIP0178-DIP0224 of *C. diphtheriae* NCTC13129. Homologous sequences were also extracted from non-toxigenic *C. diphtheriae* strains HC1 (Acc. No.: CP003212.1) and HC4 (Acc. No.: CP003215.1), *C. ulcerans* strain 809 (Acc. No.: CP002790.1) and *C. pseudotuberculosis* strain CIP52.97 (Acc. No.: CP003061.1). The location of the *tox* gene is indicated on the uppermost sequence for orientation

as a toxoid producer (Trost et al. 2012). The two copies of the phage differ only by 5 nucleotides in a genome of 36 kbp. Trost et al. (2012) identified the genomes of the toxin-producing β and ω corynephages from the published restriction maps of the prophages (Michel et al. 1982), with ω being found in *C. diphtheriae* PW8 and β corynephage being found in *C. diphtheriae* C7, CDCE8392 and a divergent version of the β phage (differing in sequence and gene content) in *C. diphtheriae* NCTC 13129. Remarkably, these strains are related to the *tox*⁻ corynephage BFK20, known from the literature and can explain some of the converting activity and recombination observed in the early studies given the high degree of homology. Comparison of the *tox*⁺corynephages showed high levels of homology, with the remarkable exception of the *tox*⁺coryneprophage in *C. diphtheriae* 31A, a strain isolated from a vaccinated adult in Rio de Janeiro, Brasil (Trost et al. 2012). This coryneprophage shows homology to the β-like phages predominantly in the *tox* containing, right-hand end of the prophage genome. Interestingly this prophage shows homology to some of the genes found in the prophage CULC22IV from *C. ulcerans* BR-AD22 (Trost et al. 2011) and similarity to a *Streptomyces* phage ϕSASD1 (http://phast.

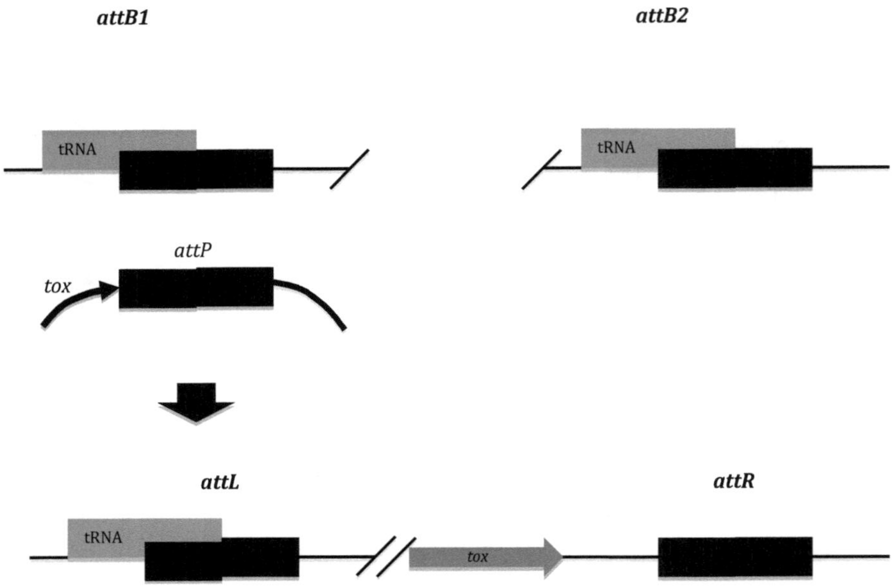

Fig. 4.2 A schematic representation of the tandem *attB* sites in the *Corynebacterium diphtheriae* genome, the formation of *attL* and *attR* following integration of the *tox*$^+$corynephage and the proximity of *tox* to the right-hand end of the corynephage genome. Note the difference in the two half-sites for *attB* representing the functional constraint on the reconstitution of a functional tRNA at *attL* upon excision

wishartlab.com/Download.html). These data suggest that *tox*$^+$ corynephages have evolved on at least two occasions in toxigenic corynebacteria or that gene shuffling occurs frequently in corynephage (Trost et al. 2012).

4.6 Integration of the Toxigenic Corynephage DNA

The integration of toxigenic corynephage in to the *C. diphtheriae* genome, as in many bacteriophages, insertion sequences and pathogenicity islands, occurs in tRNA genes, the most obvious evolutionary reason for this is the conservation and redundancy of tRNA genes in bacterial genomes. There are two distinct attachment sites found within the *C. diphtheriae* genome, *attB1* and *attB2* (Rappuoli et al. 1983). These two sites overlap with a putative $tRNA_2^{Arg}$ gene that flank a putative membrane protein and is conserved in all *C. diphtheriae* genomes (Ratti et al. 1997; Trost et al. 2012). The *attB* sites can be divided in to two segments (half-sites), the first half-site has perfect identity in both *att* sites found in the genome. The second half-site however contains several point mutations. This is likely the result of the overlap of the first half-site with the tRNA, and the constraint to resolve *attL* completely to reconstitute functionality (Ratti et al. 1997, Fig. 4.2). Earlier studies had also shown that *C. diphtheriae* could form stable single, double or triple lysogens (Rappuoli et al. 1983) with a concomitant increase in the production of toxin. Interestingly, the recent pan-genome study of Trost et al (2012) demonstrated the stable insertion of two ω^{tox+} phages in to the *attB1* and *attB2* sites in the *C. diphtheriae* PW8 genome, this indicating why this strain was probably selected as one of the main toxoid producing strains for vaccine.

Integration of the corynephage is mediated by a tyrosine-like recombinase/integrase, encoded by DIP0182 in *C. diphtheriae* NCTC 13129 (Oram et al. 2007). The mechanism of tyrosine recombinases is well studied based primarily on the similarity to the bacteriophage lambda integrase (for a detailed explanation of the integration mechanism see (Van Duyne 2005).

4.7 Evolution of Toxin Regulation by a Host Transcription Factor

The intriguing link between regulation of toxin production and iron-responsive genes suggests an intimate link between the DtxR regulon in the host bacterium and the infecting corynephage. Iron is an essential nutrient for cellular processes such as respiration and is often limited in mammalian hosts, suppressing bacterial growth; therefore pathogenic microorganisms invest heavily in acquisition and regulation of iron-scavenging and iron-responsive genes such as siderophores and proteins containing iron centres. The production of toxin in *tox*$^{+}$ strains of *C. diphtheriae* is regulated in an iron-dependent manner by DtxR, the diphtheria toxin repressor (De Zoysa et al. 2005). DtxR controls a complex network of genes involved in iron homeostasis and is highly conserved throughout the mycolata and is physiologically similar to the Fur (Ferric Uptake Regulator) protein of Gram-negative bacteria (De Zoysa et al. 2005; Wennerhold and Bott 2006). In low iron concentrations, DtxR is de-repressed from multiple promoters resulting in gene expression, including the transcription of the *tox* gene in lysogenized strains. There is also evidence of gene activation by DtxR in *C. glutamicum* (Wennerhold and Bott 2006), and given the high degree of regulon conservation it is likely that DtxR may function as an activator in other corynebacteria too. Recent comparative analysis of the pan-genome of *C. diphtheriae* revealed a high degree of conservation in the DtxR regulon across 13 strains of *C. diphtheriae* (Trost et al. 2012), with the 19 bp operator sequence being found upstream of 36 genes, including 26 that highly conserved across all strains. Interestingly the *tox* gene when carried on a corynephage is also subject to regulation by DtxR. This link between toxin production encoded by a bacteriophage and the evolution of its regulation by a host strain transcription factor is currently not well understood.

It is tempting to speculate on the evolution of this co-regulation, based on the evidence of the poorly defined non-toxigenic *C. diphtheriae* strains that carry a cryptic copy of the *tox* gene (Cianciotto and Groman 1997). It was speculated that these strains might represent a source for *tox* sequence that could be activated through recombination or mutation due to selection. Non-converting corynephage from non-toxigenic *C. diphtheriae* strains have been known since the 1950s, where *tox* was not expressed due to insertional inactivation (Groman 1955; Michel et al. 1982). Cianciotto and Groman (1997) characterised these strains and showed, through complementation tests using double lysogen strains, that the non-toxigenic phenotype was the result of the phages being *cis* dominant. This lead the authors to hypothesize that toxin-positive strains may arise by either homologous recombination between two distinct *tox*$^-$ phages or spontaneous reversion within a single mutant allele. Given the *tox* gene lies at the right-hand end of the integrated corynephages, it has been suggested that it was acquired by the corynephage by recombination (Cerdeno-Tarraga et al. 2003). This may have resulted in the gene lying in proximity to an inverted repeat (potentially homologous to the *dtxR* operator) towards the end of the corynephage genome, or that the *tox* gene was acquired with its upstream regulatory region from another strain. Recently there has been significant interest in the evolution of regulatory networks in horizontally acquired genes in closely related strains (Perez and Groisman 2009; Chen et al. 2011). The reliance on orthologous regulators (in this case DtxR) to orchestrate responses (in this case to iron limitation) conferring changes in gene expression in horizontally acquired genes (in the case *tox*) to survive deleterious conditions is an attractive scenario, especially given that diphtheria toxin causes host cell lysis, releasing a potential source of iron (Chang et al. 1989). Thus the bacteriophage-encoded promoter for a specific nutrient acquisition factor is under bacterial control (Ratti et al. 1997) offering a significant advantage to cells that have colonized a mammalian host.

4.8 Corynephage of *Corynebacterium ulcerans*

Recent genome sequencing and interest in corynebacteria has lead to the study of genomes of related organisms. Two genomes of the animal commensal and increasingly recognized zoonotic infection reservoir *C. ulcerans* was recently sequenced and have provided significant insight in to the biology of another corynebacterial strain capable of conversion by β phages (Trost et al. 2011). Both sequenced strains differ significantly in their prophage complement, with *C. ulcerans* 809 carrying a single prophage and *C. ulcerans* BR-AD22 carrying four prophage-like regions. Previous work had suggested that there was potential for *C. ulcerans* to carry β-corynephage, based on PCR assays of the *tox* gene (Mattos-Guaraldi et al. 2008), however no *tox* gene or β-corynephage sequences could be identified in the *C. ulcerans* 809 or *C. ulcerans* BR-AD22 genomes. Analysis showed that there was little synteny between the integration regions of the *C. diphtheriae* and *C. ulcerans* genomes, however remnants of a putative corynephage are present close to the tRNA^{Arg} gene and the presence of a tyrosine recombinase, homologous to that found in the β-corynephages, suggests that the *C. ulcerans* genome may have previously accommodated a β-corynephage. These data also link to the homologous genes found in the novel toxin-producing corynephage present in the genome of *C. diphtheriae* C31A (Trost et al. 2012), providing further opportunity for recombination amongst homologous phage regions.

4.9 Corynephage Resistance Mechanisms

To bacteria the constant threat of bacteriophage infection has driven the emergence and acquisition of defense genes to prevent the deleterious effects of bacteriophage infection, however it is possible that bacteriophage infection permits the acquisition of novel, possibly beneficial, genes (Hoskisson and Smith 2007). The diversity of the bacteriophage resistome in bacteria is enormous and the opportunity to acquire novel genes (such as toxin production in the case of *C. diphtheriae*) has resulted in the evolution of hypervariable resistance mechanisms.

Analysis of recently sequenced *C. diphtheriae* genomes and re-analysis of existing sequences have demonstrated that *C. diphtheriae* is a much more diverse bacterium than previously thought. While the genome of *C. diphtheriae* has a highly conserved genomic backbone there is considerable variation outside of the syntenous regions, especially around the origin of replication (Trost et al. 2012). Much of this variation across strains is due to insertion sequences (Trost et al. 2012), however it is also clear that there is a lot of variation in sequences associated with resistance to horizontal gene transfer. Mokrousov (2009) completed a detailed *in silico* analysis of the so-called CRISPR (clustered regularly interspaced short palindromic repeats) regions of the *C. diphtheriae* NCTC 13129 genome, identifying two CRISPR loci, and it is this diversity that also allowed the development of a spoligotyping method to determine the phylogenetic relationship between strains (Mokrousov et al. 2007; Mokrousov 2009). CRISPRs are a platform for acquiring resistance against phage infection, where short sequences of DNA from an infecting bacteriophage are inserted between a conserved repeated sequence within a CRISPR and thereafter confers resistance to that bacteriophage, through a mechanism similar to RNA interference, along with their associated *cas* genes which process the incoming foreign DNA (Marraffini and Sontheimer 2010). In *C. diphtheriae* NCTC 13129 four of the seven CRISPR spacer regions showed homology with hypothetical proteins from bacteriophage or plasmid/transposon encoded genes involved in integration and/or recombination (Mokrousov 2009). More recently, extensive further analysis of the CRISPR loci demonstrated the presence of three different arrays across 13 strains (Trost et al. 2012). Type I CRISPR locus is composed of three *cas* genes with between one and 28 spacers, and is the most widespread of the CRISPR arrays found in eight of the 13 strains analysed by Trost et al. (2012). Type II arrays were only found in two of the analysed strains and consist of eight *cas* genes and between four and 26 repeats. The type III array was found in five of the 13 strains and again consists of eight *cas* genes with 12–42 arrays. In addition Sangal et al. (2012a, b) indicated the presence of extensive *cas*/CRISPR arrays in *C. diphtheriae mitis* (NCTC03529) and *C. diphtheriae intermedius* (NCTC05011).

Traditionally it was thought that restriction-modification systems provided the main basis for bacteriophage resistance. Type II restriction-modification (R-M) systems have been characterised in *C. glutamicum* (Schafer et al. 1997) but few studies have been carried out other than *in silico* on the R-M systems of *C. diphtheriae.* The recent study of D'Afonseca et al. (2012) indicated that the Type I R-M system in *C. diphtheriae* NCTC 13129 maybe at least partially defective, due to the insertional inactivation of an potentially alternative modification subunit, *hsdM* (putative methylase; DIP2081), which is distant to the main genes associated with the R-M system (*hsdRSM;* DIP2312, DIP2313, DIP2314). It is possible that the inactivated methylase may play an important role in diversification of the R-M system and may result in an increase in plasticity of the genome in response to stress. Similar duplicated methylases are known in related actinomycetes such as the Phage Growth Limitation system in *Streptomyces* (Sumby and Smith 2003; Hoskisson and Smith 2007). Sangal et al. (2012b) also noted the presence of putative Type III R-M systems in the *C. diphtheriae intermedius* (NCTC05011) genome, adding to the diversity of observed R-M systems providing resistance to bacteriophages.

These data suggest that there is significant investment of genomic capacity in corynebacteria to prevent the deleterious effects of bacteriophage infection, suggesting that without these mechanisms corynebacterial genomes may be more plastic (D'Afonseca et al. 2012).

4.10 Summary

There is an intimate link between the ability to cause classical diphtheria disease symptoms by *C. diphtheriae* and the integration of a toxin carrying bacteriophage. The investment by the bacterium in bacteriophage resistance mechanisms has helped to shape the genome, but it is clear that there is a reliance on incoming genetic diversity to provide a selective advantage within a host, none more so that the provision of the diphtheria toxin to facilitate access to iron under limiting conditions. The control of a bacteriophage-encoded toxin gene by a host encoded transcription factor exemplifies the intimacy of the relationship between bacteria and their viruses and the genetic diversity that they can provide. There are still many questions to be answered in this system that will be applicable in a wider context to the evolution of bacterial pathogens. The *C. diphtheriae* tox^+ bacteriophage model system provides an ideal platform for answering questions regarding the acquisition and maintenance of horizontally transferred DNA, the regulation of horizontally transferred DNA and how this shapes the evolution of regulatory networks in pathogens. The next-generation sequencing revolution has significantly enhanced our understanding of genome structure—now is the time to begin to explore the dynamics of this structure.

Acknowledgements *Corynebacterium* work in the Hoskisson laboratory is supported by Medical Research Scotland (Grant FRG-422), the University of Strathclyde and the Society for General Microbiology. I would like to thank Florence Pethick for helpful comments on the manuscript.

References

Ackermann H-W (2009) Phage Classification and Characterization. In: Clokie MR, Kropinski AM (eds) Bacteriophages: methods and protocols, Volume 1: isolation, characterization, and interactions, vol 501. Methods in Molecular Biology. Humana Press, Totowa, pp 127–140

Bardsdale WL, Pappenheimer AM Jr (1954) Phage-host relationships in nontoxigenic and toxigenic diphtheria bacilli. J Bacteriol 67(2):220–232

Bradley DE (1967) Ultrastructure of bacteriophage and bacteriocins. Bacteriol Rev 31(4):230–314

Brussow H, Canchaya C, Hardt WD (2004) Phages and the evolution of bacterial pathogens: from genomic rearrangements to lysogenic conversion. Microbiol Mol Biol Rev 68(3):560–602

Buck G, Groman N, Falkow S (1978) Relationship between beta converting and gamma nonconverting corynebacteriophage DNA. Nature 271(5646):683–685

Bukovska G, Klucar L, Vlcek C, Adamovic J, Turna J, Timko J (2006) Complete nucleotide sequence and genome analysis of bacteriophage BFK20–a lytic phage of the industrial producer *Brevibacterium flavum*. Virology 348(1):57–71

Cerdeno-Tarraga AM, Efstratiou A, Dover LG, Holden MT, Pallen M, Bentley SD, Besra GS, Churcher C, James KD, De Zoysa A, Chillingworth T, Cronin A, Dowd L, Feltwell T, Hamlin N, Holroyd S, Jagels K, Moule S, Quail MA, Rabbinowitsch E, Rutherford KM, Thomson NR, Unwin L, Whitehead S, Barrell BG, Parkhill J (2003) The complete genome sequence and analysis of *Corynebacterium diphtheriae* NCTC13129. Nucleic Acids Res 31(22):6516–6523

Chang MP, Bramhall J, Graves S, Bonavida B, Wisnieski BJ (1989) Internucleosomal DNA cleavage precedes diphtheria toxin-induced cytolysis. Evidence that cell lysis is not a simple consequence of translation inhibition. J Biol Chem 264(26):15261–15267

Chen HD, Jewett MW, Groisman EA (2011) Ancestral genes can control the ability of horizontally acquired loci to confer new traits. PLoS Genet 7(7):e1002184

Cianciotto NP, Groman NB (1997) Characterization of bacteriophages from tox-containing, nontoxigenic isolates of *Corynebacterium diphtheriae*. Microb Pathog 22(6):343–351

D'Afonseca V, Soares SC, Ali A, Santos AR, Pinto AC, Magalhães AAC, Faria CJ, Barbosa E, Guimarães LC, Eslabão M, Almeida SS, Abreu VAC, Zerlotini A, Carneiro AR, Cerdeira LT, Ramos RTJ, Hirata Jr R, Mattos-Guaraldi AL, Trost E, Tauch A, Silva A, Schneider MP, Miyoshi A, Azevedo V (2012) Reannotation of the *Corynebacterium diphtheriae* NCTC13129 genome as a new approach to studying gene targets connected to virulence and pathogenicity in diphtheria. Open Access Bioinforma 4:1–13

De Zoysa A, Efstratiou A, Hawkey PM (2005) Molecular characterization of diphtheria toxin repressor (dtxR) genes present in nontoxigenic *Corynebacterium diphtheriae* strains isolated in the United Kingdom. J Clin Microbiol 43(1):223–228

Fouts DE (2006) Phage_Finder: automated identification and classification of prophage regions in complete bacterial genome sequences. Nucleic Acids Res 34(20):5839–5851

Freeman VJ (1951) Studies on the virulence of bacteriophage-infected strains of *Corynebacterium diphtheriae*. J Bacteriol 61(6):675–688

Freeman VJ, Morse IU (1952) Further observations on the change to virulence of bacteriophage-infected a virulent strains of *Corynebacterium diphtheria*. J Bacteriol 63(3):407–414

Frunzke J, Bramkamp M, Schweitzer JE, Bott M (2008) Population Heterogeneity in *Corynebacterium glutamicum* ATCC 13032 caused by prophage CGP3. J Bacteriol 190(14):5111–5119

Gerova M, Halgasova N, Ugorcakova J, Bukovska G (2011) Endolysin of bacteriophage BFK20: evidence of a catalytic and a cell wall binding domain. FEMS Microbiol Lett 321(2):83–91

Groman NB (1953) Evidence for the induced nature of the change from nontoxigenicity to toxigenicity in *Corynebacterium diphtheriae* as a result of exposure to specific bacteriophage. J Bacteriol 66(2):184–191

Groman NB (1955) Evidence for the active role of bacteriophage in the conversion of nontoxigenic *Corynebacterium diphtheriae* to toxin production. J Bacteriol 69(1):9–15

Groman NB (1984) Conversion by corynephages and its role in the natural history of diphtheria. J Hyg (Lond) 93(3):405–417

Halgasova N, Majtan T, Ugorcakova J, Timko J, Bukovska G (2005) Resistance of corynebacterial strains to infection and lysis by corynephage BFK 20. J Appl Microbiol 98(1):184–192

Hatfull GF, Hendrix RW (2011) Bacteriophages and their genomes. Curr Opin Virol 1(4):298–303

Holmes RK, Barksdale L (1970) Comparative studies with tox plus and tox minus corynebacteriophages. J Virol 5(6):783–784

Hoskisson PA, Smith MC (2007) Hypervariation and phase variation in the bacteriophage 'resistome'. Curr Opin Microbiol 10(4):396–400

Jain R, Rivera MC, Moore JE, Lake JA (2002) Horizontal gene transfer in microbial genome evolution. Theor Popul Biol 61(4):489–495

Kato F, Yoshimi M, Araki K, Motomura Y, Matsufune Y, Nobunaga H, Murata A (1984) Screening of Bacteriocins in Amino-Acid or Nucleic-Acid Producing Bacteria and Related Species. Agr Biol Chem 48(1):193–200

Koptides M, Barak I, Sisova M, Baloghova E, Ugorcakova J, Timko J (1992) Characterization of bacteriophage BFK20 from *Brevibacterium flavum*. J Gen Microbiol 138(7):1387–1391

Kunkle CA, Schmitt MP (2003) Analysis of the *Corynebacterium diphtheriae* DtxR regulon: identification of a putative siderophore synthesis and transport system that is similar to the *Yersinia* high-pathogenicity island-encoded yersiniabactin synthesis and uptake system. J Bacteriol 185(23):6826–6840

Kunkle CA, Schmitt MP (2005) Analysis of a DtxR-regulated iron transport and siderophore biosynthesis gene cluster in *Corynebacterium diphtheriae*. J Bacteriol 187(2):422–433

Lawrence JG, Ochman H (1997) Amelioration of bacterial genomes: rates of change and exchange. J Mol Evol 44(4):383–397

Marraffini LA, Sontheimer EJ (2010) CRISPR interference: RNA-directed adaptive immunity in bacteria and archaea. Nat Rev Genet 11(3):181–190

Mattos-Guaraldi AL, Sampaio JL, Santos CS, Pimenta FP, Pereira GA, Pacheco LG, Miyoshi A, Azevedo V, Moreira LO, Gutierrez FL, Costa JL, Costa-Filho R, Damasco PV, Camello TC, Hirata Jr R (2008) First detection of *Corynebacterium ulcerans* producing a diphtheria-like toxin in a case of human with pulmonary infection in the Rio de Janeiro metropolitan area, Brazil. Mem Inst Oswaldo Cruz 103(4):396–400

Michel JL, Rappuoli R, Murphy JR, Pappenheimer AM Jr (1982) Restriction endonuclease map of the nontoxigenic corynephage gamma c and its relationship to the toxigenic corynephage beta c. J Virol 42(2):510–518

Mokrousov I (2009) *Corynebacterium diphtheriae*: genome diversity, population structure and genotyping perspectives. Infect Genet Evol 9(1):1–15

Mokrousov I, Limeschenko E, Vyazovaya A, Narvskaya O (2007) *Corynebacterium diphtheriae* spoligotyping based on combined use of two CRISPR loci. Biotechnol J 2(7):901–906

Moreau S, Leret V, Le Marrec C, Varangot H, Ayache M, Bonnassie S, Blanco C, Trautwetter A (1995) Prophage distribution in coryneform bacteria. Res Microbiol 146(6):493–505

Oram DM, Avdalovic A, Holmes RK (2002) Construction and characterization of transposon insertion mutations in *Corynebacterium diphtheriae* that affect expression of the diphtheria toxin repressor (DtxR). J Bacteriol 184(20):5723–5732

Oram M, Woolston JE, Jacobson AD, Holmes RK, Oram DM (2007) Bacteriophage-based vectors for site-specific insertion of DNA in the chromosome of *Corynebacteria*. Gene 391(1–2):53–62

Patek M, Ludvik J, Benada O, Hochmannova J, Nesvera J, Krumphanzl V, Bucko M (1985) New bacteriophage-like particles in *Corynebacterium glutamicum*. Virology 140(2):360–363

Perez JC, Groisman EA (2009) Evolution of transcriptional regulatory circuits in bacteria. Cell 138(2):233–244

Rappuoli R, Michel JL, Murphy JR (1983) Integration of corynebacteriophages beta tox+, omega tox+, and gamma tox- into two attachment sites on the *Corynebacterium diphtheriae* chromosome. J Bacteriol 153(3):1202–1210

Ratti G, Covacci A, Rappuoli R (1997) A tRNA(2Arg) gene of *Corynebacterium diphtheriae* is the chromosomal integration site for toxinogenic bacteriophages. Mol Microbiol 25(6):1179–1181

Sangal V, Tucker NP, Burkovski A, Hoskisson PA (2012a) Draft genome of *Corynebacterium diphtheria* biovar *intermedius* NCTC 5011. J Bacteriol (in press)

Sangal V, Tucker NP, Burkovski A, Hoskisson PA (2012b) The Draft Genome Sequence of *Corynebacterium diphtheriae* bv. *mitis* NCTC 3529 Reveals Significant Diversity between the Primary Disease-Causing Biovars. J Bacteriol 194(12):3269

Schafer A, Tauch A, Droste N, Puhler A, Kalinowski J (1997) The *Corynebacterium glutamicum* cglIM gene encoding a 5-cytosine methyltransferase enzyme confers a specific DNA methylation pattern in an McrBC-deficient *Escherichia coli* strain. Gene 203(2):95–101

Simmons LE, Abbott JD, Macaulay ME, Jones AE, Ironside AG, Mandal BK, Stanbridge TN, Maximescu P (1980) Diphtheria carriers in Manchester: simultaneous infection with toxigenic and non-toxigenic *mitis* strains. Lancet 1(8163):304–305

Sonnen H, Schneider J, Kutzner HJ (1990) Corynephage Cog, a virulent bacteriophage of *Corynebacterium glutamicum*, and its relationship to phi GA1, an inducible phage particle from *Brevibacterium flavum*. J Gen Virol 71(Pt 8):1629–1633

Sumby P, Smith MC (2003) Phase variation in the phage growth limitation system of *Streptomyces coelicolor* A3(2). J Bacteriol 185(15):4558–4563

Trautwetter A, Blanco C (1988) Isolation and preliminary characterization of twenty bacteriophages infecting either *Brevibacterium* or *Arthrobacter* strains. Appl Environ Microbiol 54(6):1466–1471

Trautwetter A, Blanco C, Bonnassie S (1987) Characterization of the corynebacteriophage CG33. J Gen Microbiol 133(10):2945–2952

Trost E, Al-Dilaimi A, Papavasiliou P, Schneider J, Viehoever P, Burkovski A, Soares SC, Almeida SS, Dorella FA, Miyoshi A, Azevedo V, Schneider MP, Silva A, Santos CS, Santos LS, Sabbadini P, Dias AA, Hirata R Jr, Mattos-Guaraldi AL, Tauch A (2011) Comparative analysis of two complete *Corynebacterium ulcerans* genomes and detection of candidate virulence factors. BMC Genomics 12:383

Trost E, Blom J, de castro soares S, Huang IH, Al-Dilaimi A, Schroder J, Jaenicke S, Dorella FA, Rocha FS, Miyoshi A, Azevedo V, Schneider MP, Silva A, Camello TC, Sabbadini PS, Santos CS, Santos LS, Hirata R Jr, Mattos-Guaraldi AL, Efstratiou A, Schmitt MP, Ton-That H, Tauch A (2012) Pangenomic Study of *Corynebacterium diphtheriae* That Provides Insights into the Genomic Diversity of Pathogenic Isolates from Cases of Classical Diphtheria, Endocarditis, and Pneumonia. J Bacteriol 194(12):3199–3215

Van Duyne GD (2005) Lambda integrase: armed for recombination. Curr Biol 15(17):R658–660

Wennerhold J, Bott M (2006) The DtxR regulon of *Corynebacterium glutamicum*. J Bacteriol 188(8):2907–2918

Zhou Y, Liang Y, Lynch KH, Dennis JJ, Wishart DS (2011) PHAST: a fast phage search tool. Nucleic Acids Res 39(Web Server issue):W347–352

Chapter 5
Toxin Structure, Delivery and Action

Başak Varol, Bilge Özerman Edis and Muhammet Bektaş

Abstract Diphtheria toxin (DTx) consists of a 535 amino acids polypeptide and contains the following three domains: the amino terminal fragment A (FA or catalytic C-domain) that catalyses the transfer of an ADP-ribosyl group of NAD^+ to a post-translationally modified histidine (diphthamide) residue on eukaryotic elongation factor 2 (eEF2) and inhibits protein synthesis. Fragment B (FB) consist of the carboxy terminal receptor-binding R-domain, and the translocation (or transmembrane) T-domain. Following binding to its cell surface receptor via R-domain, DTx is internalized through the clathrin-dependent endocytosis. The acid pH created in the early endosomes triggers a conformational change in the toxin leading to the insertion of the T and C-domains in the membrane. The catalytic domain is then translocated into the cytosol across the early endosomal membrane and protein synthesis inhibition occurs. DTx-induced cytotoxicity is versatile, and it includes DNA cleavage and the depolymerisation of actin filaments. FA can interact with both G and F-actin. The binding to the latter appears to take place at the plus end of the filament blocking further polymerisation and it was concluded that G-actin has an inhibitory effect on DTx nuclease activity.

Keywords ADP-ribosylation · Diphtheria toxin · Eukaryotic elongation factor 2 · F-actin depolymerisation · Protein synthesis

M. Bektaş (✉) · B. Varol · B. Özerman Edis
Istanbul Faculty of Medicine, Department of Biophysics, Istanbul University, Istanbul, Turkey
e-mail: muhbektas@hotmail.com

B. Varol
e-mail: basakgn@yahoo.com

B. Özerman Edis
e-mail: bilgeozerman@hotmail.com

A. Burkovski (ed.), *Corynebacterium diphtheriae and Related Toxigenic Species*,
DOI 10.1007/978-94-007-7624-1_5,

5.1 Diphtheria Toxin Structure

Diphtheria toxin (DTx), a 535 amino acids containing single protein, is an example of toxins that reach their cytoplasmic targets via receptor-mediated internalization. *In vivo* by furin or *in vitro* a mild treatment with trypsin the polypeptide chain cleaves into two fragments: The N-terminal fragment A (FA) (21.167 kDa) and C-terminal fragment B (FB) (37.195 kDa) (Kantardjieff et al. 1989). FA catalyses the transfer of adenosine 5'-diphosphoribose (ADPR) group of nicotine amide dinucleotide (NAD^+) to a modified histidine (His) (diphthamide) residue on eukaryotic elongation factor 2 (eEF2) which results in the inhibition of protein synthesis. The pysiological role of diphthamide is unknown because although the mutations at His residue, or in one of the enzyme that plays a role in the modification; no abnormality is detected in cell growth. Nevertheless, the toxicity of FA is sufficiently high that one molecule of FA in the cytosol is enough to kill one cell. This process involves 3 distinct steps:

i. Binding of the toxin to the cell surface receptor the heparin binding epidermal growth factor-like precursor (HB-EGF) on the surface of sensitive cells
ii. Translocation of the catalytic domain (FA) across the endosomal membrane into the cytoplasm involves the acidic pH inside the endosome
iii. ADP-ribosylation of eEF2 and inhibition of protein synthesis (Collier 1975).

FA reaches the cytosol near the nucleus within 30 min and accumulates to intracellular concentrations in the range of 2-5 μM. Thus, DTx uptake by the cells is highly efficient ($Km = 2.2$ nM; $Vmax = 0.25$ pmol min^{-1}) (Varol et al. 2012).

A mutant form of the DTx, CRM 197, is the product of a single missense mutation (Gly^{52} to Glu) within the FA region. The mutation leads to an enzymatically inactive product, yielding a non-toxic mutant of DTx (Rönnberg and Middlebrook 1989).

The crystallographic structure of DTx homodimer has been determined to 2.5 Å resolution. The structure reveals a Y-shaped molecule consisting of three domains (C; catalytic domain, T; transmembrane domain, R; receptor binding domain) that are connected by interdomain linkers. Each domain has a typical fold and is involved in a differtent stage of cytotoxic effect. The C-domain corresponds to the A subunit (residues 1–193). The C-domain consists of eight β-strands and seven α-helices and has a role in the inhibition of protein synthesis by transfer of ADPR to a diphthamide residue of eEF2 (Choe et al. 1992). NAD^+ binding site in the catalytic domain includes Glu148, His21 and Tyr65 residues. The nucleotide adenylyl-(3,5')-uridine 3'-monophosphate (ApUp) that as an high affinity (0.3 nM as compared to ~ 8.16 μM NAD^+) and binds competively with NAD^+. FA has a single NAD-binding site (*Kd* 8.3 pM), that not only catalyzes the transfer of ADPR of eEF2 but also provides the hydrolysis of NAD^+ to ADPR, nicotinamide (NA) and H^+ with its NADase activity (Lory et al. 1980).

FB includes both the T (residues 200–387) and the R-domain (residues 387–535). The central translocation domain, also known as TM domain, includes 9 α-helices. The T-domain translocates the catalytic C domain across the endosomal membrane

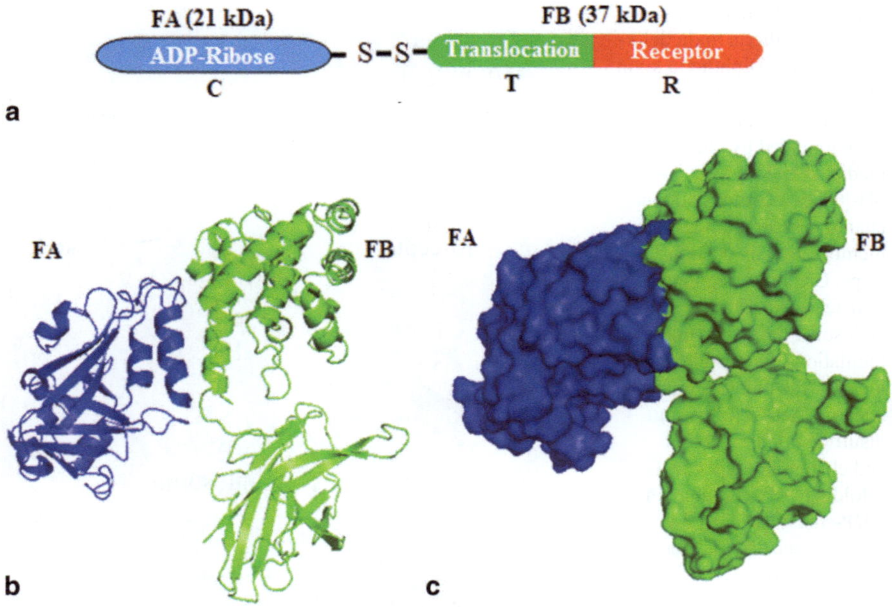

Fig. 5.1 Structure of the diphtheria toxin. **a** FA (C domain) *blue*. FB (T-domain *green* and R-domain) *red*. **b** Crystallographic structure. **c** View of the molecular surface of DTx

in response to acidification. The C-terminal receptor-binding domain is known as the R-domain, which is formed of 10 β-strands. The R-domain binds to cell surface receptor, permitting the toxin to enter the cell by receptor-mediated endocytosis (Rodnin et al. 2010) (Fig. 5.1).

5.2 Diphtheria Toxin Delivery

The delivery of DTx to the cytosol is a sequential process. The toxin binding to the cell surface is followed by receptor-mediated endocytosis. The conformational change of the toxin within endosomes initiates the translocation of the catalytic domain across the endosomal membrane. In the cytosol the catalytic fragment ADP-ribosylates the eukaryotic elongation factor 2 and initiates the depolymerisation of filamentous actin. A schematic representation of the DTx delivery is shown in Fig. 5.2.

5.2.1 Binding of Diphtheria Toxin to Cell Surface

The crucial step for the intoxication is the binding of DTx to the cell surface receptor. The precursor of heparin-binding epidermal growth factor (EGF) like growth factor, proHB-EGF, has been identified as the receptor of DTx (Naglish et al. 1992) and

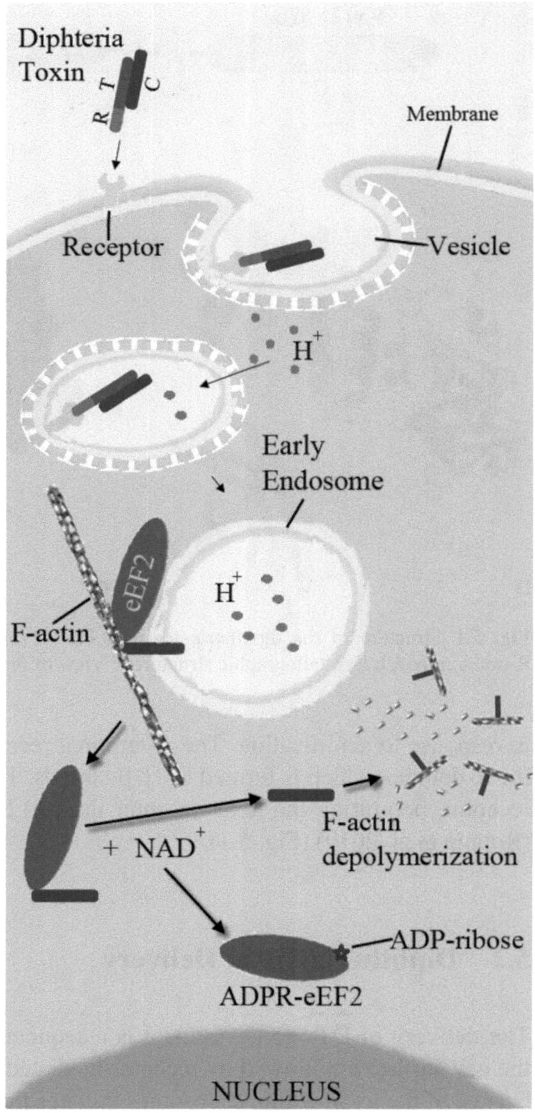

Fig. 5.2 Schematic illustration of diphtheria toxin delivery pathway. Diphtheria toxin binds to its receptor on the cell surface via R-domain and is internalized by clathrin-mediated endocytosis. Low pH-induced conformational change of T-domain initiates membrane translocation steps. Catalytic fragment transfer occurs from early endosome to cytosol under regulation of cytosolic translocation factors. C-domain becomes active following disulfide bond reduction and chaperone-dependent refolding. Toxic C-domain ADP-ribosylates eEF2 in the presence of NAD^+ and induces depolymerisation of F-actin

was first purified from monkey Vero cells (Mekada et al. 1991). The toxin binding site of proHB-EGF is the EGF-like domain (Asp106–Pro149) which is a ligand of EGF receptor of the adjacent cell (Raab and Klagsbrun 1997). This membrane anchored form of the epidermal growth factor family member affects the adjacent cell as a juxtacrine growth factor (Iwamoto and Mekada 2000). It has been reported that the interactions between DTx and proHB-EGF depend on the amino acid residue Glu141 of EGF-like domain (Louie et al. 1997). Although proHB-EGF is expressed by murine cells, due to the lack of DTx binding residues in EGF-like domain those cells are resistant to DTx (Iwamoto et al. 1994; Mitamura et al. 1995). The importance of EGF-like domain on proHB-EGF for DTx binding has been also shown, in which the inhibition of DTx binding to Vero-H cell surface has occurred by hybridoma-derived monoclonal antibodies, specific for HB-EGF (Hamaoka et al. 2010). EGF-like domain detection site of DTx has been shown to reside in the carboxyl terminal of the toxin. Within the receptor binding site, the R-domain, Lys516 and Phe530 have been determined as specific residues for the receptor recognition (Shen et al. 1994). The target cell sensitivity to DTx is also modulated by diphtheria toxin receptor-associated protein, the monkey homolog of human CD9 antigen, DAP-27/CD9 that upregulates the receptor, proHB-EGF (Iwamoto et al. 1994).

5.2.2 Internalization of Diphtheria Toxin via Clathrin-Dependent Endocytosis

The DTx internalization by receptor-mediated clathrin-dependent endocytosis was first visualized by electron microscopy on Vero cells (Morris et al. 1985). Once bound to proHB-EGF, DTx takes the logistic aid of clathrin-coated pits while entering in cells (Moya et al. 1985). Clathrin-coated pits evolve to clathrin-coated vesicles that carry the toxin to early endosomes subsequent to the loss of its lattice-like coat. During the formation of clathrin-coated pits, branched actin filaments stabilize the invaginations of membrane and support vesicles by force generation for budding and scission (McMahon and Boucrot 2011). The inhibition of clathrin-coated vesicle formation has been reported to have a protective effect against DTx intoxication (Simpson et al. 1998). When the transmembrane domain of proHB-EGF is replaced with a glycosylphosphatidylinositol (GPI), the GPI-linked DTx receptor has been shown to be internalized by clathrin-independent endocytosis and the actin cytoskeleton has been considered therefore to control the intracellular traffic of DTx bound GPI-receptor (Skretting et al. 1999). Actin-based vesicle motility has been described to have a role in deriving separated endocytic vesicles through cytosol (Qualmann et al. 2000). Besides the support of actin cytoskeleton, the Rab proteins, small GTPases in Ras superfamily, have been reported to guide endosomal vesicles, especially Rab5 protein has been reviewed for vesicular fusion with early endosomes (Somsel Rodman and Wandinger-Ness 2000). Subsequent to the fusion of DTx carrying vesicles with early endosomes, the vacuolar-type H^+-ATPase (V-ATPase) acidifies the lumen of the endosome and generates a proton gradient. The inhibition of V-ATPase with bafilomycin treatment was reported to abolish the DTx intoxication of Vero cells (Umata et al. 1990). It is accepted for long time that the acidification of the interior of vesicle triggers the unfolding of the T-domain of the DTx and its insertion into the endosomal membrane (Boquet et al. 1976).

5.2.3 Conformational Change of T-Domain in the Endosome

During the intracellular transportation, DTx undergoes conformational changes in response to the pH decrease within the last destination for the whole toxin, the early endosomal compartment. With the translocation of C-domain, the enzymatically active moiety (FA) encounters the barrier effect of the lipid bilayer of endosomal membrane due to the water-soluble properties of DTx. This obstruction has been overcome by the conformational change of the T-domain which is covalently liked to C-domain by a disulphide bond, as a result of the lowered pH in the early endosome due to activity of V-ATPase (Draper and Simon 1980; Sandvig and Olsnes 1980, Varkhoui et al. 2011). DTx-carrying early endosomes have been shown to be enlarged in size and that has been reversed under the treatment of a proton gradient reducing agent, ammonium chloride, resulting in toxin trapping (Antignani and Youle 2008). At neutral pH, globular state of DTx has been reported to be activated by furin proteolitic cleavage either on the cell surface before the internalization step or in the endosomal lumen (Gordon et al. 1995). Following the proteolysis the decrease in pH induces DTx to expose hydrophobic surfaces of the T-domain. DTx interacts with endosomal membrane through those hydrophobic surfaces formed by helices TH8 and TH9, the so-called transmembrane hairpin. The membrane insertion of the T-domain has been hypothesized to form a cation-selective channel by this transmembrane hairpin (Papini et al. 1988). Later on, channel formation has been reported not to be strictly essential (Lanzrein et al. 1996). Nevertheless, whether DTx is a fusogenic or a pore-forming peptide was debated and have been reviewed in detail (Falnes and Sandvig 2000; Parker and Feil 2005; Murphy 2011). The functional state of T-domain has been determined to be the partially folded state when the T-domain is inserted in the lipid bilayer (Man et al. 2011). The C-domain translocation has been shown to be facilitated by the pH-dependent chaperon-like activity of membrane inserted T-domain (Chassaing et al. 2011).

5.2.4 Translocation of the C-Domain to the Cytoplasm

The C-domain delivery from endosomal compartment to cytoplasm requires both disulfide bond reducing enzyme and translocation enabling factors. During the period of the membrane translocation, the catalytic fragment of DTx stays linked to the T-domain by a disulphide bond between cysteine residues which has been described to store the high potential energy and stabilize the secondary structure of the partially folded toxin (Wouters et al. 2004). The disulphide bond has been reported to be reduced by thioredoxin reductase which completes the delivery process by the aid of the cytosolic translocation factors (Ratts et al. 2003). The necessity of ATP and cytosolic factors such as β-COP for C-domain release was first determined in 1997 (Lemichez et al. 1997). Recent evidence indicates that actin filaments may provide a structural framework for translocation factors. The heat shock protein Hsp90, an actin-binding protein, has been shown to act as a chaperon in translocation and refolding of C-domain (Ratts et al. 2003). The strong avidity of V-ATPase for filamentous actin (F-actin) is a sign of the interactions between F-actin and DTx loaded early endosomes (Carnell et al. 2011). F-actin has been reported to interact also with catalytic fragment of DTx and eEF2 (Bektaş et al. 1994; 2009). Shortly after the delivery of C-domain from early endosome, the amount of the release has

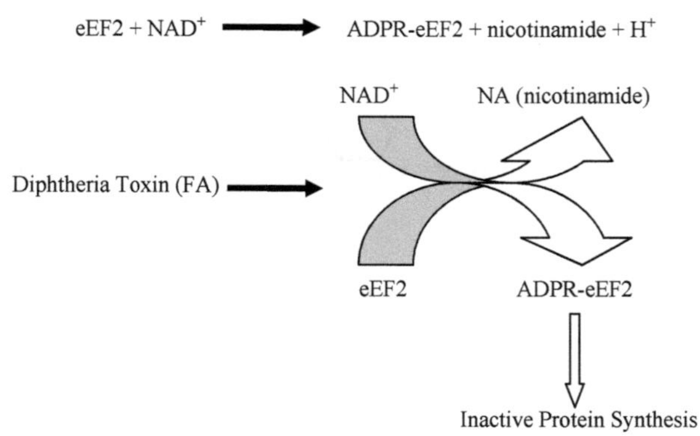

Fig. 5.3 ADP-ribosylation of eukaryotic elongation factor 2 (eEF2). Adenosine 5'-diphosphoribose (ADPR), nicotinamide adenine dinucleotide (NAD^+), fragment A (FA)

been reported to be controlled by cytosolic factors and it has been hypothesized that ADP-ribosylated eEF2 emerges as a new candidate for blocking further entry of catalytic fragment (Bektaş et al. 2011).

5.3 Cytotoxicity of Diphtheria Toxin

5.3.1 ADP-Ribosylation of eEF2 by Diphtheria Toxin

DTx catalyzes the ADP-ribosylation of a unique post-translationally modified histidine (His715), termed diphthamide residue, in eukaryotic elongation factor 2 (eEF2) (Collier 1967; Honjo et al. 1968; Van Ness et al. 1980; Collier 2001). eEF2 promotes ribosomal translocation as one of the three protein factors involved in eukaryotic polypeptide chain elongation (Bermek 1978). eEF2 is a GTP-binding protein that catalyzes the translocation of the peptidyl-tRNA and mRNA from the 80S ribosome A site to the P site, and therefore is essential for protein biosynthesis (Jorgensen et al. 2005). Subunit FA is the N-terminal component of the toxin and contains the catalytic domain for the nicotinamide adenine dinucleotide (NAD^+) -dependent ADP-ribosylation of eEF2, which inhibits protein synthesis by eukaryotic cells (Fig. 5.3). This reaction forms a covalent bond between adenosine 5'-diphosphoribose (ADPR) and eEF2. The reaction blocks the functional site which interacts with RNA in translation, and consequentially prevents protein synthesis (Jorgensen et al. 2004, 2005; Taylor et al. 2007).

Diphthamide is a post-translational modification conserved in all archaea and eukaryotes. Diphthamide modification is present in all eukaryotic organisms, in which it is restricted to a single protein and amino acid, a histidine residue of eEF2 (position H^{699} in yeast and H^{715} in mammals). There is also evidence of a diphthamide modification, or a precursor of diphthamide, in the analogous bacterial elongation factor (EF-G) (Pappenheimer et al. 1983). The function of diphthamide is not entirely clear. Within the eEF2 structure, the diphthamide is located at the tip of a domain loop that is proposed to mimic the anticodon loop of tRNA (Ortiz et al. 2006). It has

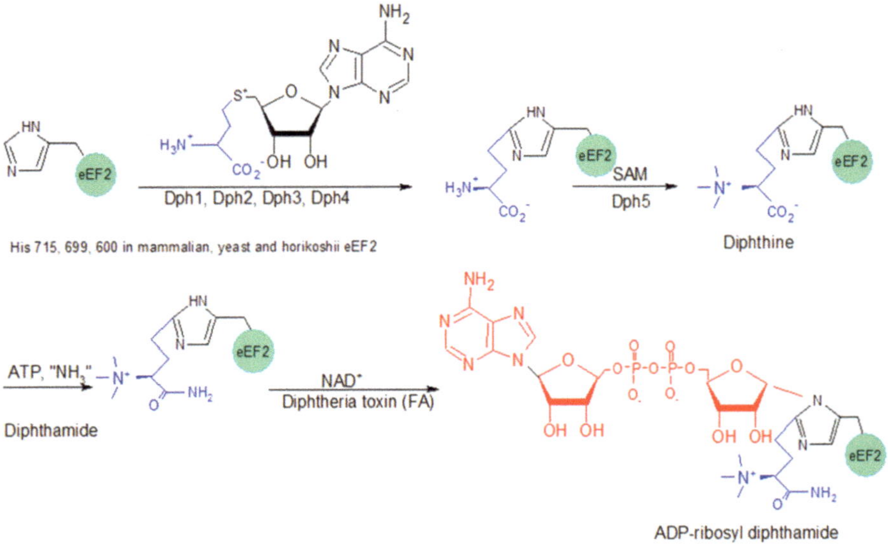

Fig. 5.4 The proposed biosynthesis pathway for diphthamide and its ADP-ribosylation by diphtheria toxin. Catalytic domain (FA) catalyses the ADP-ribose group of NAD^+ to the diphtamide moiety of eEF2

been suggested that this loop, bearing the diphthamide modification, stabilizes the mRNA-tRNA (codon-anticodon) interaction and is necessary to protect the translation reading frame. Diphthamide biosynthesis has been well characterized in yeast and has been shown genetically to require the action of five proteins (Dph1-Dph5) (Liu et al. 2004). Diphthamide-deficient yeast strains have allowed analysis of the pathway, which proved to be a complex, multi step process (Fichtner et al. 2003; Liu et al. 2004). The first step involves the transfer of a 3-amino-3-carboxypropyl moiety from S-adenosylmethionine (SAM or AdoMet) to the C-2 imidazole of the target histidine to produce an intermediate structure. This requires Dph1-Dph4. Subsequently, trimethylation of the intermediate compound by an AdoMet-dependent methyltransferase, a process requiring Dph5, produces diphthine (Mattheakis et al. 1992; Jorgensen et al. 2006). The final step the carboxyl group (–COOH) of the diphthine by an ATP-dependent enzyme, results in diphthamide (Fig. 5.4).

5.3.2 Deoxyribonuclease Activity of Diphtheria Toxin

Besides inhibition of protein synthesis, diphtheria toxin gives rise to breakdown of cytoskeleton, internucleosomal DNA cleavage and cytolysis (Chang et al. 1989; Komatsu et al. 1998). DTx catalytic domain (FA) has been repeatedly shown to possess deoxyribonuclease activity (Lessnick et al. 1992; Lee et al. 2005; Varol et al. 2012). In addition to DTx, we discovered that CRM197, the ADPrT-deficient G52E-mutant form of DTx, possesses endonuclease activity (Bruce et al. 1990) and its specific activity is four times higher than that of bovine pancreatic DNase I and 20 times higher than that of DTx.

5.3.3 Effect of Diphtheria Toxin on F-Actin Stability

Diphtheria toxin induced cytotoxicity is yet versatile and includes DNA cleavage, increases in caspase-3 type proteolytic activity as well as depolymerisation of actin filaments (F-actin). These events follow the onset of inhibition of protein synthesis by a lag of several hours and correspond to those associated with programmed cell death type I (Bras et al. 2005). Catalytic domain (FA) was found to bind to F-actin (filamentous) in a stoichiometric manner (Bektaş et al. 2009; Varol et al. 2012). This binding was inhibited by gelsolin and globular actin monomer (G-actin), but not by DNase I. The binding of FA inhibited polymerisation of G-actin induced a time-dependent breakdown of F-actin under polymerisation conditions. The study indicates that FA of DTx can interact with both G-actin and F-actin. The binding to the latter appears to take place at the positive end of the filament, blocking its further polymerisation (Bektaş et al. 2009). Inhibition of the ADP-ribosyltransferase (ADPrT) activity did not have any effect on the interactions of FA with actin (Bektaş et al. 2004). Moreover, the role of F-actin and eEF2 in the endocytic trafficking of DTx are also determined (Bektaş et al. 2011).

References

Antignani A, Youle RJ (2008) Endosome fusion induced by diphtheria toxin translocation domain. Proc Natl Acad Sci U S A 105(23):8020–8025

Bektaş M, Nurten R, Gürel Z, Sayers Z, Bermek E (1994) Interactions of eukaryotic elongation factor 2 with actin: a possible link between protein synthetic machinery and cytoskeleton. FEBS Lett 356(1):89–93

Bektaş M, Günçer M, Güven C, Nurten R, Bermek E (2004) Actin—an inhibitor of eukaryotic elongation factor activities. Biochem Biophys Res Commun 317(4):1061–1066

Bektaş M, Varol B, Nurten R, Bermek E (2009) Interaction of diphtheria toxin (fragment A) with actin. Cell Biochem Funct 27(7):430–439

Bektaş M, Hacıosmanoğlu E, Özerman B, Varol B, Nurten R, Bermek E (2011) On diphtheria toxin fragment A release into the cytosol-Cytochalasin D effect and involvement of actin filaments and eukaryotic elongation factor 2. Int J Biochem Cell Biol 43(9):1365–1372

Bermek E (1978) Mechanisms in Polypeptide Chain Elongation on Ribosomes. Proc Natl Acad Sci U S A 21:63–100

Boquet P, Silverman MS, Pappenheimer AM Jr, Vernon WB (1976) Binding of triton X-100 to diphtheria toxin, crossreacting material 45, and their fragments. Proc Natl Acad Sci U S A 73(12):4449–4453

Bras M, Queenan B, Susin SA (2005) Programmed cell death via mitochondria: Different modes of dying. Biochemistry (Mosc) 70(2):231–239

Bruce C, Baldwin RL, Lessnick SL, Wisnieski BJ (1990) Diphtheria toxin and its ADP-ribosyltransferase-defective homologue CRM197 possess deoxyribonuclease activity. Proc Natl Acad Sci U S A 87(8):2995–2998

Carnell M, Zech T, Calaminus SD, Ura S, Hagedorn M, Johnston SA, May RC, Soldati T, Machesky LM, Insall RH (2011) Actin polymerization driven by WASH causes V-ATPase retrieval and vesicle neutralization before exocytosis. J Cell Biol 193(5):831–839

Chang MP, Baldwin RL, Bruce C, Wisnieski BJ (1989) Second cytotoxic pathway of diphtheria toxin suggested by nuclease activity. Science 246(4934):1165–1168

Chassaing A, Pichard S, Araye-Guet A, Barbier J, Forge V, Gillet D (2011) Solution and membrane bound chaperone activity of the diphtheria toxin translocation domain towards the catalytic domain. FEBS J 278(23):4516–4525

Choe S, Bennett MJ, Fujii G, Curmi PM, Kantardjieff KA, Collier RJ, Eisenberg D (1992) The crystal structure of diphtheria toxin. Nature 357(6375):216–222

Collier RJ (1967) Effect of diphtheria toxin on protein synthesis: inactivation of one of the transfer factors. J Biol Chem 25(1):83–98

Collier RJ (1975) Diphtheria toxin: mode of action and structure. Bacteriol Rev 39(1):54–85

Collier RJ (2001) Understanding the mode of action of diphtheria toxin: a perspective on progress during the 20th century. Toxicon 39(11):1793–1803

Draper RK, Simon MI (1980) The entry of diphtheria toxin into the mammalian cell cytoplasm: evidence for lysosomal involvement. J Cell Biol 87(3):849–854

Falnes PO, Sandvig K (2000) Penetration of protein toxins into cells. Curr Opin Cell Biol 12(4):407–413

Fichtner L, Jablonowski D, Schierhorn A, Kitamoto HK, Stark MJ, Schaffrath R (2003) Elongator's toxin-target (TOT) function is nuclear localization sequence dependent and suppressed by post-translational modification. Mol Microbiol 49(5):1297–1307

Gordon VM, Klimpel KR, Arora N, Henderson MA, Leppla SH (1995) Proteolytic activation of bacterial toxins by eukaryotic cells is performed by furin and by additional cellular proteases. Infect Immun 63(1):82–87

Hamaoka M, Chinen I, Murata T, Takashima S, Iwamoto R, Mekada E (2010) Anti-human HB-EGF monoclonal antibodies inhibiting ectodomain shedding of HB-EGF and diphtheria toxin binding. J Biochem 148(1):55–69

Honjo T, Nishizuka Y, Hayaishi O (1968) Diphtheria toxin-dependent adenosine diphosphate ribosylation of aminoacyl transferase II and inhibition of protein synthesis. J Biol Chem 243(12):3553–3555

Iwamoto R, Mekada E (2000) Heparin-binding EGF-like growth factor: a juxtacrine growth factor. Cytokine Growth Factor Rev 11(4):335–344

Iwamoto R, Higashiyama S, Mitamura T, Taniguchi N, Klagsbrun M, Mekada E (1994) Heparin-binding EGF-like growth factor, which acts as the diphtheria toxin receptor, forms a complex with membrane protein DRAP27/CD9, which up-regulates functional receptors and diphtheria toxin sensitivity. EMBO J 13(10):2322–2330

Jorgensen R, Yates SP, Teal DJ, Nilsson J, Prentice GA, Merrill AR, Rom Andersen GR (2004) Crystal structure of ADP-ribosylated sibosomal translocase from *Saccharomyces cerevisiae*. J Biol Chem 279(44):45919–45925

Jorgensen R, Merrill AR, Yates SP, Marquez VE, Schwan AL, Boesen T, Andersen GR (2005) Exotoxin A–eEF2 complex structure indicates ADP ribosylation by ribosome mimicry. Nature 436(7053):979–984

Jorgensen R, Merrill AR, Andersen GR (2006) The life and death of translation elongation factor 2. Biochem Soc Trans 34:1–6

Kantardjieff K, Collier RJ, Eisenberg D (1989) X-ray grade crystals of the enzymatic fragment of diphtheria toxin. J Biol Chem 264(18):10402–10404

Komatsu N, Oda T, Muramatsu T (1998) Involvement of both caspase-like proteases and serine proteases in apoptotic cell death induced by ricin, modeccin, diphtheria toxin, and pseudomonas toxin. J Biochem 124(5):1038–1044

Lanzrein M, Sand O, Olsnes S (1996) GPI-anchored diphtheria toxin receptor allows membrane translocation of the toxin without detectable ion channel activity. EMBO J 15(4):725–734

Lee JW, Nakamura LT, Chang MP, Wisnieski BJ (2005) Mechanistic aspects of the deoxyribonuclease activity of diphtheria toxin. Biochim Biophys Acta 1747(1):121–131

Lemichez E, Bomsel M, Devilliers G, vanderSpek J, Murphy JR, Lukianov EV, Olsnes S, Boquet P (1997) Membrane translocation of diphtheria toxin fragment A exploits early to late endosome trafficking machinery. Mol Microbiol 23(3):445–447

Lessnick SL, Lyczak JB, Bruce C, Lewis DG, Kim PS, Stolowitz ML, Hood L, Wisnieski BJ (1992) Localization of diphtheria toxin nuclease activity to fragment A. J Bacteriol 174(6):2032–2038

Liu S, Milne GT, Jeffrey G, Kuremsky JG, Fink GR, Leppla SH (2004) Identification of the proteins required for biosynthesis of diphthamide, the target of bacterial ADP-ribosylating toxins on translation elongation factor 2. Mol Cell Biol 24(21):9487–9497

Lory S, Carroll SF, Bernard PD, Collier RJ (1980) Ligand interactions of diphtheria toxin I. Binding and hydrolysis of NAD. J Biol Chem 255(24):12011–12015

Louie GV, Yang W, Bowman ME, Choe S (1997) Crystal structure of the complex of diphtheria toxin with an extracellular fragment of its receptor. Mol Cell 1(1):67–78

Man P, Montagner C, Vitrac H, Kavan D, Pichard S, Gillet D, Forest E, Forge V (2011) Accessibility changes within diphtheria toxin T domain upon membrane penetration probed by hydrogen exchange and mass spectrometry. J Mol Biol 414(1):123–134

Mattheakis LC, Shen WH, Collier RJ (1992) DPH5, a methyltransferase gene required for diphthamide biosynthesis in *Saccharomyces cerevisiae*. Mol Cell Biol 12(9):4026–4037

McMahon HT, Boucrot E (2011) Molecular mechanism and physiological functions of clathrin-mediated endocytosis. Nat Rev Mol Cell Biol 12(8):517–533

Mekada E, Senoh H, Iwamoto R, Okada Y, Uchida T (1991) Purification of diphtheria toxin receptor from Vero cells. J Biol Chem 266(30):20457–20462

Mitamura T, Higashiyama S, Taniguchi N, Klagsbrun M, Mekada E (1995) Diphtheria toxin binds to the epidermal growth factor (EGF)-like domain of human heparin-binding EGF-like growth factor/diphtheria toxin receptor and inhibits specifically its mitogenic activity. J Biol Chem 270(3):1015–1019

Morris RE, Gerstein AS, Bonventre PF, Saelinger CB (1985) Receptor-mediated entry of diphtheria toxin into monkey kidney (Vero) cells: electron microscopic evaluation. Infect Immun 50(3):721–727

Moya M, Dautry-Varsat A, Goud B, Louvard D, Boquet P (1985) Inhibition of coated pit formation in Hep2 cells blocks the cytotoxicity of diphtheria toxin but not that of ricin toxin. J Cell Biol 101(2):548–559

Murphy JR (2011) Mechanism of diphtheria toxin catalytic domain delivery to the eukaryotic cell cytosol and the cellular factors that directly participate in the process. Toxins (Basel) 3(3):294–308

Naglish JG, Metherall JE, Russell DW, Eidels L (1992) Expression cloning of a diphtheria toxin receptor: identity with a heparin-binding EGF-like growth factor precursor. Cell 69(6):1051–1061

Ortiz PA, Ulloque R, Kihara GK, Zheng H, Kinzy TG (2006) Translation elongation factor 2 anticodon mimicry domain mutants affect fidelity and diphtheria toxin resistance. J Biol Chem 281(43):32639–32648

Papini E, Sandoná D, Rappuoli R, Montecucco C (1988) On the membrane translocation of diphtheria toxin: at low pH the toxin induces ion channels on cells. EMBO J 7(11):3353–3359

Pappenheimer AM Jr, Dunlop PC, Adolph KW, Bodley JW (1983) Occurrence of diphthamide in archaebacteria. J Bacteriol 153(3):1342–1347

Parker MW, Feil SC (2005) Pore-forming protein toxins: from structure to function. Prog Biophys Mol Biol May 88(1):91–142

Qualmann B, Kessels MM, Kelly RB (2000) Molecular links between endocytosis and the actin cytoskeleton. J Cell Biol 150(5):111–116

Raab G, Klagsbrun M (1997) Heparin-binding EGF-like growth factor. Biochim Biophys Acta 1333(3):F179–199

Ratts R, Zeng H, Berg EA, Blue C, McComb ME, Costello CE, vanderSpek JC, Murphy JR (2003) The cytosolic entry of diphtheria toxin catalytic domain requires a host cell cytosolic translocation factor complex. J Cell Biol 160(7):1139–1150

Rodnin MV, Kyrychenko A, Kienker P, Sharmo O, Posokhov YO, Collier RJ, Finkelstein A, Ladokhin AS (2010) Conformational switching of the diphtheria toxin T domain. J Mol Biol 402(1):1–7

Rönnberg BJ, Middlebrook JL (1989) Cellular regulation of diphtheria toxin cell surface receptors. Toxicon 27(12):1377–1388

Sandvig K, Olsnes S (1980) Diphtheria toxin entry into cells is facilitated by low pH. J Cell Biol 87(3):828–832

Shen WH, Choe S, Eisenberg D, Collier RJ (1994) Participation of lysine 516 and phenylalanine 530 of diphtheria toxin in receptor recognition. J Biol Chem 269(46):29077–29084

Simpson JC, Smith DC, Roberts LM, Lord JM (1998) Expression of mutant dynamin protects cells against diphtheria toxin but not against ricin. Exp Cell Res 239(2):293–300

Skretting G, Torgersen ML, van Deurs B, Sandvig K (1999) Endocytic mechanisms responsible for uptake of GPI-linked diphtheria toxin receptor. J Cell Sci 112(22):3899–3909

Somsel Rodman J, Wandinger-Ness A (2000) Rab GTPases coordinate endocytosis. J Cell Sci 113(2):183–192

Taylor DJ, Nilsson J, Merrill AR, Andersen GR, Nissen P, Frank J (2007) Structures of modified eEF280S ribosome complexes reveal the role of GTP hydrolysis in translocation. EMBO J 26(9):2421–2431

Umata T, Moriyama Y, Futai M, Mekada E (1990) The cytotoxic action of diphtheria toxin and its degradation in intact Vero cells are inhibited by bafilomycin A1, a specific inhibitor of vacuolar-type H(+)-ATPase. J Biol Chem 265(35):21940–21945

Van Ness BG, Howard JB, Bodley JW (1980) ADP-ribosylation of elongation factor 2 by diphtheria toxin. NMR spectra and proposed structures of ribosyl-diphthamide and its hydrolysis products. J Biol Chem 255(22):10710–10716

Varkouhi AK, Scholte M, Storm G, Haisma HJ (2011) Endosomal escape pathways for delivery of biologicals. J Control Release 151(3):220–228

Varol B, Bektaş M, Nurten R, Bermek E (2012) The cytotoxic effect of diphtheria toxin on actin cytoskeleton. Cell Mol Biol Lett 17(1):49–61

Wouters MA, Lau KK, Hogg PJ (2004) Cross-strand disulphides in cell entry proteins: poised to act. Bioessays 26(1):73–79

Chapter 6
Iron Acquisition and Iron-Dependent Gene Expression in *Corynebacterium diphtheriae*

Michael P. Schmitt

Abstract The ability of bacterial pathogens to acquire iron during infection of mammalian hosts is often an essential component of the disease process. The human pathogen *Corynebacterium diphtheriae* utilizes a variety of iron sources, including both host iron compounds and inorganic iron for growth in iron depleted conditions. Recent studies have shown that the upper respiratory tract of humans, a region colonized by *C. diphtheriae*, is low in available iron, which suggests that mechanisms for acquiring iron from the host environment are essential for survival. To support growth in low iron medium, *C. diphtheriae* produces the siderophore corynebactin as well as systems involved in the use of heme and hemoglobin. These iron uptake systems utilize ATP-binding cassette- (ABC) type transporters and associated solute binding lipoproteins. Heme-iron acquisition also requires membrane anchored hemin- and hemoglobin-binding proteins, as well as the intracellular heme degrading enzyme HmuO. Expression of all of these iron and heme acquisition systems is coordinately regulated with that of diphtheria toxin (DT), a major virulence determinant of *C. diphtheriae*. Transcription of the genes encoding DT and the iron uptake systems is controlled by the Diphtheria Toxin Repressor, DtxR; a global iron-dependent regulatory factor. Expression of HmuO is modulated by the novel heme-responsive signal transduction systems, ChrAS and HrrAS.

Keywords DtxR · Heme transport · Hemoglobin · HmuO · Iron transport

M. P. Schmitt (✉)
Laboratory of Respiratory and Special Pathogens, Division of Bacterial, Parasitic, and Allergenic Products, Center for Biologics Evaluation and Research Food and Drug Administration, 20892 Bethesda, MD, USA
e-mail: michael.schmitt@fda.hhs.gov

A. Burkovski (ed.), *Corynebacterium diphtheriae and Related Toxigenic Species*, DOI 10.1007/978-94-007-7624-1_6,

6.1 Introduction

Iron is essential for the growth of almost all organisms including most bacteria species, many of which are pathogenic to humans (Cornelis and Andrews 2010; Braun and Hantke 2011). The importance of iron in disease has been well established and has been extensively reviewed (Litwin and Calderwood 1993; Crosa et al. 2004; Wyckoff et al. 2007; Honsa and Maresso 2011). The host environment inhabited by many bacterial pathogens is frequently low in available iron, and much of the iron in the host is sequestered in the extracellular environment by the high molecular weight glycoproteins transferrin and lactoferrin, which are present in serum and in various secretions on mucous membranes, respectively. Intracellular iron in mammalian hosts is predominately bound by heme, which is associated with protein, primarily hemoglobin (Honsa and Maresso 2011). In the last few decades, studies have shown that the expression of numerous bacterial virulence determinants is controlled by the iron content in the growth medium (Skaar 2010). Transcriptional regulation of many of these virulence systems is repressed in high iron environments and de-repressed in iron depleted medium. The findings suggest that the low-iron environment of the host functions as a signal to many bacterial pathogens to activate a variety of virulence determinants (Skaar 2010). Many of these iron-regulated factors, which have been characterized in both gram positive and gram negative bacteria, are involved in the uptake of extracellular iron. Iron acquisition in bacteria involves a variety of mechanisms including high affinity siderophore uptake systems (Chu et al. 2010) and various membrane or cell-wall anchored binding-proteins that sequester iron-compounds at the bacterial surface and subsequently transport the iron source into the cytosol (Wyckoff et al. 2007; Hammer and Skaar 2011). Siderophores, which are ubiquitous in bacteria, are low-molecular weight iron chelating compounds that are secreted by bacteria into the extracellular environment where they bind ferric iron with high affinity (Winkelmann 2002; Chu et al. 2010). The ferric-siderophore initially interacts with specifc receptors on the bacterial surface and this iron complex is subsequently transported into the bacterial cytosol where the iron is removed by either a reductive process or through degradation of the siderophore. Iron uptake systems that utilize surface exposed binding-proteins are common in bacteria, especially in pathogenic species, and they are frequently involved in the acquisition of iron from host sources including transferrin, lactoferrin and various heme-containing proteins (Yu and Schryvers 2002; Ouattara et al. 2010; Noinaj et al. 2012). Studies in gram negative bacteria have identified outer membrane proteins that specifically bind transferrin, lactoferrin, and hemin or hemoproteins such as hemoglobin (Mey and Payne 2001; Yu and Schryvers 2002; Mazmanian et al. 2003; Wyckoff et al. 2007; Noinaj et al. 2012). The uptake of hemin in gram negative bacteria is very similar to the transport of ferric siderophores. Outer membrane hemin binding proteins in gram negative species are exposed on the bacterial surface and are proposed to extract the bound hemin from hemoglobin, where the hemin is subsequently mobilized into the bacterial cytosol via an ATP-binding cassette (ABC) transporter and its cognate periplasmic solute binding protein. Hemin uptake in gram positive bacteria has also recently been described for several human pathogens (Mazmanian et al. 2003; Allen and Schmitt 2009; Ouattara et al. 2010; Nobles and Maresso 2011). Since gram positive bacteria lack an outer membrane, heme or hemoproteins are proposed to bind to cell wall anchored receptors, where heme is initially extracted from the hemoprotein and subsequently transported into the cytoplasm through an ABC-type transporter. Once in the cytosol, the removal of iron from heme is not well understood in many bacterial species, however, heme degrading enzymes, such as heme oxygenases are important for the utilization of heme iron in several well characterized systems (Schmitt 1997b; Zhu et al. 2000; Ratliff et al. 2001; Skaar et al. 2004).

Hemophores, which are low-molecular weight heme-binding proteins that are secreted into the extracellular medium, were originally characterized in *Serratia* species and in a few other gram negative organisms (Cescau et al. 2007), but recently this class of heme-binding proteins have been identified in the gram positive bacterium *Bacillus anthracis* (Maresso et al. 2008). Hemophores are proposed to remove heme from hemoproteins such as hemoglobin or hemopexin, and transport the heme to the bacterial surface where the hemin is ultimately taken-up by the bacteria through a binding-protein dependent process.

Corynebacterium diphtheriae, the cause of the severe upper respiratory tract disease diphtheria, is a gram positive bacterium that secretes diphtheria toxin (DT), an iron-regulated 58-kDa exotoxin that is the primary cause of the morbidity and mortality associated with human infection by this organism (Barksdale 1970; Hadfield et al. 2000; Holmes 2000; Collier 2001). *C. diphtheriae* is also associated with skin ulcerations, and in recent years, both toxinogenic and non-toxigenic strains have in rare instances been linked to severe infections including endocarditis (Muttaiyah et al. 2011). DT was the first bacterial virulence determinant shown to be regulated by the iron content of the growth medium. Studies from the 1930s demonstrated that DT production was repressed under conditions of excess iron and that removal of iron from the growth medium resulted in increased toxin synthesis (Pappenheimer and Johnson 1936). Many of these early studies from 1930s and 1940s that examined the effects of iron on DT expression were primarily concerned with optimizing bacterial growth conditions to maximize DT synthesis for vaccine production, and only in recent decades has attention focused on the effects of iron on the broader physiology of *C. diphtheriae.*

6.2 Siderophore Synthesis and Transport

Interest in iron transport and iron-regulated gene expression in *C. diphtheriae* originated from years of research examining the expression and synthesis of DT (Pappenheimer 1977; Collier 2001). The mucous membranes and epithelial cell surfaces that are colonized by *C. diphtheriae* in the human upper respiratory tract are predicted to be relatively low in available iron, since DT, which *in vitro* is produced only under low iron conditions, is expressed during human infection. While the specific iron sources available to *C. diphtheriae* in the host are not known, the associated necrosis, and cell lysis at the site of *C. diphtheriae* colonization likely produces an environmental milieu that contains various host iron sources (Hadfield et al. 2000). These iron-containing compounds may include transferrin that is present in serum, lactoferrin found on mucous membranes, ferritin, and various heme-containing proteins, such as hemoglobin.

Early studies that examined the iron-dependent regulation of DT identified chemically-derived mutants that expressed toxin in high iron conditions. These mutations were shown to reside both on the bacterial chromosome and on the phage genome, which encodes the *tox* gene, the structural gene for DT. These studies identified three classes of mutations that resulted in the loss of iron-dependent repression of *tox*: one group of mutants was predicted to carry mutations that inactivated a chromosomally encoded regulatory protein that controls *tox* expression (one of these mutants was later shown to have a mutation in *dtxR*, which is discussed below). A second class of mutants contains mutations that reside on the phage genome and were predicted to alter sequences upstream of the *tox* gene that interfere with the binding of a regulatory protein (Murphy et al. 1976; Uchida et al. 1977; Welkos and Holmes 1981), while the third class of mutants were presumed to be defective in factors involved in iron uptake, which resulted in reduced intracellular iron levels during growth in high iron medium (Cryz et al. 1983). A reduced intracellular iron level was predicted to result in increased toxin production, since previous studies had demonstrated that optimal DT production occurs in iron-depleted medium. These presumed iron uptake mutants, isolated in *C. diphtheriae* strain C7 (β) (Freeman 1951), were designated HC1, HC3, HC4 and HC5, and were severely defective in iron uptake and grew poorly in low-iron medium (Cryz et al. 1983). Subsequent studies that examined iron uptake in *C. diphtheriae* C7 (β) showed that iron transport was an energy dependent process and that an iron chelating compound was secreted into the extracellular medium during growth in low iron conditions (Russell and Holmes 1983; Russell et al. 1984). This extracellular iron chelating activity was initially described as a corynebacterial siderophore and was later named corynebactin (Russell and Holmes 1985). A chemically derived mutant (HC6) that failed to produce the siderophore was isolated and used in bioassays to demonstrate that corynebactin is able to stimulate the growth of the mutant in low iron conditions. Several bacterial and fungal siderophores, including enterobactin, DHBA, rhodotorulic acid, ferrichrome A, desferrioxamine B, and vibriobactin, are unable to stimulate growth of the HC6 mutant; however, aerobactin, a hydroxamate type siderophore produced by various gram negative enteric bacteria, stimulated growth of HC6 in low-iron medium. Biochemical analysis with partially purified corynebactin gave negative results in both the Arnow test for phenolates and in the Csaky assay for hydroxamate type siderophores, suggesting a novel structure for corynebactin (Russell et al. 1984). *C. diphtheriae* strain PW8, which is used in the production of DT for vaccine development, is also defective for siderophore synthesis, however, corynebactin as well as aerobactin enhance the growth of PW8 in low iron conditions (Russell and Holmes 1985). The genetic defects that are responsible for the failure to produce siderophore in the HC6 and PW8 strains are not known.

It was subsequently shown that corynebactin could be detected by the Chrome Azurol S (CAS) assay, which provided a convenient method to quantitate siderophore production in *C. diphtheriae* (Tai et al. 1990). Both corynebactin and DT production are repressed in high iron medium, and the iron dependent regulation of these compounds was later shown to be mediated by DtxR (Schmitt and Holmes 1991). In 2002, Qian et al. identified a DtxR-regulated operon in *C. diphtheriae* C7 that was proposed to be essential for siderophore uptake. The iron-regulated promoter for this operon designated IRP6 was identified using a SELEX-like system that specifically identified DNA sequences that bound DtxR. The *irp6* operon, located downstream from the IRP6 promoter, included three genes, designated *irp6A*, *irp6B*, and *irp6C* that are predicted to encode the permease (Irp6B) and ATPase (Irp6C) of an ABC transporter, and a solute binding protein, Irp6A, that is homologous to periplasmic solute binding proteins found in gram negative bacteria. The predicted products share significant sequence similarity to iron uptake systems found in other bacteria. A clone carrying the complete *irp6ABC* operon is able to complement the constitutive corynebactin production phenotype of three of the previously described *C. diphtheriae* HC mutants (HC1, HC4, and HC5). These chemically-derived HC mutants were previously shown to have diminished ability to transport iron and also produced siderophore and DT constitutively (Cryz et al. 1983; Qian et al. 2002). Reduced ability to transport iron and the overproduction of siderophore in high iron conditions is a phenotype associated with mutations in siderophore transport systems. DNA sequence analysis of the chromosomally encoded *irp6ABC* genes in the HC mutants revealed that they carried various point mutations that could be complemented *in-trans* by specific *irp6ABC* genes. The ability to repress production of siderophore in the HC mutants in high iron conditions by the cloned *irp6ABC* genes suggested that the Irp6ABC iron transport system is involved in the uptake of corynebactin; however, it was noted that the HC mutants are chemically derived and that some of the strains are known to harbor multiple mutations. Additionally, defined mutations in the *irp6ABC* genes were not constructed in this study, and therefore, the impact of other random mutations in these strains was not assessed.

In 2005, a DtxR and iron regulated genetic locus, which contains seven genes (*ciuA-G*), five of which are predicted to encode factors involved in corynebactin synthesis and uptake, was identified in *C. diphtheriae* strain C7 and in the clinical isolate 1737 (Kunkle and Schmitt 2005). The genes were identified through a search of the recently completed *C. diphtheriae* genome for proteins with sequence similarity to siderophore biosynthetic enzymes (Cerdeno-Tarraga et al. 2003). Genes involved in siderophore transport are located on a four gene operon, *ciuABCD*, and are predicted to encode an ABC transporter that includes the permease proteins CiuB and CiuC, an ATPase, CiuD, and a putative lipoprotein receptor CiuA, which has homology to gram negative periplasmic solute binding proteins. Downstream of the transport operon is a second DtxR and iron regulated operon that contains three genes including a putative siderophore biosynthetic gene, *ciuE*, which is predicted to encode a product that shares sequence similarity to the aerobactin biosynthetic enzymes IucC and IucA. Two additional genes of unknown function, *ciuF* and

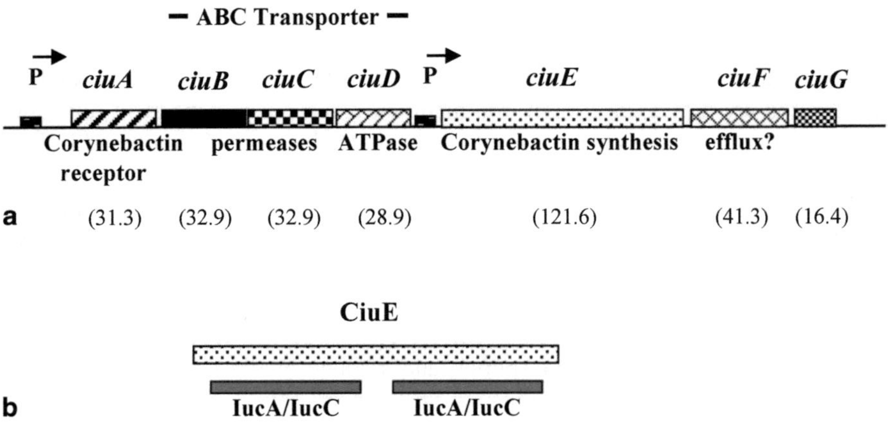

Fig. 6.1 **a** Genetic map of the *ciu* locus in *C. diphtheriae*. CiuA-D are essential for transport of corynebactin and CiuE is critical for siderophore synthesis. **P** indicates the presence of DtxR-regulated promoters and the arrows show the direction of transcription. Numbers below gene description indicate size of predicted products in kDa. **b** Structural features of CiuE show the two regions of sequence similarity to the aerobactin biosynthetic enzymes IucA and IucC

ciuG, are also encoded in this operon (Fig. 6.1a). Pfam analysis shows that CiuF has sequence similarity to proteins in the major facilitator superfamily of membrane transporters. CiuE with a predicted mass of 121.6 kDa is approximately twice the size of the IucC and IucA proteins, and contains two discreet regions of sequence similarity to IucC/IucA (Fig. 6.1b). A non-polar deletion in the *ciuE* gene resulted in a severe defect in siderophore production in both the C7 and 1737 strains, indicating that the *ciuE* gene is essential for corynebactin synthesis. The *ciuE* gene is the only gene in this cluster that encodes a product that shares sequence similarity to proteins known to be required for siderophore biosynthesis, and it is not known if additional genes are needed for corynebactin production.

Deletion of the *ciuA* gene in *C. diphtheriae* has no effect on siderophore synthesis, but results in a severe defect in iron uptake and in the ability to use corynebactin as an iron source for growth in low iron medium (Kunkle and Schmitt 2005). The *ciuA* mutant also exhibits reduced growth in low-iron medium relative to the parent strain, and a similar growth profile is observed with the *ciuE* mutant. It was noted that mutations in either the *ciuA* or *ciuE* gene do not fully abolish growth in low-iron medium, which suggests that additional iron uptake systems are present in *C. diphtheriae*. It was previously shown that the cloned *irp6A* gene was able to complement the siderophore production phenotype in the chemically-derived mutant HC1, which contains a point mutation in the *irp6A* gene and is severely defective in iron uptake (Cryz et al. 1983; Qian et al. 2002). In contrast, a defined deletion in the *irp6A* gene in both C7 and in 1737 showed no defect in iron uptake or in the ability to use corynebactin as an iron source (Kunkle and Schmitt 2005). The C7 *irp6A* deletion mutant exhibits a slight growth defect in low-iron medium, and is unable to utilize aerobactin as an iron source, suggesting the function for the Irp6ABC system is most likely aerobactin utilization and not corynebactin transport. The reason for the severe iron uptake defect in the HC strains is not known, but it seems likely that mutations in addition to those found in the *irp6* operon contribute to the phenotype for the HC mutants. Studies with the Ciu system definitively demonstrate that the CiuABCD transporter is essential for corynebactin uptake, and that CiuE has a critical role in the synthesis of the corynebactin siderophore (Kunkle and Schmitt 2005).

A previous report (Kunkle and Schmitt 2003) described a second DtxR and iron regulated operon that included four genes, two of which were predicted to encode enzymes involved in siderophore biosynthesis, designated *sidA* and *sidB*, and two additional genes encode products involved in siderophore transport, CdtP and CdtQ. The predicted gene products share sequence similarity to proteins involved in the synthesis and uptake of yersiniabactin, a siderophore produced by *Yersinia pestis* (Perry and Fetherston 2011). The yersiniabactin gene cluster is contained within a high pathogenicity island on the *Y. pestis* chromosome and is adjacent to phage genes and an IS element. A similar genetic configuration is observed at the *isdAB/cdtPQ* gene cluster in some *C. diphtheriae* strains, including the 1737 strain and other isolates from the recent diphtheria epidemic in the former Soviet Union (Popovic et al. 1996; Dittmann et al. 2000; Kunkle and Schmitt 2003). Mutations in *sidA* or *sidB* have no effect on siderophore production in the 1737 strain, and most of the *sidA* and *sidB* coding region is deleted in the C7 strain, however, the iron regulated *sidA* promoter as well as the *cdtPQ* transport genes are present in C7 (Fig. 6.2). A *sidA/ciuE* double mutant in 1737 exhibits the identical siderophore production defect as that of the *ciuE* single mutant, providing further evidence that the *sidA* and *sidB* genes do not contribute to siderophore production in the 1737 strain (Kunkle and Schmitt 2005). The function of the *sidA* operon in the *C. diphtheriae* clinical isolates remains to be determined.

The structure for the siderophore corynebactin from *C. diphtheriae* C7 (β) was recently determined, and biochemical analysis revealed a novel structure in which a single lysine amino acid is substituted at both amino groups with two molecules of citrate (Fig. 6.3) (Zajdowicz et al. 2012). This structure is similar to that of staphyloferrin A, which contains an ornithine in replace of lysine. Purified corynebactin is able to enhance iron uptake and stimulate growth of a *C. diphtheriae* siderophore biosynthetic mutant (*ciuE*). It was further noted that a conserved region in the corynebactin biosynthetic enzyme, CiuE, shares homology to FhuF, a ferric iron reductase involved in ferrioxamine B transport in *E. coli*. While the mechanism of iron removal from ferric-corynebactin is not known, it was suggest that this conserved motif in CiuE could potentially function to reduce the ferric iron in the ferric-corynebactin complex, thus allowing the re-cycling of corynebactin (Zajdowicz et al. 2012).

6.2.1 C. diphtheriae *Siderophore Nomenclature*

In recent years there has been some confusion regarding the name corynebactin in regards to the *C. diphtheriae* siderophore. The siderophore produced by *C. diphtheriae* was originally described in 1984 (Russell et al. 1984) and was subsequently given the name corynebactin in a report published in 1985 (Russell and Holmes

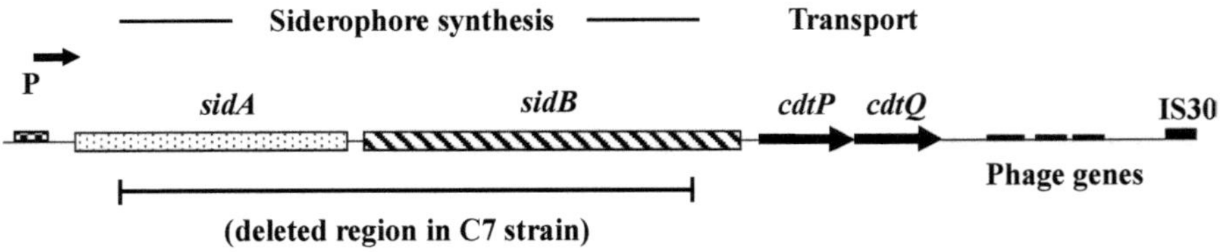

Fig. 6.2 Map of the *sidAB/cdtPQ* gene cluster is shown and the region deleted in the C7 strain is indicated. The presence of phage genes and an IS element (IS30) suggests this region is part of a genomic island in NCTC13129. Although the genes present in the *sidAB* operon are predicted to encode products involved in siderophore transport and synthesis, no siderophore is known to be associated with this genetic region. **P** indicates the presence of the DtxR-regulated promoter and the arrow shows the direction of transcription

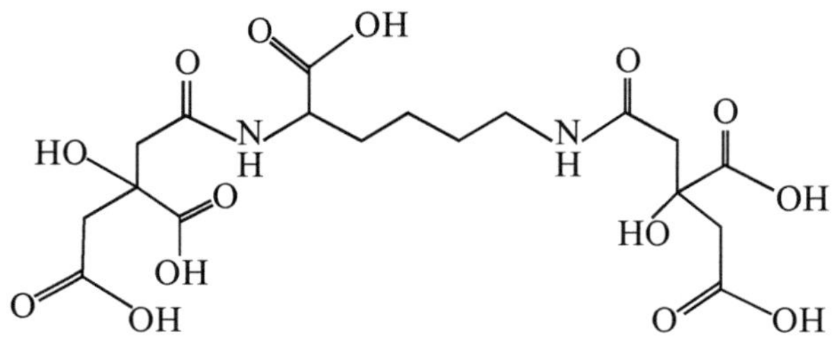

Fig. 6.3 Corynebactin structure is depicted. The siderophore is composed of a lysine residue substituted at both amino groups with two molecules of citrate

1985). A study in 1997 (Budzikiewicz et al. 1997) described the purification and structure of a siderophore purportedly isolated from *C. glutamicum* that was also designated corynebactin. The *C. glutamicum* siderophore described in the 1997 report was a catechol-type siderophore that was identical to the siderophore bacillibactin produced by *Bacillus subtilis*. In a subsequent study in 2006 (Dertz et al. 2006), investigators were unable to identify a siderophore produced by *C. glutamicum*. A BLAST search of the *C. glutamicum* genome failed to identify genes with significant similarity to the bacillibactin biosynthetic genes, and the authors concluded that *C. glutamicum* does not produce a siderophore and they suggested that the original usage of the name corynebactin, referring to the siderophore of *C. diphtheriae*, should be restored. With the determination of the structure of corynebactin from *C. diphtheriae,* the nomenclature issue is now resolved.

6.3 Additional Iron Transporters

In 1994, a search for DtxR and iron regulated promoters in *C. diphtheriae* C7 identified two promoter/operator sequences that were designated IRP1 and IRP2 (Schmitt and Holmes 1994). Analysis of the sequence downstream from the IRP1 promoter revealed the presence of a gene (*irp1A*) that encodes a predicted product that shares significant sequence similarity to the periplasmic binding component of iron transport systems found in gram negative bacteria and also with the FhuD ferrichrome receptor from *B. subtilis*. Irp1A is proposed to be a lipoprotein anchored to the cytoplasmic membrane and a possible receptor for an as yet identified siderophore (Schmitt et al. 1997). Subsequent studies showed that the *irp1A* promoter controls the expression of a four gene operon, *irp1ABCD*, which is predicted to encode the permease (Irp1B/Irp1C) and ATPase (Irp1D) components of an ABC iron transporter (Qian et al. 2002). The cloned *irp1ABCD* genes were unable to complement the siderophore dependent iron uptake defect in the various HC mutants, and defined mutations in the *irp1* genes have not been constructed. While it is assumed that the Irp1ABCD transporter is involved in iron acquisition, a specifc function for this system has not been reported. A recent report (Trost et al. 2012a) that describes the genomic sequence of 13 different *C. diphtheriae* strains, including C7 (β), revealed that the *irp1ABCD* operon is not present in the genome in several strains including NCTC13129, which is a clinical isolate from the Russian epidemic and the first *C. diphtheriae* strain in which the genomic sequence was reported (Cerdeno-Tarraga et al. 2003). The effect of the *irp1ABCD* deletion on iron transport in the NCTC13129 strain has not been investigated.

No open reading frame was described downstream of the IRP2 promoter in the original description of this cloned region from *C. diphtheriae* C7 (Schmitt and Holmes 1994), and no sequences homologous to the IRP2 promoter region or downstream sequences are present in the genomic strain NCTC13129. However, analysis of the completed C7 (β) genome sequence revealed the presence of a nine gene operon downstream from the IRP2 promoter, designated *irp2A-I* (Trost et al. 2012a). The predicted product of several of these genes share sequence similarity to enzymes involved in siderophore biosynthesis, including *irp2C* and *irp2F*, which encode products with similarities to non-ribosomal peptide synthetases. The *irp2I* gene is predicted to encode the ATPase component of a metal-type ABC transporter, but other components that are typically associated with gram positive iron uptake systems were not identified in this operon nor were they present in an adjacent operon (*irp2J-N*), which encoded two putative permease components of an ABC transporter. The *irp2A-I* operon and the adjacent sequences are flanked by genes predicted to encode transposase enzymes and the entire region is proposed to be a putative genomic island in C7 (β).

Over the last 20 years studies using genetic screens or proteomic analysis have identified factors in *C. diphtheriae* that are part of the DtxR regulon, and many of these proteins are proposed to function in iron transport or in iron regulation (Tai and Zhu 1995; Lee et al. 1997; Kunkle and Schmitt 2003; Yellaboina et al. 2004; Hansmeier et al. 2006; Trost et al. 2012a). In 2003, Kunkle and Schmitt developed a repressor titration assay (DRTA) using the cloned *dtxR* gene to identify *C. diphtheriae* DNA fragments that harbored DtxR binding sites. This genetic screen was modeled after a similar system developed using the *fur* gene from *E. coli* (FURTA) (Stojiljkovic et al. 1994). Ten previously unknown DtxR binding sites were identified using DRTA, and the associated promoter regions were examined in expression studies. Genes downstream from seven of these DtxR-regulated promoters were predicted to encode proteins associated with iron or heme transport systems, including the previously described *sidAB-cdtPQ* genes. Other putative iron transport systems identified from the search included two ABC iron transport systems; SufB-D, which is associated with Fe-S cluster assembly, and the FrgA-D iron uptake system, which shares similarity to siderophore transporters. The FrgA-D system is composed of an ABC transporter (FrgA-C) and a solute binding lipoprotein receptor, FrgD. A system that shares sequence similarity to FrgA-D (FagA-D) was characterized in *C. pseudotuberculosis*, a close relative of *C. diphtheriae* (Billington et al. 2002). Mutations in the *fagBC* genes had no effect on iron uptake *in-vitro* in *C. pseudotuberculosis*; however, the mutant showed reduced virulence in a goat model of caseous lymphadenitis, suggesting that the FagABCD iron uptake system may have a role in the pathogenesis for this animal pathogen. *C. diphtheriae frgA-D* mutants have not been reported, and the role of this system in iron uptake or virulence is not known. A recent analysis of the proteome of *C. diphtheriae* identified FrgD as the most abundant of the ABC iron transport solute binding proteins on the bacterial surface (Hansmeier et al. 2006).

The recent completion of the genomic sequence of 13 *C. diphtheriae* strains has identified additional genes encoding putative ABC-type iron transporters including the *iutA-E* system, and *iusA-E*, which is deleted from the genome sequence in NCTC13129, C7 (β) and five other strains, but present in PW8 (Trost et al. 2012a, b). These two genetic systems were identified based on the presence of a putative DtxR binding site located upstream of these operons. While a function for either of these transporters has not been determined, the IutA-E system is conserved in all 13 of the *C. diphtheriae* strains: the iron uptake systems CiuABCD (corynebactin transporter), Irp6ABC (aerobactin uptake in C7), and FrgABCD are also conserved in all 13 strains, suggesting these uptake systems may have a critical function in *C. diphtheriae* iron acquisition (Table 6.1). A more detailed analysis of the IutA-E system revealed that in addition to harboring components for an ABC transporter (IutBCD), this system also contains two putative solute binding proteins, IutA and IutE. The IutA and IutE proteins share almost 60 % amino acid sequence identity, and they both show significant sequence similarity to the MntA solute binding protein in *C. diphtheriae,* which is associated with an ABC-type Mn transporter (Schmitt 2002). The similarity to MntA, suggests that the Iut system may primarily function in the acquisition of Mn or perhaps has a dual function in Mn and Fe uptake. Typically, ABC-type metal transporters contain a single solute binding protein, which in gram positive bacteria is a lipoprotein that functions as a surface receptor for a ferric-siderophore complex or as a receptor for a specific metal or metal containing compound. The reason for two solute binding proteins for the Iut system remains to be determined.

Table 6.1 *C. diphtheriae* genes involved in iron or heme transport and regulation

Gene/operon	Strain[a]	Function/homology	Regulation[b]
irp6ABC	C7, PW8, NC	siderophore (aerobactin) transport	Fe/DtxR
ciuA-D	C7, PW8, NC	corynebactin transport (ABC)	Fe/DtxR
ciuEFG	C7, PW8, NC	corynebactin synthesis (CiuE)	Fe/DtxR
sidA/B	PW8, NC	siderophore synthesis	Fe/DtxR
cdtP/Q	C7, PW8, NC	Fe transport	Fe/DtxR
irp1A-D	C7, PW8	ABC-type Fe transporter	Fe/DtxR
irp2A-I	PW8, NC	siderophore biosynthesis/ Fe-transport	Fe/DtxR
frgA-D	C7, PW8, NC	ABC-type Fe transporter	Fe/DtxR
iutA-E	C7, PW8, NC	ABC-type Mn/Fe-transport	Fe/DtxR
iusA-E	PW8	ABC-type Fe transporter	Fe/DtxR
htaA-hmuTUV	C7, PW8, NC	hemin-transport	Fe/DtxR
htaB	C7, PW8, NC	hemin binding (transport?)	Fe/DtxR
htaC	C7, PW8, NC	hemin transport?	Fe/DtxR
chtA/chtB	NC	hemin transport?	Fe/DtxR
dip0522	NC	hemin transport?	Fe/DtxR
hmuO	C7, PW8, NC	heme oxygenase	Fe/DtxR/heme ChrAS/HrrAS
chrAS	C7, PW8, NC	heme-responsive regulator	unknown
hrrAS	C7, PW8, NC	heme-responsive regulator	heme/HrrA[c]
hemA	C7, PW8, NC	heme synthesis	heme/ChrAS/ HrrAS
hrtAB	C7, PW8, NC	resistance to heme toxicity	heme/ChrAS
dtxR	C7, PW8, NC	Fe-repressor	unknown
ripA (irp3)	C7, PW8, NC	Repressor of Fe-proteins?	Fe/DtxR
mntR	C7, PW8, NC	Mn-repressor (DtxR homolog)	Mn/MntR

[a] Indicates whether gene (s) is present in *C. diphtheriae* strains C7, PW8 or NCTC13129 (NC)
[b] Regulatory protein that controls expression is indicated and metal/heme co-factor
[c] *hrrA* expression is activated by heme and requires HrrA

6.4 Heme Transport

While the acquisition of iron is important for many bacterial pathogens to cause disease, much of the iron in the human host is associated with heme, which is bound to hemoproteins such as hemoglobin. To acquire heme-iron sources, bacterial pathogens, such as *C. diphtheriae*, have developed mechanisms for the uptake of heme and the acquisition of heme from heme containing proteins. A study in 1997 (Schmitt 1997b) demonstrated that *C. diphtheriae* C7 is able to use heme and hemoglobin as iron sources for growth in low iron medium. Chemically derived mutants in *C. diphtheriae* and *C. ulcerans* that failed to use heme and hemoglobin as an iron source were isolated, and the heme utilization defects were complemented using a genomic library of cloned *C. diphtheriae* DNA (Schmitt 1997b; Drazek et al. 2000). Two groups of complementing clones were identified; one group contained the *hmuO* gene, which encodes a heme degrading heme oxygenase enzyme (discussed below) (Schmitt 1997b), while the second group encoded an ABC hemin transporter, HmuUV, and its associated solute binding protein, HmuT (Drazek et al. 2000). This was the first report of an ABC-type hemin transport system in a gram positive bacterium. These studies showed that HmuT is a lipoprotein and is able bind to a hemin agarose resin, suggesting that HmuT binds hemin. Moreover, cell fractionation studies show that HmuT is localized to the cytoplasmic membrane, which indicates that HmuT may function as a cell surface hemin receptor in *C. diphtheriae*. A related HmuTUV hemin transporter was identified in *C. ulcerans*, and a defined mutation in the *hmuT* gene in this species abolishes the ability to use both hemin and hemoglobin as iron sources (Schmitt and Drazek 2001). A similar mutation in the *hmuT* gene in *C. diphtheriae* C7 had no effect on the ability to use heme-iron, however, later studies (discussed below) using a more sensitive heme utilization assay show that mutations in *hmuT* in *C. diphtheriae* result in a defect in the use of heme and hemoglobin as iron sources (Allen and Schmitt 2009).

Results from the DRTA study along with data derived from the completion of the *C. diphtheriae* genome sequence, allowed a more detailed analysis of the heme transport gene cluster, designated *hmu*. The *hmu* genetic region contains six genes that are transcribed from three separate DtxR and iron regulated promoters (Allen and Schmitt 2009). One operon includes *htaA* and the *hmuTUV* genes, while two additional genes, designated *htaB* and *htaC*, are each transcribed from independent promoters (Fig. 6.4). HtaA and HtaB share amino acid sequence similarity with a variety of proteins of unknown function in species closely related to *C. diphtheriae*. HtaA contains an N-terminal region of approximately 150 amino acids, designated CR1 (conserved region 1), that shares almost 50 % similarity to a second CR domain in its C-terminus (CR2) (Fig. 6.4). A region with significant sequence similarity to the CR domain is also present in single copy in HtaB. In addition to the sequence similarity that HtaA and HtaB share at the CR domains, the proteins exhibit extensive structural similarities. Both HtaA and HtaB contain putative signal peptidase I leader peptides, indicating they are secreted; however, both proteins possess a C-terminal hydrophobic region followed by a positively charged sequence, which suggests the proteins may be tethered to the cytoplasmic membrane at their C-termini. HtaC is predicted to be a membrane protein and exhibits only minimal sequence similarity to proteins of unknown function. *C. diphtheriae* 1737 mutants that harbor deletions in the *htaA*, *hmuTUV* or the complete *hmu* locus exhibit reduced ability to use both heme and hemoglobin as iron sources, indicating that the product of these genes are involved in the utilization of heme-iron and likely function in heme uptake (Allen and Schmitt 2009). Mutations in *htaB* have no defect in the ability to use heme as an iron source, and a specific function for HtaB has not been determined. All of these mutants that showed a diminished efficiency in heme-iron use, including the *hmu* deletion, still maintained the ability to use heme-iron sources, suggesting that alternate mechanisms for heme iron utilization are present in *C. diphtheriae*. The presence of multiple hemin-iron utilization systems has been described or proposed for several species of both gram negative and gram positive organisms (Stojiljkovic and Hantke 1994; Occhino et al. 1998;

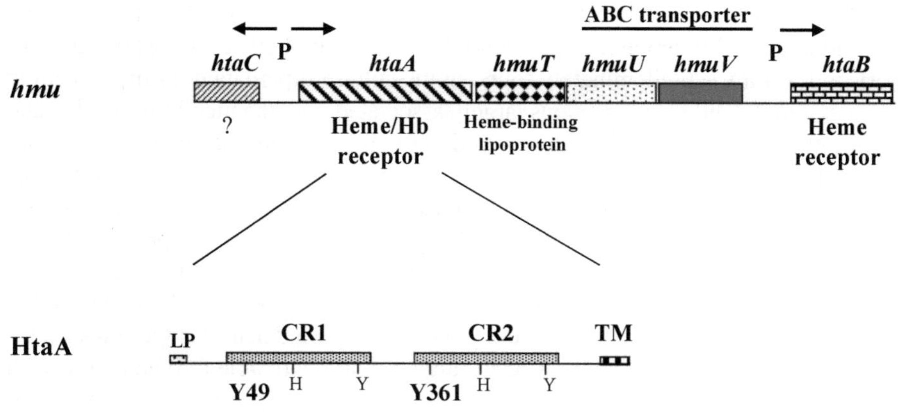

Fig. 6.4 Genetic map of the *hmu* locus is shown. **P** indicates the presence of DtxR-regulated promoters and the arrows show the direction of transcription. HtaA is a surface anchored hemin and hemoglobin binding receptor that is involved in hemin transport. Structural features of HtaA are shown below the genetic map: LP, leader peptide; CR1 and CR2, conserved regions; TM, trans-membrane region. Conserved tyrosine (Y) and histidine (H) residues in the CR domains are indicated; Y49 and Y361 are critical for hemin and hemoglobin binding by the CR domains

Bates et al. 2003). In many of these bacteria, it was reported that mutations in the hemin-specific ABC transporter resulted in only a minimal or no defect in hemin uptake, suggesting these bacteria possess alternate ABC-type hemin transport systems. Analysis of the genome of *C. diphtheriae* has revealed several putative ABC-type iron or siderophore transporters, some of which exhibit sequence similarity to hemin transport systems (Table 6.1) (Trost et al. 2012a). However, the HmuTUV system shows the highest sequence similarity to previously described hemin uptake systems, and it is not known if other ABC-type iron transporters in *C. diphtheriae* are capable of hemin transport. Additionally, genes that are predicted to encode proteins with sequence similarity to the surface proteins HtaA (DIP0522 and ChtA) and HtaB (ChtB) are also present in the *C. diphtheriae* genome, and each of these factors contain putative N-terminal secretion signals, a C-terminal transmembrane region, and a CR domain. However, a function for DIP0522, ChtA or ChtB in hemin transport has not been described.

UV-visible spectroscopy was used to demonstrate that both HtaA and HtaB are able to bind hemin (Allen and Schmitt 2009). Both proteins produce a unique soret band at 406 nm in the presence of varying concentrations of hemin, a spectral property consistent with hemin binding proteins. Cellular localization studies show that HtaA and HtaB are present in the cytoplasmic membrane in bacteria grown in an iron-depleted semi-defined minimal medium, suggesting these proteins are most likely anchored to the membrane through their C-terminal hydrophobic tail regions. A similar anchoring mechanism is believed to occur with the heme binding Shr and Shp proteins in *Streptococcus pyogenes* (Bates et al. 2003; Fisher et al. 2008); however, the heme binding Isd proteins described in *S. aureus* (Mazmanian et al.

2003) and *B. anthracis* (Honsa and Maresso 2011) are anchored to the cell wall peptidoglycan through a covalent bond that is mediated by sortase enzymes. Surprisingly, HtaA is predominately found in the culture supernatant during growth in rich medium, while HtaB remains largely associated with the membrane (Allen and Schmitt 2009). The reason for the secretion of HtaA in nutrient rich medium is not known and the physiological relevance of this observation is not clear, since soluble HtaA containing bound hemin is not used as an iron source for *C. diphtheriae*, suggesting that the secreted form of HtaA does not function as a hemophore. Protease studies with *C. diphtheriae* whole cells grown in minimal medium indicate that the membrane anchored forms of HtaA and HtaB are exposed on the surface of *C. diphtheriae*, which suggests that these proteins likely bind hemin at the cell surface.

The two CR domains in HtaA, CR1 and CR2, are both able to bind hemin with similar affinities, indicating that the hemin binding ability ascribed to HtaA is conferred by these two 150 amino acid conserved domains (Allen and Schmitt 2011). It was further shown by use of an ELISA procedure that the CR domains in HtaA are able to bind hemoglobin, with the CR2 domain exhibiting significantly stronger binding than the CR1 domain, though neither CR domain bound hemoglobin as strongly as the full length HtaA protein. The binding of hemoglobin by HtaA is specific and requires hemoglobin to be associated with hemin since the apo form of hemoglobin fails to bind HtaA. HtaB shows no ability to bind hemoglobin in an ELISA system, and its binding capability appears to be specific to hemin.

Sequence alignment of the CR domains from HtaA, HtaB and from various HtaA homologs shows that while the overall sequence similarity is low, it was observed that two tyrosine residues and a histidine are conserved in all of the CR sequences (Allen and Schmitt 2011). Tyrosine and histidine residues are known to coordinate the iron atom of heme in numerous well characterized hemin binding proteins (Grigg et al. 2007; Pilpa et al. 2009). Site directed mutagenesis studies that targeted the conserved tyrosine and histidine residues within the CR domains confirmed the importance of these amino acids in hemin and hemoglobin binding. Amino acid substitutions in a tyrosine residue at the N-terminus of the CR domains in HtaA (Y49A in CR1, and Y361A in CR2) exhibited the sharpest decrease in absorbance at 406 nm, indicating that this tyrosine is critical for hemin binding and potentially may function as the ligand involved in coordinating the heme iron, although more detailed biochemical analysis will be required to confirm the identity of the iron coordinating ligand in the CR domains (Fig. 6.4). The Y361A substitution in the CR2 domain of HtaA not only causes a strong reduction in hemin binding, but decreases hemoglobin binding by 90%, which suggests that the binding site for hemin and hemoglobin may share common sequences. It was further observed that the Y361A substitution abolishes the ability of HtaA to utilize heme or hemoglobin as iron sources, since the cloned *htaA* gene carrying this mutation is unable to complement the heme-iron utilization phenotype of a *C. diphtheriae htaA* deletion mutant (Allen and Schmitt 2011).

Studies using purified proteins showed that HtaA is able to acquire heme from hemoglobin, and that heme-loaded HtaA is able to transfer heme to HtaB (Allen and Schmitt 2011). Based on these findings, a model for the utilization of heme-iron

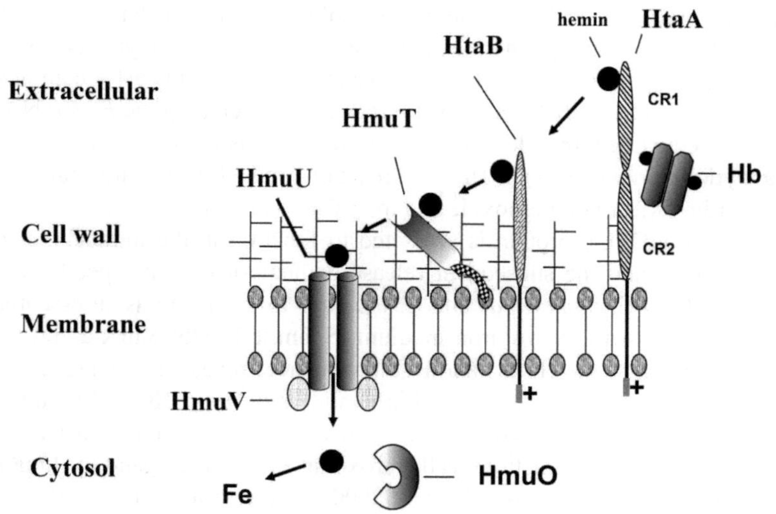

Fig. 6.5 A model for hemin transport in *C. diphtheriae* is shown. Hemin and hemoglobin (Hb) are proposed to bind to the surface anchored HtaA protein at one of the two CR domains. Hemin is extracted from Hb by an unknown mechanism and is transferred to either another CR domain on HtaA or to HtaB. Hemin is then passaged to HmuT, a heme binding lipoprotein that is anchored to the cytoplasmic membrane through an N-terminal lipid anchor. Hemin is mobilized into the cytosol by the ABC-type hemin transporter HmuUV, where heme is degraded by the heme oxygenase HmuO, which results in the release of the heme bound iron

from hemoglobin was described. This model proposes that hemoglobin binds to one or both of the HtaA CR domains on the surface of *C. diphtheriae*, where hemin is subsequently extracted from hemoglobin by an unknown mechanism. Hemin is then transferred to HtaB (or possibly to another CR domain on HtaA) and then to HmuT, where hemin is ultimately transported into the cytosol through the ABC uptake system, HmuUV. Once in the cytosol, heme is thought to be degraded by the action of the heme oxygenase, HmuO, with the subsequent releases of iron (Fig. 6.5).

6.5 HmuO

Heme oxygenases are enzymes involved in the degradation of heme and have a critical function in heme homeostasis (Maines 1992). Although heme oxygenase enzymes have been well characterized in eukaryotes for many years, this group of heme degradation proteins was initially described in prokaryotes with the identification and characterization of the *C. diphtheriae hmuO* gene and its product (Schmitt 1997b; Wilks and Schmitt 1998). The cloned *hmuO* gene, which was identified from a *C. diphtheriae* C7 genomic library, is able to complement a *C. ulcerans* mutant that is defective in the use of heme and hemoglobin as iron sources.

HmuO is a predicted 24.1-kDa soluble cytosolic protein, and shares 33 % identity and almost 70 % similarity with the human heme oxygensase enzyme HO-1. HmuO purified from *E. coli* is a catalytically active protein that forms a 1:1 complex with heme (Wilks and Schmitt 1998). Several reductases, including the *E. coli* NADPH-dependent reductase, are able to effectively function as redox partners to promote the catalytic activity of HmuO. The final reaction products of HmuO bound to heme are biliverdin IX, carbon monoxide and iron. Based on both the *in-vitro* and *in-vivo* activity of HmuO, the protein is predicted to function in the intracellular degradation of heme and the subsequent release of heme-iron. This specific function of HmuO allows *C. diphtheriae* and *C. ulcerans* to use heme as an essential iron source during growth in low iron medium (Schmitt 1997b; Kunkle and Schmitt 2007). Alignment of the HmuO and HO-1 sequences reveals that residue H25, the HO-1 axial heme ligand, is conserved in HmuO at position 20 (H20). Subsequent mutagenesis studies confirmed that H20 is essential for the catalytic activity of the protein and this residue functions as the proximal axial heme ligand in HmuO (Chu et al. 2000; Wilks and Moenne-Loccoz 2000). The crystal structure of the ferric and ferrous forms of the HmuO-heme complex were solved in 2004, and provided additional information on the mechanism of HmuO activity (Hirotsu et al. 2004).

Heme oxygenases were subsequently identified in other bacterial species, including *Neisseria* (Zhu et al. 2000) and *Pseudomonas* (Ratliff et al. 2001), where the enzymes function in heme-iron utilization in a similar manner to that described for HmuO in *Corynebacterium*. A novel class of heme oxygenases, designated IsdG and IsdI, were recently reported in several organisms including the gram positive pathogens *S. aureus* (Wu et al. 2005) and *B. anthracis* (Skaar et al. 2006) as well as in the gram negative bacterium *Bradyrhizobium japonicum*, HmuQ/D (Puri and O'Brian 2006). However, heme oxygenase genes have not been identified in several well characterized bacterial pathogens that are able to use heme as an iron source, including *Shigella dysenteriae* and *Vibrio cholerae*. The *hutZ* gene in *V. cholerae* is encoded in a hemin transport operon and is required for efficient heme utilization (Wyckoff et al. 2007). A clone carrying the *hutZ* gene is able to complement the heme-iron utilization defect in *C. diphtheriae* and *C. ulcerans hmuO* mutants (Wyckoff et al. 2004). The product of *hutZ* is a cytosolic protein that binds heme, but has no detectable catalytic activity and no sequence similarity to heme oxygenases. The function of HutZ has not been determined and the mechanism as to how it complements an *hmuO* mutant is not known (Wyckoff et al. 2007).

An analysis of HmuO activity in strains of *C. diphtheriae* and *C. ulcerans* reveals that the importance of HmuO in heme-iron utilization varies considerably among different strains and species. A deletion of the *hmuO* gene in *C. ulcerans* fully abolishes the ability to use hemin or hemoglobin as iron sources, while *hmuO* mutations in various *C. diphtheriae* strains results in only a partial defect in heme-iron use (Kunkle and Schmitt 2007). Moreover, in the *C. diphtheriae* clinical isolate, 1737, a deletion of *hmuO* has no measurable effect on the ability to use heme as an iron source. These findings suggest that *C. diphtheriae* utilize alternate systems for heme-iron use that substitute for the function of HmuO. An ortholog of the *S. aureus isdG/I* genes is present on the *C. diphtheriae* chromosome (*dip1841*), however, analysis of mutations in *dip1841*, or in a *hmuO-dip1841* double mutant, suggests that DIP1841 is not involved in the use of hemin as an iron source (Kunkle and Schmitt 2007).

6.6 Heme-Dependent Gene Expression

The initial analysis of the *hmuO* gene and the upstream promoter sequences identified a putative DtxR binding site that overlaps the promoter region (Schmitt 1997b). A subsequent study that examined the transcriptional regulation of *hmuO* showed that expression of *hmuO* is repressed by DtxR in high iron conditions and de-repressed in low iron medium (Schmitt 1997a). However, significant levels of transcription from the *hmuO* promoter are only observed in the presence of a heme source, such as hemin or hemoglobin, with optimal levels of expression occurring under low iron conditions in the presence of heme. It was subsequently reported that the heme responsive expression observed at the *hmuO* promoter is controlled by a two-component signal transduction system, which is composed of the response regulator ChrA and the sensor kinase ChrS (Schmitt 1999). The heme-responsive regulation at the *hmuO* promoter was reconstituted in an *E. coli* system using the cloned *chrA/S* genes and the *S. dysenteriae* heme transport system provided *in-trans*. This was the first description of a bacterial two-component system that regulates gene expression by a heme-responsive mechanism.

The *chrA* and *chrS* genes are organized as a two gene operon on the *C. diphtheriae* chromosome and are expressed from a constitutive promoter that is specific for the *chrSA* genes (Bibb et al. 2005). Upstream from *chrS* is the zinc and Zur-regulated *cmrA* gene that is predicted to encode a high molecular weight surface protein of unknown function that is anchored to the cell wall by the action of a sortase (Fig. 6.6a). Deletion of the *chrA* or *chrS* genes in *C. diphtheriae* results in significant reduction in the heme-responsive expression of *hmuO*; however, these mutations do not fully abolish heme activation of *hmuO* transcription. A search of the *C. diphtheriae* genome for additional heme-dependent regulators identified the two-component system HrrA and HrrS (Bibb et al. 2007). The HrrA response regulator and the sensor kinase HrrS show significant sequence similarity to ChrA and ChrS, respectively. Deletion of both two component systems (*chrAS* and *hrrAS*) from the *C. diphtheriae* chromosome results in the complete loss of heme-dependent expression from the *hmuO* promoter, which confirms that both signal transduction systems are involved in the heme-responsive transcription of *hmuO*. It was further shown that both the ChrAS and HrrAS systems are involved in the heme and hemoglobin-dependent repression at the promoter for *hemA*, the promoter proximal gene in an operon that encodes factors involved in heme biosynthesis. This study also provides evidence for "cross-talk" between the ChrAS and HrrAS systems.

High levels of heme or iron is toxic to many bacteria, including *C. diphtheriae*, and the expression of genetic systems involved in the uptake and utilization of iron or iron containing compounds such as heme is often tightly regulated in bacteria. *C. diphtheriae* strains harboring mutations in the *chrAS* genes exhibit increase

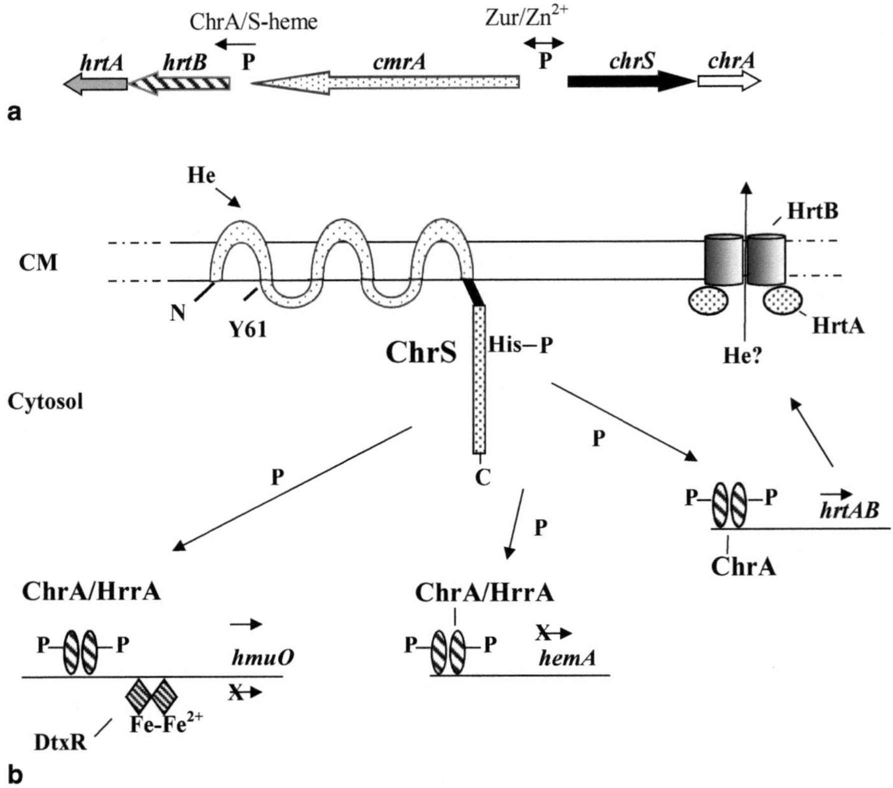

Fig. 6.6 **a** Genetic map of the *hrtAB-chrSA* locus in *C. diphtheriae* is shown. **P** indicates the location of known promoters in *C. diphtheriae* and the direction of transcription. The *hrtAB* genes are positively regulated by ChrAS in the presence of hemin, while *cmrA* is repressed in high levels of Zn^{2+} by the Zn-responsive repressor, Zur. Transcription from the *chrS* promoter occurs at a low constitutive level and expression is not affected by either hemin or iron levels. **b** Proposed model of ChrSA regulation at the *hrtAB, hmuO and hemA* promoters is depicted. In the presence of a heme source (**He**), hemin or heme associated factors are proposed to interact with the membrane anchored N-terminus of ChrS, which results in autophosphorylation (**P**) at a conserved histidine (His) and the subsequent trans-phosphorylation of ChrA. HrrA is presumed to be phosphorylated by HrrS in a similar manner. Phosphorylated ChrA (shown as a dimer) binds upstream of the *hmuO* and *hrtB* promoters to activate transcription, and ChrA along with HrrA are proposed to repress transcription at the *hemA* promoter. HrrA also contributes to the activation of *hmuO* in the presence of heme, while DtxR represses transcription at *hmuO* in high iron conditions. The products of *hrtAB* are predicted to be components of an ABC transporter that may function as an efflux pump to export hemin or factors produced in response to hemin exposure. The location of a conserved tyrosine reside that is critical for ChrS function (Y61) is indicated in the ChrS sensor domain. CM; cytoplasmic membrane

sensitivity to high levels of hemin (Bibb et al. 2005). The reduced expression of *hmuO,* which occurs in the *chrAS* mutant, is not the cause of the hemin sensitivity, since an *hmuO* deletion mutant shows no greater sensitivity to hemin than the wild type strain. The cause of the increase in hemin sensitivity in the *chrAS* mutants is due to the reduced expression of the *hrtAB* genes, which are predicted to encode an ABC transporter involved in the protection of *C. diptheriae* from toxic levels of hemin (Bibb and Schmitt 2010). While the mechanism of protection from hemin toxicity by HrtAB is not known, it is proposed that this system functions to export excess hemin from the cytosol. The *hrtAB* genes are located downstreram of *cmrA* and transcription of *hrtAB* is activated in the presence of hemin or hemoglobin, which is mediated by the ChrAS system (Fig. 6.6a). Heme-dependent expression of *hrtAB* is independent of the HrrAS system and the genes are not repressed by iron; however, expression of *hrtAB* is increased in a *dtxR* mutant, suggesting that DtxR has a positive affect on expression. The mechanism for the *dtxR* regulation is not known, although it is likely indirect, since a DtxR binding site is not present upstream of the *hrtAB* genes. Systems with similar functions to ChrAS and HrtAB are present in *S. aureus* (Stauff and Skaar 2009b) and *B. anthracis* (Stauff and Skaar 2009a), and are designated, HssRS/HrtAB. An ortholog to the HrrAS system was recently described in *C. glutamicum* (Frunzke et al. 2011). A model depicting the heme-responsive regulation of *hmuO*, *hemA* and *hrtAB* by ChrAS and HrrAS is presented in Fig. 6.6b.

Analysis of the ChrS sensor kinase shows that it contains a C-terminal kinase domain and an N-terminal region, which includes six transmembrane regions (Bibb and Schmitt 2010). Like other senor kinases, it is proposed that the N-terminal membrane spanning region is involved in the detection and response to environmental stimuli, which for ChrS is presumed to be heme or factors produced in response to heme exposure. Amino acid substitutions in the C-terminal region identified several residues that were critical for function including a highly conserved tyrosine residue that is predicted to reside in a cytosolic region of ChrS (Fig. 6.6b). Although tyrosine residues are frequently associated with heme binding by proteins, the significance of this conserved tyrosine in the function of ChrS has not been determined, and it is not known if ChrS binds heme. An *in-vitro* study recently showed that ChrS present in liposomes has kinase activity that is responsive to heme, but the mechanism of heme activation was not demonstrated (Ito et al. 2009).

Studies with purified proteins confirmed the kinase function of ChrS and showed that ChrS was able to activate the DNA binding ability of ChrA through phosphorylation at a conserved aspartate residue. Phosphorylated ChrA was able to bind to specific regions upstream of the *hmuO* and *hrtAB* promoters, providing evidence that ChrA activates gene expression directly at these promoters (Burgos and Schmitt 2012).

6.7 Iron Regulation of Gene Expression

As early as the 1930s it was shown that toxinogenic strains of *C. diphtheriae* repressed DT production when grown in medium that contained high levels of iron, and that optimal expression of DT is observed only under iron-depleted conditions. Studies that attempted to understand the mechanism of how iron controlled the expression of DT synthesis led to the identification of *C. diphtheriae* mutants that are unable to repress DT production in high iron conditions (Kanei et al. 1977). Several of theses mutants were mapped to the chromosome of *C. diphtheriae* and it was proposed that these mutations disrupted a gene encoding a regulatory factor involved in the iron-dependent repression of DT. In 1990, the *dtxR* gene from *C. diphtheriae* C7 was identified and shown to mediate the iron-dependent repression at the *tox* promoter (Boyd et al. 1990). It was subsequently shown (Schmitt and Holmes 1991) that the cloned *dtxR* gene is able to complement a *C. diphtheriae dtxR* point mutant, C7βhm723. This study demonstrated that DT and siderophore production are coordinately regulated by iron, which is mediated by DtxR. DNA binding studies showed that Purified DtxR binds to a 19-bp sequence that overlaps the *tox* promoter (Schmitt et al. 1992; Tao and Murphy 1992). DtxR binding is enhanced in the presence of various divalent cations, including Fe^{2+}, Mn^{2+}, Ni^{2+}, Co^{2+}, Cd^{2+} and Zn^{2+}, although it is believed that Fe^{2+} is the primary metal responsible for activation of DtxR *in-vivo*. DtxR is a global iron dependent regulatory factor, and in the last 20 years numerous DtxR binding sites were identified on the *C. diphtheriae* chromosome, and a 19-bp DtxR consensus binding site was derived, 5'-TTAGGTTAGCCTAACCTAA-3' (Tao and Murphy 1994; Kunkle and Schmitt 2003). Analysis of the genomic sequence of 13 *C. diphtheriae* strains identified a total of 36 DtxR binding sites among all of the genomic sequences (Trost et al. 2012a). Approximately half of the DtxR binding sites are located upstream of genes encoding systems involved in iron or heme transport. One of the genes predicted to be controlled by DtxR encodes an ortholog of the *C. glutamicum* transcriptional regulator RipA (regulator of iron proteins A), which in *C. glutamicum* is repressed by DtxR and iron, and is involved in the repression of several iron-containing enzymes (Wennerhold et al. 2005). The *ripA* gene in *C. diphtheriae* was previously identified downstream from the DtxR-regulated IRP3 promoter (Lee et al. 1997). Based on the similarities between the RipA proteins in these *Corynebacterium* species, it seems likely that *C. diphtheriae* RipA may have a similar function to the ortholog in *C. glutamicum*.

6.7.1 DtxR Structure

DtxR is a 226 amino acid (26-kDa) protein and is a member of a super-family of metalloregulatory factors that control gene expression by primarily acting as transcriptional repressors, but occasionally as activators (Feese et al. 2001; Pennella and Giedroc 2005). DtxR-like proteins are found in both gram positive and gram negative bacteria and their DNA binding function is activated *in-vivo* by either Fe^{2+} or Mn^{2+}. The *C. diphtheriae* MntR protein is a homolog of DtxR and controls the expression of a putative Mn transport operon by a Mn-responsive mechanism (Schmitt 2002). The primary amino acid sequence of DtxR has no significant similarity to the functionally related Fur protein, a well characterized metalloregulatory factor present in numerous gram negative and gram positive bacteria (Lee and Helmann 2007). *C. diphtheriae* encodes a single Fur-like protein, Zur, which is responsive to elevated levels of zinc and controls the expression of at least three promoters on the *C. diphtheriae* chromosome (Smith et al. 2009).

The crystal structure for DtxR was reported independently by two separate research groups (Qiu et al. 1995; Schiering et al. 1995). Crystallographic analysis indicated that DtxR is a dimer with each monomer containing three domains (Qiu et al. 1995). A separate study described the DtxR monomer as composed of two major structural domains connected by a 23 amino acid flexible linker (Schiering et al. 1995; Love et al. 2004). Amino acids 1–73 of DtxR contain a helix-turn-helix DNA binding motif, while residues 74–140 include sequences critical for dimerization and metal binding. The remaining C-terminal region of DtxR contains a *src* homology 3-like fold (SH3), which is unusual in bacteria, and is a motif that is associated with protein-protein interaction (Qiu et al. 1996; Wang et al. 1999). While the function of the SH3-like domain is not fully understood, it was proposed that this region may serve to modulate repressor activity through interaction at a metal binding site and at other regions near the junction between the N- and C-terminal domains (Love et al. 2004). Two unique metal binding sites were identified in DtxR; site 1, which is composed of residues H79, E83, H98, E170 and Q173, is considered an ancillary site, and although important for full activity, it is not essential for the repressor function of DtxR. Site 2 contains residues C102, E105, H106 and M10 and occupancy of this site by divalent cations is critical for the repressor activity of DtxR. The crystal structure of DtxR bound to DNA shows that two dimers of DtxR bind to opposite faces of the DNA helix, with no contact between the dimer pairs (White et al. 1998; Pohl et al. 1999a). A similar mode of DNA binding was reported for the Fur protein from *P. aueruginosa* (Pohl et al. 2003). In *Mycobacterium tuberculosis*, the iron-dependent regulator, IdeR, is a homolog to DtxR and shares 88 % amino acid similarity. The cloned *ideR* gene is able to repress expression of the *C. diphtheriae tox* gene (Schmitt et al. 1995), which demonstrates the high level of functional similarity between these global iron repressors. The structure of IdeR is also similar to DtxR especially in the N-terminal region that contains the metal binding, dimerization and DNA binding regions (Pohl et al. 1999b).

References

Allen CE, Schmitt MP (2009) HtaA is an iron-regulated hemin binding protein involved in the utilization of heme iron in *Corynebacterium diphtheriae*. J Bacteriol 191(8):2638–2648

Allen CE, Schmitt MP (2011) Novel hemin binding domains in the *Corynebacterium diphtheriae* HtaA protein interact with hemoglobin and are critical for heme iron utilization by HtaA. J Bacteriol 193(19):5374–5385

Barksdale L (1970) *Corynebacterium diphtheriae* and its relatives. Bacteriol Rev 34(4):378–422

Bates CS, Montanez GE, Woods CR, Vincent RM, Eichenbaum Z (2003) Identification and characterization of a *Streptococcus pyogenes* operon involved in binding of hemoproteins and acquisition of iron. Infect Immun 71(3):1042–1055

Bibb LA, Schmitt MP (2010) The ABC transporter HrtAB confers resistance to hemin toxicity and is regulated in a hemin-dependent manner by the ChrAS two-component system in *Corynebacterium diphtheriae*. J Bacteriol 192(18):4606–4617

Bibb LA, King ND, Kunkle CA, Schmitt MP (2005) Analysis of a heme-dependent signal transduction system in *Corynebacterium diphtheriae*: deletion of the *chrAS* genes results in heme sensitivity and diminished heme-dependent activation of the *hmuO* promoter. Infect Immun 73(11):7406–7412

Bibb LA, Kunkle CA, Schmitt MP (2007) The ChrA-ChrS and HrrA-HrrS signal transduction systems are required for activation of the *hmuO* promoter and repression of the *hemA* promoter in *Corynebacterium diphtheriae*. Infect Immun 75(5):2421–2431

Billington SJ, Esmay PA, Songer JG, Jost BH (2002) Identification and role in virulence of putative iron acquisition genes from *Corynebacterium pseudotuberculosis*. FEMS Microbiol Lett 208(1):41–45

Boyd J, Oza MN, Murphy JR (1990) Molecular cloning and DNA sequence analysis of a diphtheria tox iron-dependent regulatory element (*dtxR*) from *Corynebacterium diphtheriae*. Proc Natl Acad Sci U S A 87(15):5968–5972

Braun V, Hantke K (2011) Recent insights into iron import by bacteria. Curr Opin Chem Biol 15(2):328–334

Budzikiewicz H, Bossenkamp A, Taraz K, Pandey A, Meyer JM (1997) corynebactin, a cyclic catecholate siderophore from *Corynebacterium glutamicum* ATCC 14067. Z Naturforsch C Biosci 52:551–554

Burgos JM, Schmitt MP (2012) The ChrA Response Regulator in *Corynebacterium diphtheriae* Controls Hemin-Regulated Gene Expression through Binding to the *hmuO* and *hrtAB* Promoter Regions. J Bacteriol 194(7):1717–1729

Cerdeno-Tarraga AM, Efstratiou A, Dover LG, Holden MT, Pallen M, Bentley SD, Besra GS, Churcher C, James KD, De Zoysa A, Chillingworth T, Cronin A, Dowd L, Feltwell T, Hamlin N, Holroyd S, Jagels K, Moule S, Quail MA, Rabbinowitsch E, Rutherford KM, Thomson NR, Unwin L, Whitehead S, Barrell BG, Parkhill J (2003) The complete genome sequence and analysis of *Corynebacterium diphtheriae* NCTC13129. Nucleic Acids Res 31(22):6516–6523

Cescau S, Cwerman H, Letoffe S, Delepelaire P, Wandersman C, Biville F (2007) Heme acquisition by hemophores. Biometals 20(3–4):603–613

Chu BC, Garcia-Herrero A, Johanson TH, Krewulak KD, Lau CK, Peacock RS, Slavinskaya Z, Vogel HJ (2010) Siderophore uptake in bacteria and the battle for iron with the host; a bird's eye view. Biometals 23(4):601–611

Chu GC, Katakura K, Tomita T, Zhang X, Sun D, Sato M, Sasahara M, Kayama T, Ikeda-Saito M, Yoshida T (2000) Histidine 20, the crucial proximal axial heme ligand of bacterial heme oxygenase Hmu O from *Corynebacterium diphtheriae*. J Biol Chem 275(23):17494–17500

Collier RJ (2001) Understanding the mode of action of diphtheria toxin: a perspective on progress during the 20th century. Toxicon 39(11):1793–1803

Cornelis P, Andrews SC (eds) (2010) Iron Uptake and Homeostasis in Microorganisms. Caister Academic Press, Norfolk

Crosa JH, Mey AR, Payne SM (eds) (2004) Iron Transport in Bacteria. ASM Press, Washington DC

Cryz SJ Jr, Russell LM, Holmes RK (1983) Regulation of toxinogenesis in *Corynebacterium diphtheriae*: mutations in the bacterial genome that alter the effects of iron on toxin production. J Bacteriol 154(1):245–252

Dertz EA, Stintzi A, Raymond KN (2006) Siderophore-mediated iron transport in *Bacillus subtilis* and *Corynebacterium glutamicum*. J Biol Inorg Chem 11(8):1087–1097

Dittmann S, Wharton M, Vitek C, Ciotti M, Galazka A, Guichard S, Hardy I, Kartoglu U, Koyama S, Kreysler J, Martin B, Mercer D, Ronne T, Roure C, Steinglass R, Strebel P, Sutter R, Trostle M (2000) Successful control of epidemic diphtheria in the states of the Former Union of Soviet Socialist Republics: lessons learned. J Infect Dis 181(Suppl 1):S10–22

Drazek ES, Hammack CA, Schmitt MP (2000) *Corynebacterium diphtheriae* genes required for acquisition of iron from haemin and haemoglobin are homologous to ABC haemin transporters. Mol Microbiol 36(1):68–84

Feese MD, Pohl E, Holmes RK, Hol WGJ (2001) Iron-dependent regulators. In: Messerschmidt A, Huber R, Poulos T, Wieghardt K (eds) Handbook of metalloproteins. Wiley, Chichester, pp 85–863

Fisher M, Huang YS, Li X, McIver KS, Toukoki C, Eichenbaum Z (2008) Shr is a broad-spectrum surface receptor that contributes to adherence and virulence in group A streptococcus. Infect Immun 76(11):5006–5015

Freeman VJ (1951) Studies on the virulence of bacteriophage-infected strains of *Corynebacterium diphtheriae*. J Bacteriol 61(6):675–688

Frunzke J, Gätgens C, Brocker M, Bott M (2011) Control of heme homeostasis in *Corynebacterium glutamicum* by the two-component system HrrSA. J Bacteriol 193(5):1212–1221

Grigg JC, Vermeiren CL, Heinrichs DE, Murphy ME (2007) Haem recognition by a *Staphylococcus aureus* NEAT domain. Mol Microbiol 63(1):139–149

Hadfield TL, McEvoy P, Polotsky Y, Tzinserling VA, Yakovlev AA (2000) The pathology of diphtheria. J Infect Dis 181(Suppl 1):S116–120

Hammer ND, Skaar EP (2011) Molecular mechanisms of *Staphylococcus aureus* iron acquisition. Annu Rev Microbiol 65:129–147

Hansmeier N, Chao TC, Pühler A, Tauch A, Kalinowski J (2006) The cytosolic, cell surface and extracellular proteomes of the biotechnologically important soil bacterium *Corynebacterium efficiens* YS-314 in comparison to those of *Corynebacterium glutamicum* ATCC 13032. Proteomics 6(1):233–250

Hirotsu S, Chu GC, Unno M, Lee DS, Yoshida T, Park SY, Shiro Y, Ikeda-Saito M (2004) The crystal structures of the ferric and ferrous forms of the heme complex of HmuO, a heme oxygenase of *Corynebacterium diphtheriae*. J Biol Chem 279(12):11937–11947

Holmes RK (2000) Biology and molecular epidemiology of diphtheria toxin and the *tox* gene. J Infect Dis 181(Suppl 1):S156–167

Honsa ES, Maresso AW (2011) Mechanisms of iron import in anthrax. Biometals 24(3):533–545

Ito Y, Nakagawa S, Komagata A, Ikeda-Saito M, Shiro Y, Nakamura H (2009) Heme-dependent autophosphorylation of a heme sensor kinase, ChrS, from *Corynebacterium diphtheriae* reconstituted in proteoliposomes. FEBS Lett 583(13):2244–2248

Kanei C, Uchida T, Yoneda M (1977) Isolation from *Corynebacterium diphtheriae* C7(beta) of bacterial mutants that produce toxin in medium with excess iron. Infect Immun 18(1):203–209

Kunkle CA, Schmitt MP (2003) Analysis of the *Corynebacterium diphtheriae* DtxR regulon: identification of a putative siderophore synthesis and transport system that is similar to the *Yersinia* high-pathogenicity island-encoded yersiniabactin synthesis and uptake system. J Bacteriol 185(23):6826–6840

Kunkle CA, Schmitt MP (2005) Analysis of a DtxR-regulated iron transport and siderophore biosynthesis gene cluster in *Corynebacterium diphtheriae*. J Bacteriol 187(2):422–433

Kunkle CA, Schmitt MP (2007) Comparative analysis of *hmuO* function and expression in *Corynebacterium* species. J Bacteriol 189(9):3650–3654

Lee JH, Wang T, Ault K, Liu J, Schmitt MP, Holmes RK (1997) Identification and characterization of three new promoter/operators from *Corynebacterium diphtheriae* that are regulated by the diphtheria toxin repressor (DtxR) and iron. Infect Immun 65(10):4273–4280

Lee JW, Helmann JD (2007) Functional specialization within the Fur family of metalloregulators. Biometals 20(3–4):485–499

Litwin CM, Calderwood SB (1993) Role of iron in regulation of virulence genes. Clin Microbiol Rev 6(2):137–149

Love JF, vanderSpek JC, Marin V, Guerrero L, Logan TM, Murphy JR (2004) Genetic and biophysical studies of diphtheria toxin repressor (DtxR) and the hyperactive mutant DtxR(E175K) support a multistep model of activation. Proc Natl Acad Sci U S A 101(8):2506–2511

Characterization and regulation of heme oxygenase isozymes. In: Heme oxygenase: clinical applicationsand function. CRC Press, Boca Raton, pp 109–144

Maresso AW, Garufi G, Schneewind O (2008) *Bacillus anthracis* secretes proteins that mediate heme acquisition from hemoglobin. PLoS Pathog 4(8):e1000132

Mazmanian SK, Skaar EP, Gaspar AH, Humayun M, Gornicki P, Jelenska J, Joachmiak A, Missiakas DM, Schneewind O (2003) Passage of heme-iron across the envelope of *Staphylococcus aureus*. Science 299(5608):906–909

Mey AR, Payne SM (2001) Haem utilization in *Vibrio cholerae* involves multiple TonB-dependent haem receptors. Mol Microbiol 42(3):835–849

Murphy JR, Skiver J, McBride G (1976) Isolation and partial characterization of a corynebacteriophage beta, *tox* operator constitutive-like mutant lysogen of *Corynebacterium diphtheriae*. J Virol 18(1):235–244

Muttaiyah S, Best EJ, Freeman JT, Taylor SL, Morris AJ, Roberts SA (2011) *Corynebacterium diphtheriae* endocarditis: a case series and review of the treatment approach. Int J Infect Dis 15(9):e584–588

Nobles CL, Maresso AW (2011) The theft of host heme by Gram-positive pathogenic bacteria. Metallomics 3(8):788–796

Noinaj N, Easley NC, Oke M, Mizuno N, Gumbart J, Boura E, Steere AN, Zak O, Aisen P, Tajkhorshid E, Evans RW, Gorringe AR, Mason AB, Steven AC, Buchanan SK (2012) Structural basis for iron piracy by pathogenic *Neisseria*. Nature 483(7387):53–58

Occhino DA, Wyckoff EE, Henderson DP, Wrona TJ, Payne SM (1998) *Vibrio cholerae* iron transport: haem transport genes are linked to one of two sets of *tonB*, *exbB*, *exbD* genes. Mol Microbiol 29(6):1493–1507

Ouattara M, Cunha EB, Li X, Huang YS, Dixon D, Eichenbaum Z (2010) Shr of group A *streptococcus* is a new type of composite NEAT protein involved in sequestering haem from methaemoglobin. Mol Microbiol 78(3):739–756

Pappenheimer AM Jr (1977) Diphtheria toxin. Annu Rev Biochem 46:69–94

Pappenheimer AM Jr, Johnson S (1936) Studies on diphtheria toxin production: the effect of iron and copper to the medium. Br J Exp Pathol 17:335–341

Pennella MA, Giedroc DP (2005) Structural determinants of metal selectivity in prokaryotic metal-responsive transcriptional regulators. Biometals 18(4):413–428

Perry RD, Fetherston JD (2011) Yersiniabactin iron uptake: mechanisms and role in *Yersinia pestis* pathogenesis. Microbes Infect 13(10):808–817

Pilpa RM, Robson SA, Villareal VA, Wong ML, Phillips M, Clubb RT (2009) Functionally distinct NEAT (NEAr Transporter) domains within the *Staphylococcus aureus* IsdH/HarA protein extract heme from methemoglobin. J Biol Chem 284(2):1166–1176

Pohl E, Holmes RK, Hol WG (1999a) Crystal structure of a cobalt-activated diphtheria toxin repressor-DNA complex reveals a metal-binding SH3-like domain. J Mol Biol 292(3):653–667

Pohl E, Holmes RK, Hol WG (1999b) Crystal structure of the iron-dependent regulator (IdeR) from *Mycobacterium tuberculosis* shows both metal binding sites fully occupied. J Mol Biol 285(3):1145–1156

Pohl E, Haller JC, Mijovilovich A, Meyer-Klaucke W, Garman E, Vasil ML (2003) Architecture of a protein central to iron homeostasis: crystal structure and spectroscopic analysis of the ferric uptake regulator. Mol Microbiol 47(4):903–915

Popovic T, Kombarova SY, Reeves MW, Nakao H, Mazurova IK, Wharton M, Wachsmuth IK, Wenger JD (1996) Molecular epidemiology of diphtheria in Russia, 1985–1994. J Infect Dis 174(5):1064–1072

Puri S, O Brian MR (2006) The *hmuQ* and *hmuD* genes from *Bradyrhizobium japonicum* encode heme-degrading enzymes. J Bacteriol 188(18):6476–6482

Qian Y, Lee JH, Holmes RK (2002) Identification of a DtxR-regulated operon that is essential for siderophore-dependent iron uptake in *Corynebacterium diphtheriae*. J Bacteriol 184(17):4846–4856

Qiu X, Verlinde CL, Zhang S, Schmitt MP, Holmes RK, Hol WG (1995) Three-dimensional structure of the diphtheria toxin repressor in complex with divalent cation co-repressors. Structure 3(1):87–100

Qiu X, Pohl E, Holmes RK, Hol WG (1996) High-resolution structure of the diphtheria toxin repressor complexed with cobalt and manganese reveals an SH3-like third domain and suggests a possible role of phosphate as co-corepressor. Biochemistry 35(38):12292–12302

Ratliff M, Zhu W, Deshmukh R, Wilks A, Stojiljkovic I (2001) Homologues of neisserial heme oxygenase in gram-negative bacteria: degradation of heme by the product of the *pigA* gene of *Pseudomonas aeruginosa*. J Bacteriol 183(21):6394–6403

Russell LM, Holmes RK (1983) Initial characterization of the ferric iron transport system of *Corynebacterium diphtheriae*. J Bacteriol 155(3):1439–1442

Russell LM, Holmes RK (1985) Highly toxinogenic but avirulent Park-Williams 8 strain of *Corynebacterium diphtheriae* does not produce siderophore. Infect Immun 47(2):575–578

Russell LM, Cryz SJ Jr, Holmes RK (1984) Genetic and biochemical evidence for a siderophore-dependent iron transport system in *Corynebacterium diphtheriae*. Infect Immun 45(1):143–149

Schiering N, Tao X, Zeng H, Murphy JR, Petsko GA, Ringe D (1995) Structures of the apo- and the metal ion-activated forms of the diphtheria tox repressor from *Corynebacterium diphtheriae*. Proc Natl Acad Sci U S A 92(21):9843–9850

Schmitt MP (1997a) Transcription of the *Corynebacterium diphtheriae hmuO* gene is regulated by iron and heme. Infect Immun 65(11):4634–4641

Schmitt MP (1997b) Utilization of host iron sources by *Corynebacterium diphtheriae*: identification of a gene whose product is homologous to eukaryotic heme oxygenases and is required for acquisition of iron from heme and hemoglobin. J Bacteriol 179(3):838–845

Schmitt MP (1999) Identification of a two-component signal transduction system from *Corynebacterium diphtheriae* that activates gene expression in response to the presence of heme and hemoglobin. J Bacteriol 181(17):5330–5340

Schmitt MP (2002) Analysis of a DtxR-like metalloregulatory protein, MntR, from *Corynebacterium diphtheriae* that controls expression of an ABC metal transporter by an Mn(2+)-dependent mechanism. J Bacteriol 184(24):6882–6892

Schmitt MP, Holmes RK (1991) Iron-dependent regulation of diphtheria toxin and siderophore expression by the cloned *Corynebacterium diphtheriae* repressor gene *dtxR* in *C. diphtheriae* C7 strains. Infect Immun 59(6):1899–1904

Schmitt MP, Holmes RK (1994) Cloning, sequence, and footprint analysis of two promoter/operators from *Corynebacterium diphtheriae* that are regulated by the diphtheria toxin repressor (DtxR) and iron. J Bacteriol 176(4):1141–1149

Schmitt MP, Drazek ES (2001) Construction and consequences of directed mutations affecting the hemin receptor in pathogenic *Corynebacterium* species. J Bacteriol 183(4):1476–1481

Schmitt MP, Twiddy EM, Holmes RK (1992) Purification and characterization of the diphtheria toxin repressor. Proc Natl Acad Sci U S A 89(16):7576–7580

Schmitt MP, Predich M, Doukhan L, Smith I, Holmes RK (1995) Characterization of an iron-dependent regulatory protein (IdeR) of *Mycobacterium tuberculosis* as a functional homolog of the diphtheria toxin repressor (DtxR) from *Corynebacterium diphtheriae*. Infect Immun 63(11):4284–4289

Schmitt MP, Talley BG, Holmes RK (1997) Characterization of lipoprotein IRP1 from *Corynebacterium diphtheriae*, which is regulated by the diphtheria toxin repressor (DtxR) and iron. Infect Immun 65(12):5364–5367

Skaar EP (2010) The battle for iron between bacterial pathogens and their vertebrate hosts. PLoS Pathog 6(8):e1000949

Skaar EP, Gaspar AH, Schneewind O (2004) IsdG and IsdI, heme-degrading enzymes in the cytoplasm of *Staphylococcus aureus*. J Biol Chem 279(1):436–443

Skaar EP, Gaspar AH, Schneewind O (2006) *Bacillus anthracis* IsdG, a heme-degrading monooxygenase. J Bacteriol 188(3):1071–1080

Smith KF, Bibb LA, Schmitt MP, Oram DM (2009) Regulation and activity of a zinc uptake regulator, Zur, in *Corynebacterium diphtheriae*. J Bacteriol 191(5):1595–1603

Stauff DL, Skaar EP (2009a) *Bacillus anthracis* HssRS signalling to HrtAB regulates haem resistance during infection. Mol Microbiol 72(3):763–778

Stauff DL, Skaar EP (2009b) The heme sensor system of *Staphylococcus aureus*. Contrib Microbiol 16:120–135

Stojiljkovic I, Hantke K (1994) Transport of haemin across the cytoplasmic membrane through a haemin-specific periplasmic binding-protein-dependent transport system in *Yersinia enterocolitica*. Mol Microbiol 13(4):719–732

Stojiljkovic I, Bäumler AJ, Hantke K (1994) Fur regulon in gram-negative bacteria. Identification and characterization of new iron-regulated *Escherichia coli* genes by a fur titration assay. J Mol Biol 236(2):531–545

Tai SS, Zhu YY (1995) Cloning of a *Corynebacterium diphtheriae* iron-repressible gene that shares sequence homology with the AhpC subunit of alkyl hydroperoxide reductase of *Salmonella typhimurium*. J Bacteriol 177(12):3512–3517

Tai SP, Krafft AE, Nootheti P, Holmes RK (1990) Coordinate regulation of siderophore and diphtheria toxin production by iron in *Corynebacterium diphtheriae*. Microb Pathog 9(4):267–273

Tao X, Murphy JR (1992) Binding of the metalloregulatory protein DtxR to the diphtheria tox operator requires a divalent heavy metal ion and protects the palindromic sequence from DNase I digestion. J Biol Chem 267(30):21761–21764

Tao X, Murphy JR (1994) Determination of the minimal essential nucleotide sequence for diphtheria tox repressor binding by in vitro affinity selection. Proc Natl Acad Sci U S A 91(20):9646–9650

Trost E, Blom J, de Castro Soares S, Huang I, Al-Dilaimi A, Schröder J, Dorella FA, Rocha FS, Miyoshi A, Azevedo V, Schneider MP, Silva A, Camello TC, Sabbadini PS, Santos CS, Santos LS, Hirata R, Mattos-Guaraldi AL, Efstratiou A, Schmitt MP, Ton-That H, Tauch A (2012a) Pan-genomics of *Corynebacterium diphtheriae*: insights into the genomic diversity of pathogenic isolates from cases of classical diphtheria, endocarditis and pneumonia. J Bacteriol 194(12):3199–3215

Trost E, Blom J, de Castro Soares S, Huang IH, Al-Dilaimi A, Schröder J, Jaenicke S, Dorella FA, Rocha FS, Miyoshi A, Azevedo V, Schneider MP, Silva A, Camello TC, Sabbadini PS, Santos CS, Santos LS, Hirata R Jr, Mattos-Guaraldi AL, Efstratiou A, Schmitt MP, Ton-That H, Tauch A (2012b) Pangenomic Study of *Corynebacterium diphtheriae* That Provides Insights into the Genomic Diversity of Pathogenic Isolates from Cases of Classical Diphtheria, Endocarditis, and Pneumonia. J Bacteriol 194(12):3199–3215

Uchida T, Kanei C, Yoneda M (1977) Mutations of corynephage beta that affect the yield of diphtheria toxin. Virology 77(2):876–879

Wang G, Wylie GP, Twigg PD, Caspar DL, Murphy JR, Logan TM (1999) Solution structure and peptide binding studies of the C-terminal src homology 3-like domain of the diphtheria toxin repressor protein. Proc Natl Acad Sci U S A 96(11):6119–6124

Welkos SL, Holmes RK (1981) Regulation of toxinogenesis in *Corynebacterium diphtheriae*. II. Genetic mapping of a tox regulatory mutation in bacteriophage beta. J Virol 37(3):946–954

Wennerhold J, Krug A, Bott M (2005) The AraC-type regulator RipA represses aconitase and other iron proteins from *Corynebacterium* under iron limitation and is itself repressed by DtxR. J Biol Chem 280(49):40500–40508

White A, Ding X, vanderSpek JC, Murphy JR, Ringe D (1998) Structure of the metal-ion-activated diphtheria toxin repressor/tox operator complex. Nature 394(6692):502–506

Wilks A, Moenne-Loccoz P (2000) Identification of the proximal ligand His-20 in heme oxygenase (Hmu O) from *Corynebacterium diphtheriae*. Oxidative cleavage of the heme macrocycle does not require the proximal histidine. J Biol Chem 275(16):11686–11692

Wilks A, Schmitt MP (1998) Expression and characterization of a heme oxygenase (Hmu O) from *Corynebacterium diphtheriae*. Iron acquisition requires oxidative cleavage of the heme macrocycle. J Biol Chem 273(2):837–841

Winkelmann G (2002) Microbial siderophore-mediated transport. Biochem Soc Trans 30(4):691–696

Wu R, Skaar EP, Zhang R, Joachimiak G, Gornicki P, Schneewind O, Joachimiak A (2005) *Staphylococcus aureus* IsdG and IsdI, heme-degrading enzymes with structural similarity to monooxygenases. J Biol Chem 280(4):2840–2846

Wyckoff EE, Schmitt M, Wilks A, Payne SM (2004) HutZ is required for efficient heme utilization in *Vibrio cholerae*. J Bacteriol 186(13):4142–4151

Wyckoff EE, Mey AR, Payne SM (2007) Iron acquisition in *Vibrio cholerae*. Biometals 20(3–4):405–416

Yellaboina S, Ranjan S, Chakhaiyar P, Hasnain SE, Ranjan A (2004) Prediction of DtxR regulon: identification of binding sites and operons controlled by Diphtheria toxin repressor in *Corynebacterium diphtheriae*. BMC Microbiol 4:38

Yu RH, Schryvers AB (2002) Bacterial lactoferrin receptors: insights from characterizing the *Moraxella bovis* receptors. Biochem Cell Biol 80(1):81–90

Zajdowicz s, Haller JC, Krafft AE, Hunsucker SW, Mant CT, Duncan MW, Hodges RS, Jones DNM, Holmes RK (2012) Purification and structural characterization of siderophore (corynebactin) from *Corynebacterium diphtheriae*. PLoS ONE 7(4):e34591

Zhu W, Wilks A, Stojiljkovic I (2000) Degradation of heme in gram-negative bacteria: the product of the *hemO* gene of *Neisseriae* is a heme oxygenase. J Bacteriol 182(23):6783–6790

Chapter 7
Assembly and Function of *Corynebacterium diphtheriae* Pili

Melissa E. Reardon-Robinson and Hung Ton-That

Abstract Pili or fimbriae are known virulence factors that facilitate bacterial colonization of specific host tissues and pathogenesis. *Corynebacterium diphtheriae*, the causative agent of pharyngeal diphtheria, harbors three pilus gene clusters encoding the heterotrimeric SpaA-, SpaD-, and SpaH-type pili. The current model for Gram-positive pilus assembly is based on the studies of the SpaA-type pili, which may serve as a major virulence determinant for *C. diphtheriae* because they are specifically required for adherence to pharyngeal epithelial cells. The SpaA-type pilus is comprised of the shaft pilin SpaA and minor proteins SpaB and SpaC, which make up the base and tip of the surface structure, respectively. Pilus biogenesis requires tandem sortase enzymes that covalently link SpaABC monomers and subsequently anchor the resulting heterotrimeric polymer to the bacterial peptidoglycan. Here, a decade of research aimed to reveal the assembly, structure, and role of *C. diphtheriae* pili in pathogenesis is discussed.

Keywords *Corynebacterium diphtheriae* · Pili · Pilin motif · Pilus assembly · Sortase

7.1 Introduction

To successfully colonize a host, a bacterium must first adhere to target epithelial tissues. The repulsive force generated by similar charges present on both the bacterium and target cell surfaces, however, presents a challenge (Proft and Baker 2009).

H. Ton-That (✉) · M. E. Reardon-Robinson
Department of Microbiology & Molecular Genetics,
University of Texas Health Science Center, Houston, TX, USA
e-mail: Ton-That.Hung@uth.tmc.edu

H. Ton-That
Department of Microbiology and Molecular Genetics,
The University of Texas Medical School at Houston,
6431 Fannin Street, MSE R213, Houston, TX 77030, USA

A. Burkovski (ed.), *Corynebacterium diphtheriae and Related Toxigenic Species*,
DOI 10.1007/978-94-007-7624-1_7, © Springer Science+Business Media Dordrecht (outside the USA) 2014

To overcome this, prokaryotic organisms assemble proteinaceous structures, known as pili or fimbriae, on the cell surface that recognize cognate host receptors. Pili are polymeric virulence factors required for bacterial colonization and pathogenesis. Their surface expression has been demonstrated to foster progression of infectious diseases that affect periodontal, urinary, gastrointestinal, and respiratory tissue (Sauer et al. 2000). Their importance to pathogenesis makes the surface structures attractive candidates for the development of vaccines and antimicrobial therapies (Maione et al. 2005; Soriani and Telford 2010).

Both Gram-positive and Gram-negative species of bacteria produce pili, but utilize remarkably different mechanisms of assembly (Thanassi et al. 1998; Ton-That and Schneewind 2004). Gram-negative pili are multimeric structures comprised of non-covalently associated subunits that are fastened to the outer-membrane. The chaperone-usher pathway, responsible for the display of uropathogenic *Escherichia coli* type I and Pap pili on the surface is the best characterized example of Gram-negative surface pilus adhesins among others including type IV pili and curli (Kline et al. 2010). To construct these adhesins, precursors translocated into the periplasmic space first associate with a designated chaperone charged with preventing protein aggregation, and the delivery of substrates to an outer membrane usher which catalyzes both polymerization of subunits and secretion of resulting complexes to the cell surface (Phan et al. 2011; Remaut et al. 2008). Pilin precursors are joined by a process known as 'donor strand exchange' in which the N-terminal extension of a preceding subunit fills a gap within the incomplete immunoglobulin fold of an incoming pilin (Sauer et al. 1999, 2002).

The study of pili expressed by Gram-positive bacteria is a relatively recent endeavor. These prokaryotes, unlike their Gram-negative counterparts, lack outer membranes. Their cell envelopes are comprised of a single inner membrane surrounded by a dense layer of peptidoglycan. Sjoquist and coworkers (Sjoquist et al. 1972) first recognized that Gram-positives utilize their cell wall as an organelle to anchor proteins upon observing the *Staphylococcus aureus* protein A, an MSCRAMM protein known for its ability to bind to the constant region of immunoglobulins, became soluble only after cells were treated with lysostaphin, a peptidoglycan hydrolase. Decades later, Schneewind and colleagues demonstrated that cell wall anchoring of protein A is dependent on a C-terminal cell wall sorting signal (CWSS) comprised of an LPXTG motif, a hydrophobic domain, and a positively charged tail (Schneewind et al. 1992). Soon afterwards, a unique transpeptidase named sortase SrtA in *S. aureus* was discovered and credited as the catalysis for a transpeptidation reaction that links CWSS-containing substrates like protein A to the cell wall peptidoglycan (Mazmanian et al. 1999, 2001). Finally, the finding that sortase mediates pilus polymerization in Gram-positive bacteria was first demonstrated in *Corynebacterium diphtheriae* (Ton-That and Schneewind 2003), the causative agent of diphtheria (Klebs 1883).

This important study has launched subsequent investigations of pili expressed by *Streptococcus agalactiae, Streptococcus pyogenes, Streptococcus pneumoniae, Enterococcus faecalis, Bacillus cereus*, and *Actinomyces oris* (Lauer et al. 2005; Mora et al. 2005; Barocchi et al. 2006; LeMieux et al. 2006; Dramsi et al. 2006;

Rosini et al. 2006; Nallapareddy et al. 2006; Budzik et al. 2007; Mishra et al. 2007), establishing *C. diphtheriae* as a model system for the biogenesis of Gram-positive pili (Mandlik et al. 2008b). This chapter will review the most current knowledge available regarding pilus expression, structure, and sortase-mediated assembly in *C. diphtheriae*, as well as the role of pili in establishing disease. Pilus assembly is a worthy research endeavor as it will expand our understanding of the infection process potentially leading to the development of new antibiotics or vaccines.

7.2 Pili and Pilus Gene Clusters of *Corynebacterium diphtheriae*

Gram-positive pili were first observed in 1968 on the surface of *Corynebacterium renale* using electron microscopy (Yanagawa et al. 1968). Subsequent protein analysis revealed that these structures were remarkably different compared to those expressed by Gram-negative bacteria. The integrity of pili isolated from the cell surface, for example, was highly resistant to heat boiling (Kumazawa and Yanagawa 1972). Pili were also observed in many Gram-positive bacteria, such as *Actinomyces viscosus* and *Actinomyces naeslundii* (Girard and Jacius 1974). These pili are required for interbacterial adhesion and bacterial attachment to host surfaces (Yeung 1999). Despite a flurry of research activities in this area for decades, the mechanism of pilus assembly was not revealed until recently.

In 2003, Ton-That and Schneewind surveyed the available genome of *C. diphtheriae* NCTC13129 (Cerdeno-Tarraga et al. 2003), a toxin-producing clinical isolate, for open reading frames (ORFs) encoding sortase homologs and CWSS-containing proteins (Ton-That and Schneewind 2003). They revealed three gene clusters encoding a total of nine surface proteins with the LPXTG motif, termed SpaA-I (Spa for *s*ortase-mediated *p*ilus *a*ssembly), and five sortases (SrtA-E). In addition, a sixth sortase encoding gene named *srtF* was found elsewhere in the genome (Fig. 7.1). Polyclonal antibodies raised against purified Spa proteins were used to label immobilized cells, followed by staining with IgG-conjugated gold particles, and subsequent detection by electron microscopy. This methodology revealed three distinct pilus structures assembled on the cell surface of *C. diphtheriae*, designated as SpaA-, SpaD-, and SpaH-type pili based on the major pilin (Ton-That and Schneewind 2003; Gaspar and Ton-That 2006; Swierczynski and Ton-That 2006). Each pilus is composed of a shaft pilin that displays a tip pilin and is interspersed by a minor pilin also forming the pilus base; for example, the SpaA-type pilus is made of the shaft pilin SpaA, the tip pilin SpaC and the pilus base SpaB (Ton-That and Schneewind 2003; Mandlik et al. 2008a). Pili observed were typically smaller than those seen on the surface of Gram-negative bacteria, ranging from 1–2 μm in length and 1–2 nm in width (Ton-That and Schneewind 2003). Unlike Gram-negative pili, these pili are covalently linked, evident by their resistance to hot SDS and formic acid treatment (Ton-That and Schneewind 2003). Overexpression of shaft pilins resulted in exceedingly long

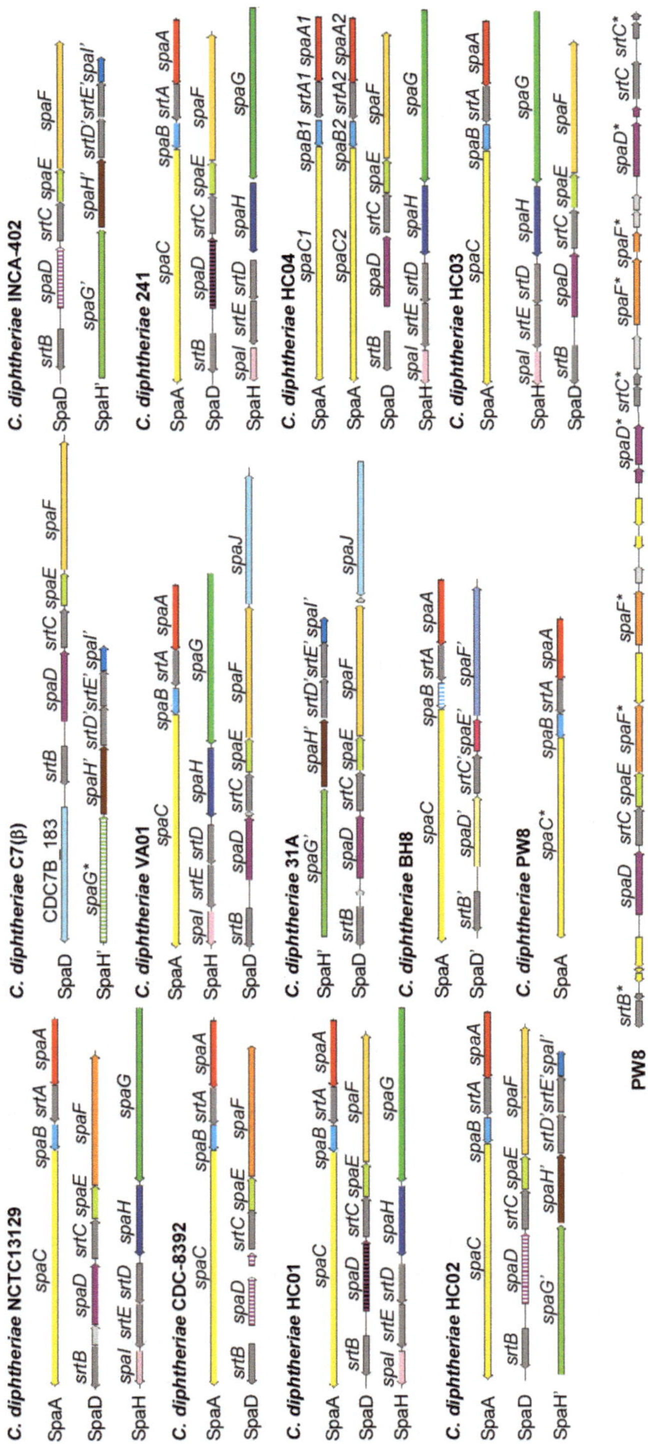

Fig. 7.1 Pilus gene clusters of *Corynebacterium diphtheriae* NCTC13129 and recently sequenced clinical isolates. Homology of pilin and sortase genes is indicated by color; with genes similar to SpaD and SpaH types denoted with primes. Fragmented genes are indicated by asterisks and hatched arrows. (Reprinted from (Trost et al. 2012) with permission)

pili (Swierczynski and Ton-That 2006; Ton-That and Schneewind 2003), indicating pilus polymerization is a stoichiometric process. Deletion of a sortase gene within a pilus gene cluster, i.e. *srtA* within the SpaA-type locus, resulted in the abrogation of pilus polymers, thus designating pilin-specific sortases responsible for assembling specific pilin subunits (Mandlik et al. 2008b).

A recent pangenomic study of 13 *C. diphtheriae* clinical isolates revealed heterogeneity within pilus gene clusters (Trost et al. 2012). All strains were found to harbor at least two distinct pilus forms with the SpaA-type pili being the most common followed by SpaD-type. Few isolates expressed SpaH-type pili. Interestingly, genes within clusters appeared to have co-evolved. Among all strains tested, the genes encoding SpaABC pilins and their respective sortase were found to be conserved-type specific sortase and pilins (Fig. 7.2), while components of the SpaD-type and SpaH-type were found to be highly divergent in comparison. One strain in particular, the vaccine strain Park-William No. 8 (PW8) contains a *spaD* cluster with multiple intact and disrupted *spaD*, *spaE*, *spaF*, *srtE*, and *srtB* genes, indicating that it was the target of mobile DNA elements. The maintenance of the *spa* locus throughout evolution underscores its importance for the bacterium's ability to cause disease. SpaA-type pili specifically facilitate corynebacterial adherence to human pharyngeal epithelial cells, the major site for corynebacterial infection (Mandlik et al. 2007). This specific adherence is attributed to the two minor pilins of SpaA-type pili, SpaB and SpaC (Mandlik et al. 2007). Maintenance of these genes is important since mutations affecting SpaA/B/C or SrtA would compromise the fitness of these bacteria. PW8 appears to be an exception as it harbors a frame-shift mutation within *spaC* preventing the tip pilin from being fully translated (Trost et al. 2012). Interestingly, this correlates with overall decreased virulence compared to other isolates (Iwaki et al. 2010). Low conservation of the *spaD* and *spaH* loci suggests that mutations in these regions are better tolerated because they are less important to establishing *C. diphtheriae* infection. Indeed, adherence assays have demonstrated that while SpaA-type pili target pharyngeal epithelial cells, SpaD and SpaH pili preferentially mediate binding to laryngeal and lung epithelial cells (Mandlik et al. 2007).

7.3 The Archetype SpaA-type Pili: Conserved Pilin Elements and the Mechanism of Sortase-Mediated Pilus Assembly

The SpaA-type pilus, encoded by the gene locus *spaA-srtA-spaB-spaC* (Ton-That and Schneewind 2003), serves as a model of pilus assembly in Gram-positive bacteria. As mentioned above, immuno-electron microscopic analysis, in combination with genetics and biochemical methods, revealed that SpaA forms the pilus shaft, with SpaC located at the tip. SpaB was observed along the pilus structures (Ton-That and Schneewind 2003) and also at the pilus base (Mandlik et al. 2008a).

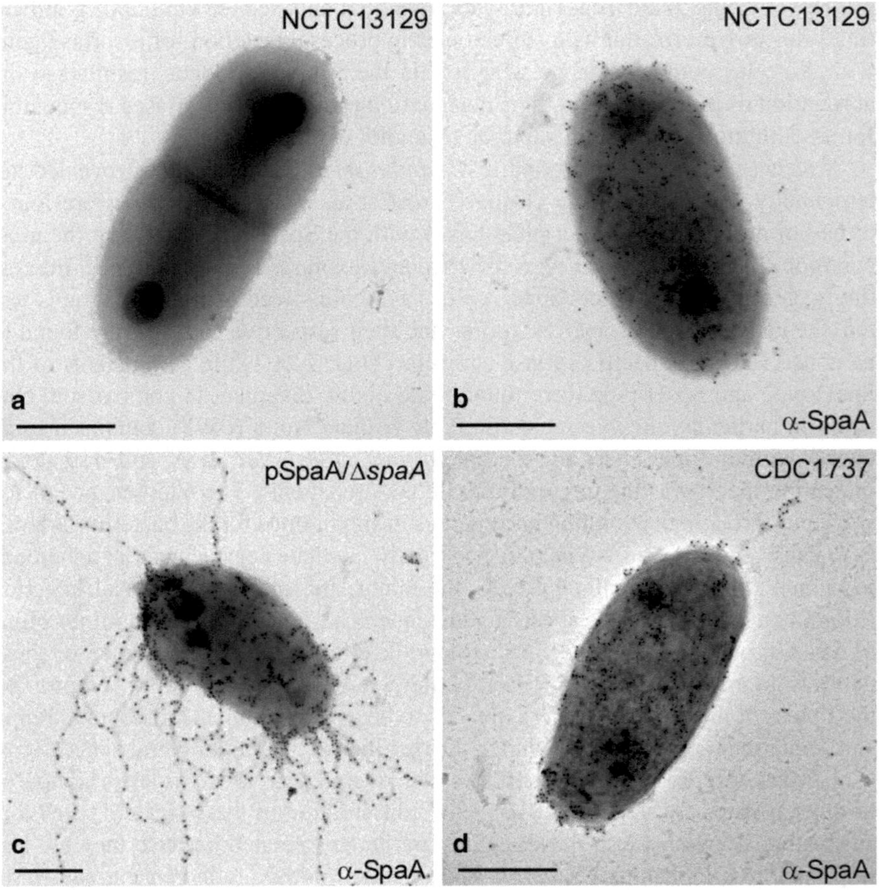

Fig. 7.2 Electron micrographs of *Corynebacterium diphtheriae* NCTC13129 and a toxigenic strain. a Corynebacterial cells were immobilized on nickel carbon-coated grids and viewed by an electron microscopy after staining with 1 % uranyl acetate. **b–d** Cells of *C. diphtheriae* NCTC13129 **b**, its isogenic strain over-expressing the pilin shaft SpaA **c**, and a clinical isolate from Russia (CDC1737; (Popovic et al. 1996)) **d** were immobilized on nickel carbon-coated grids, stained with antibodies against SpaA (α-SpaA), followed by goat anti-rabbit IgG conjugated to 18 nm gold particles, and stained with 1 % uranyl acetate before electron microscopy. Bars indicate 0.5 μm

Consistent with the role of sortase in pilus assembly, a mutant lacking *srtA* failed to assemble SpaA pili on the cell surface (Ton-That and Schneewind 2003).

Like all other pilins, SpaA/B/C contain an N-terminal signal peptide and a C-terminal CWSS for sortase recognition and cleavage (Ton-That and Schneewind 2003). Sequence comparison between corynebacterial major pilin proteins, i.e. SpaA, SpaD and SpaH, with the putative shaft pilins of many Gram-positive bacteria revealed several conserved elements, including the pilin motif WxxxVxVYP**K** and the E-box LX**E**T, which harbor conserved lysine and glutamic acid residues (bold), respectively (Ton-That and Schneewind 2003; Ton-That et al. 2004b). Mutations altering lysine (K190) in the SpaA pilin motif to alanine or arginine completely abrogated pilus assembly (Ton-That and Schneewind 2003), whereas similar mutations of the glutamic acid residue (E446) in the E-box did not dramatically affect pilus polymerization. Instead, these mutations resulted in fragmented SpaA polymers and failed incorporation of SpaB into the SpaA pili (Ton-That et al. 2004b). While it was apparent that the pilin motif is involved in pilus polymerization, it was not clear how the E-box participates in pilus assembly.

Compelling evidence that the pilin motif is essential for pilus polymerization and that both the pilin motif and the LPXTG motif are sufficient and necessary for this transpeptidation, was provided by an experiment in which sortase-mediated polymerization of a hybrid protein was studied (Ton-That et al. 2004b). A region containing the N-terminal signal peptide and the pilin motif of SpaA was fused to the mature domain of staphylococcal enterotoxin B (SEB) at its amino-terminus, whereas its C-terminus was linked to the SpaA-CWSS. This fusion protein was expressed in strains lacking *spaA* or *srtA*. Remarkably, SEB polymers were observed by immunoblotting with the Δ*spaA* mutant but not with the Δ*srtA* mutant. A similar phenotype was also observed in a strain harboring a mutation in the lysine residue of the SpaA pilin motif, i.e. K190A. The results have established the specific role of pilin-specific sortase and critical elements for pilus assembly in Gram-positive bacteria (see below).

7.3.1 A Biphasic Model of Sortase-Mediated Pilus Assembly

Using a classic mutagenesis approach in which *S. aureus* cells failed to "sort" protein A to the bacterial cell wall were selected, Mazmanian and colleagues identified the first sortase gene termed *srtA* (Mazmanian et al. 1999), which encodes a transpeptidase enzyme that catalyzes cell wall anchoring of staphylococcal surface proteins (Ton-That et al. 1999). Further analysis of available bacterial genomes revealed that sortase enzymes were conserved among Gram-positive bacteria (excluding *Mycobacterium* and *Microplasma*) (Comfort and Clubb 2004; Dramsi et al. 2005), revealing that the display of cell surface proteins in these organisms relied on a universal mechanism (Ton-That et al. 2004a).

Homologs of *S. aureus* SrtA are grouped into class A sortases (often collectively referred to as housekeeping sortases), whereas sortases involved in pilus polymerization, termed pilin-specific sortases like SrtA of *C. diphtheriae*, belong to class C (Comfort and Clubb 2004; Dramsi et al. 2005). There are additional sortase groups termed B, D, E and F. Class B sortases are involved in iron acquisition, but the others are less well understood (Spirig et al. 2011). It was first determined in *C. diphtheriae* that the housekeeping sortase SrtF plays an essential role in surface display of pilus polymers generated by a pilin-specific sortase (Ton-That and Schneewind 2003; Swaminathan et al. 2007). By fractionating corynebacterial cells into extracelluar milieu and cell wall compartments, Swaminathan and colleagues showed that deletion of *srtF* resulted in abundant secretion of SpaA pilus polymers into the culture medium, a phenotype similar to a multiple deletion mutant expressing only SrtA; furthermore, a mutant strain expressing only SrtF failed to produce SpaA polymers (Swaminathan et al. 2007). This work has established a principle for a biphasic model of pilus assembly in Gram-positive bacteria, whereby pilus polymerization catalyzed by a pilin-specific sortase is terminated by cell wall anchoring of the resulting pilus polymers that is catalyzed by non-polymerizing sortase or the housekeeping sortase (Mandlik et al. 2008b).

According to this model using *C. diphtheriae* SpaABC pili as an example (Fig. 7.3), Spa pilin precursors are targeted to the general secretion machinery (Sec) via the N-terminal signal peptide for translocation into the extracellular space. Subunits move across the membrane until they are tethered in place by the C-terminal CWSS. Within the exoplasm, the pilin-specific sortase SrtA recognizes the LPXTG motif of each pilin and cleaves between threonine (T) and glycine (G), resulting in formation of an acyl-enzyme intermediate between the catalytic cysteine residue of the sortase enzyme and the threonine residue of the Spa pilins. Polymerization of SpaA or the cross-linking of subunits SpaC to SpaA occurs by a nucleophilic attack of the pilin motif lysine of a neighboring SpaA-SrtA acyl-enzyme intermediate to the thioester linkage of the next. This process occurs repeatedly to construct the SpaA shaft; pilus assembly, thus, is catalyzed in a bottom-up fashion with SpaC-SpaA linkage as the first transpeptidase reaction. Pilus polymerization is terminated when SpaB tethered the pilus base is transferred to the housekeeping sortase, which catalyzes cell wall anchoring of SpaB, hence the pilus polymers.

This model has been supported with several lines of evidence. First, as discussed above the mutational analysis of the pilin motif and the SEB fusion protein provided strong evidence for the transpeptidation reaction that crosslinks pilin subunits. Indeed, mass spectrometry analysis of native SpaA pili revealed the predicted isopeptide bond between SpaA subunits, which is formed between the lysine residue K190 of the SpaA pilin motif and the threonine residue within the LPXTG motif of an adjacent subunit (Kang et al. 2009). Similarly, this linkage has been also demonstrated in *B. cereus* pili using mass spectrometry (Budzik et al. 2008). Presumably, SrtA catalyzes cross-linking to SpaC by a similar mechanism, as this was shown by Budzik and colleagues for the linkage between the tip pilin BcpB and the pilin shaft BcpA (Budzik et al. 2009). Second, the presence of an acyl-enzyme intermediate between pilin and sortase was implicated by mutational analysis of the pilin-specific sortase SrtA of *C. diphtheriae* (Guttilla et al. 2009). In this study, several truncations of the membrane anchor domain of SrtA were generated, and the affect of these mutations were analyzed by western blotting with antibodies against SpaA and SrtA. It was found that SrtA mutants loosely bound to the membrane were secreted into the culture medium in complex with SpaA polymers. Third, the two-step pilus assembly, i.e. pilus polymerization by pilin-specific sortase preceding cell wall anchoring by non-polymerizing sortase, is evident by the study of Swaminathan and colleagues with the housekeeping sortase SrtF, as described above. Finally, in agreement with this mode of assembly Mandlik and coworkers elegantly showed that the pilus base

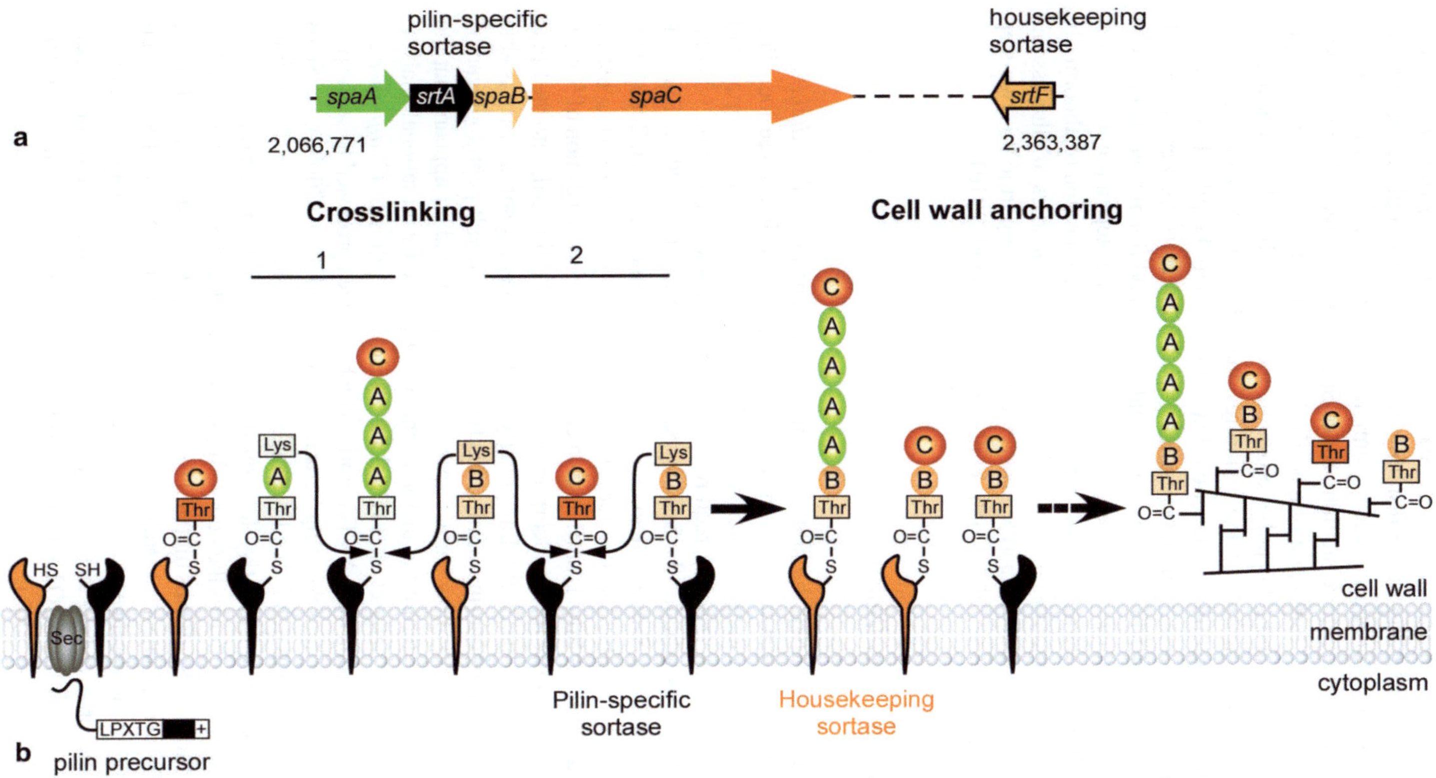

Fig. 7.3 Biphasic model of pilus assembly with *C. diphtheriae* SpaA pili. Shown are the gene locus *spaA-srtA-spaB-spaC* that encodes the prototype SpaA-type pilus and the housekeeping sortase gene located elsewhere **a** Pilus assembly on the cell surface occurs by a two-step mechanism **b** Pilin-specific sortase (*black*) catalyzes polymerization of Spa pilins with SpaC at the tip and SpaA forming the shaft. SpaB enters the pilus base, which terminates pilus polymerization and triggers cell wall anchoring by the housekeeping sortase (*orange*) (see text for detail). Heterodimers of SpaB and SpaC are also depicted. Modified from (Chang et al. 2011) with permission

SpaB functions as a molecular switch that terminates pilus polymerization, leading to cell wall anchoring of the resulting pilus polymer (Mandlik et al. 2008a). Specifically, deletion of *spaB* caused a similar phenotype as deletion of *srtF*, i.e. pilus secretion into the culture medium. Since SpaB is a preferred substrate of SrtF (Mandlik et al. 2008a), formation of a SpaB-SrtF acyl-enzyme intermediate might serve as the signal to end pilus polymerization and begin cell wall anchoring. For this to occur, it was shown that the lysine residue K139 of SpaB functions as a nucleotide for a transpeptidation reaction that links SpaB to the pilus base, similar to the transpeptidation reaction that crosslinks SpaA pilins. SrtF then catalyzes cell wall anchoring of SpaB-linked polymers. SpaB, thus, additionally contributes to control the pilus length because it serves as the rate-limiting step in polymerization termination. Indeed, deletion of *spaB* is associated with the secretion of abnormally long SpaA polymers into the culture medium. Significantly, studies of pilus assembly in *B. cereus*, *S. agalactiae* and *S. pyogenes* lend further support to this two-step mechanism (Budzik et al. 2007; Nobbs et al. 2008; Smith et al. 2010).

7.3.2 Sortase Specificity

As aforementioned, unlike the housekeeping sortase gene *srtF* of *C. diphtheriae* NCTC 13129 other pilin-specific sortase genes are clustered into three loci. Previous work has shown there is no cross-activity between pilin-specific sortases of different loci, i.e. SrtA is solely required for the assembly of SpaA pili, whereas SrtB and SrtC are specific for the SpaD-type pili (Ton-That et al. 2004b; Ton-That and Schneewind 2003; Swaminathan et al. 2007; Gaspar and Ton-That 2006). However, specificity of pilin-specific sortases within a pilus type is promiscuous; for example, either SrtB or SrtC is sufficient to catalyze pilus polymerization of SpaDF pilin, but SrtB is specific for the incorporation of the minor pilin SpaE into the SpaD pili (Gaspar and Ton-That 2006). Intriguingly, the housekeeping sortase SrtF is able to catalyze cell wall anchoring of all pilus types to the cell wall peptidoglycan. A long standing question is how sortase specificity is determined? It appears that the LPXTG motif is one determining factor, as this is supported by mutational analysis of the SpaB CWSS (Chang et al. 2011). Like all the base pilins of *C. diphtheriae*, the SpaB CWSS contains the LAXTG motif, which has been proposed to be the preferred substrate of the housekeeping sortase SrtF (Mandlik et al. 2008a). As SpaB serves as a molecular switch for cell wall anchoring, it would have a greater affinity for SrtF than SrtA. Consistently, deletion of *srtF* led to increased expression of SpaB that failed to anchor to the cell wall (Mandlik et al. 2008a). However, when the LAFTG motif of SpaB was mutated to the SpaA LPXTG motif, i.e. LPLTG, the mutant SpaB now became a substrate of SrtA, as shown by the ability of SrtA to anchor SpaA pili in the absence of *srtF* (Chang et al. 2011). This is in agreement with previous studies in *S. aureus* that show SrtA enzymes recognize the five amino acid peptide LPETG, not the NPQTN peptide, which is the substrate of class B sortase SrtB (Mazmanian et al. 2002).

The LPXTG motif is not the only determining factor for sortase specificity. Evidently, both SpaA and SpaH pilins contain the LPLTG motif; however, SrtA cannot polymerize SpaH, and neither SrtD nor SrtE catalyzed pilus polymerization of SpaA (Mandlik et al. 2007; Swaminathan et al. 2007). Intriguingly, corynebacterial SrtD is able to polymerize FimA, the major pilin shaft of *A. oris* type 2 fimbriae (Mishra et al. 2007), when FimA is expressed in *C. diphtheriae* (Ton-That et al. 2004b). Phylogenetic analysis of FimA and SpaH sequences shows both pilins are closely related (Mishra et al. 2007). While both have the same LPLTG motif, several conserved elements are revealed. It is thus appealing to know if these elements contribute to sortase specificity. Given the proximity of pilin substrates and pilin-specific sortases on the membrane (Guttilla et al. 2009), we speculate that other domains of the CWSS sequence, i.e. hydrophobic domain and positively charged tail, may also be important as they are required for substrate retention within the bacterial membrane (Mazmanian et al. 2001; Mandlik et al. 2008a).

7.3.3 Pilusosome: A Pilus Assembly Center

Sortases are membrane-bound enzymes working on the outer leaflet of the cytoplasmic membrane (Mazmanian et al. 1999, 2001). Both sortases and pilin substrates harbor an N-terminal signal peptide; thus, they are subjected to translocation across the membrane by the Sec machinery. Based upon the requirements of protein secretion, the close proximity of sortases and their cognate substrates, specificity of sortase enzymes, and highly organized manner of assembly, it is hypothesized that secretion and pilus assembly machineries are found in close proximity and coupled. A hint for this conjecture came from the study of the pilus base SpaB (Mandlik et al. 2008a). In this study, the CWSS of SpaB was removed, hence preventing it from being inserted into the bacterial membrane. However, SpaB was found to be incorporated into the SpaA pilus structures, although these polymers were secreted into the extracellular milieu. This indicated that SrtA catalyzes the attachment of SpaB to the pilus base before completion of protein translocation suggesting that sortase and the secretion machinery are in close proximity and coupled. In support of this model, immuno-electron microscopic studies have revealed that SrtA, its cognate pilin substrates, and the translocation motor SecA are co-localized (Guttilla et al. 2009). This observation was corroborated by another study that examined co-localization of *E. faecalis* sortase and SecA (Kline et al. 2009). Interestingly, co-localization was shown to be dependent on a positively charged cytoplasmic domain within the sortase enzyme, suggesting that a retention signal may be responsible for pilusosome maintenance. An outstanding problem is how the orderly pilus assembly is orchestrated within the pilusosome. More experimental work is necessary to elucidate the apparent complex pathway of pilus assembly in Gram-positive bacteria.

7.4 A Structural View of Pilus Assembly

7.4.1 Three-Dimensional Structures of Pilins

The first crystal structure of Gram-positive pilins was solved using the minor pilin GBS52 of *Streptococcus agalactiae* (Krishnan et al. 2007). This pilin is comprised of two IgG-rev (CnaB) domains termed N1 and N2 that are joined by a short linker region. An *in vitro* experiment with recombinant GBS52 protein conjugated to fluorescently labeled beads revealed that the IgG-rev (CnaB) regions of the protein were crucial for binding to host target tissues. This was interesting because the *S. aureus* Cna B repeat region with IgG-rev fold exhibits no adhesive properties (Rich et al. 1998), suggestive of additional functions of the IgG-rev fold.

Shortly after, a crystallization study by the Baker group revealed similar IgG-like domains in the shaft pilin Spy0128 of Group A *Streptococcus* (GAS) pili (Kang et al. 2007). More important is the discovery of intramolecular isopeptide bonds in Spy0128, a feature not previously reported in any organism (Kang et al. 2007). The crystal structure of Spy0128 is characterized by two IgG-like domains both containing an isopeptide bond formed between lysine (Lys) and aspargine (Asn) residues within hydrophobic regions of the structure. These linkages, confirmed by mass spectrometry, form autocatalytically with assistance from a nearby Asp or glutamate residues. These residues act as proton shuttles permitting Lys to nucleophilicly attack the carbonyl carbon of the Asn R-group resulting in isopeptide bond formation. Significantly, the thermal stability and proteolytic stability of Spy0128 were defected in Spy0128 mutants that are devoid of the isopeptide bonds (Kang and Baker 2009). As more and more structures of Gram-positive pilins become available, it has become clear that the folding patterns of crystallized pilus proteins are remarkably similar, i.e. IgG-like fold, despite low amino acid identity among known Gram-positive pilins (see (Krishnan and Narayana 2011; Vengadesan and Narayana 2011) for an in-depth analysis).

A crystal structure of the *C. diphtheriae* shaft pilin SpaA has been also solved to a resolution of 1.6 A (Kang et al. 2009) (Fig. 7.4). Although the structure exhibits common features, like the IgG-like fold and isopeptide linkages, it has some unique features that are absent from other Gram-positive pilins. The SpaA structure is comprised of three tandem Ig-like domains termed N-terminal (N-domain), middle (M-domain) and C-termimal (C-domain). Both the N and C domains exhibited an IgG-rev fold, while the M-domain is characterized as DEv-IgG fold or CnaA-type fold. Lys-199 and Asn-321 in the M-domain and Lys-363 and Asn-482 in the C-domain form the two intramolecular isopeptide bonds with assistance from acidic residues Asp-241 and Glu-446, respectively. Interestingly, Glu-446 is the conserved residue of the E-Box that involves formation of the Lys-363–Asn-462 intramolecular bond. Furthermore, given that the mutations of Glu-446 severely affect the incorporation of SpaB into the pilus structure (Ton-That et al. 2004b) and the E-Box is in close proximity of the SpaA LPXTG motif, it has been speculated that stability of the C-domain conferred by the intramolecular linkage is required for SpaB

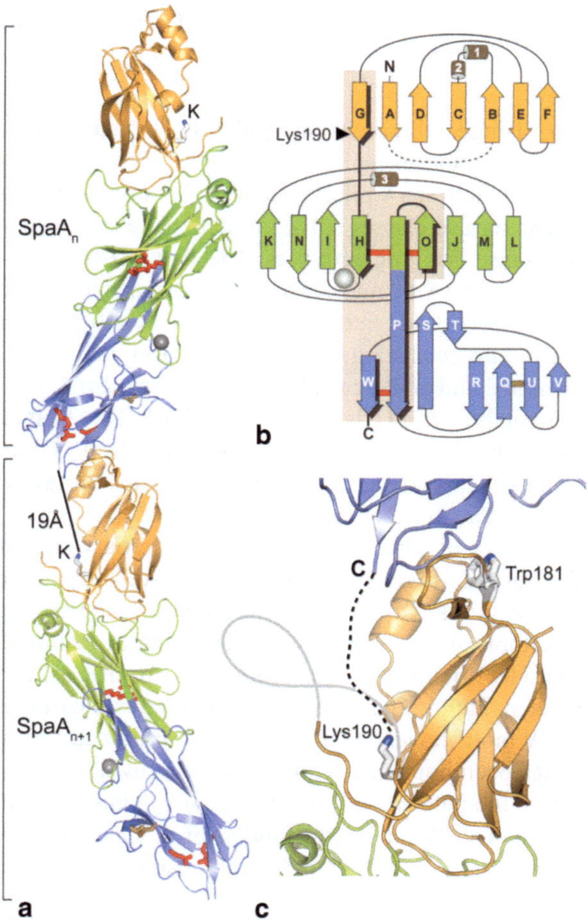

Fig. 7.4 Crystal structure of the *Corynebacterium diphtheriae* shaft pilin SpaA. **a** The SpaA molecules are stacked end-to-end, in which the C-domain (*blue*) of one SpaA molecule packs against the N-domain (*gold*) of the next. The middle domain (M; *green*) contains a Ca^{2+} ion (*grey* sphere). Residues forming isopeptide bonds are shown in red. The lysine pilin motif is labeled K. **b** Schematic representation of IgG-like folds in each domain with isopeptide bonds showing as a red bar and a disulfide bond as a grey bar. **c** Shown is the enlargement of end-to-end linkage between two SpaA molecules. The 10 missing residues of the C-terminus is shown with a broken line. Positions of lysine and tryptophan residues in the pilin motif are indicated. (Reprinted from (Kang et al. 2009) with permission)

incorporation (Kang et al. 2009). Within the crystal, SpaA molecules are arranged in a head-to-toe manner in which the lysine pilin motif abuts the C-terminus of a neighboring protein. SpaA positioning, remarkably, is reminiscent of the proposed ordered assembly proposed to occur in *vivo*.

SpaA exhibits a number of distinguishing features compared to other known pilins. First, the M-domain displays a calcium-binding site that was demonstrated to show high affinity as high levels of chelating agent EDTA could not remove it. The importance of this calcium binding site to the overall SpaA structure has not been explored. In addition, a disulfide bond joining two neighboring β-strands is present within the C-terminus, and is believed to provide additional stability to the protein (Kang et al. 2009). This bond, interestingly, appears to be unique to actinobacterial pili as similar bonds have only been detected in the major shaft pilin FimA of *A. oris* (Mishra et al. 2011). It remains to be seen if these unique features contribute to the stability and function of SpaA pilins.

7.4.2 Three-Dimensional Structures of Pilin-Specific Sortase Enzymes

Surprisingly, three-dimensional (3D) structures of neither pilin-specific sortase nor the housekeeping sortase of *C. diphtheriae* are available, considering the well-studied SpaA pilus system in this organism. Nonetheless, structures of different sortase classes solved to date have provided some insights into the mode of sortase catalytic activities. The first 3D structure of sortase was solved with *S. aureus* SrtA by nuclear magnetic resonance (NMR) (Ilangovan et al. 2001). *S. aureus* SrtA folds into an eight-stranded anti-parallel β-barrel structure with the active site consisting of His120, Cys184 and Arg197. Based on structural, biochemical and genetic analyses of *S. aureus* SrtA, sortase activity is dependent on this catalytic triad (Ilangovan et al. 2001; Ton-That et al. 2002; Zong et al. 2004). While the enzymatic activity of *S. aureus* SrtA is drastically increased by the presence of Ca^{2+} (Ilangovan et al. 2001), other sortase enzymes tested so far do not require Ca^{2+} for their activity.

Classes of sortase vary in terms of tertiary structure, but the overall shape of the transpeptidase enzymes is conserved. They all contain the core β-barrel and have a similar configuration of the active site (Spirig et al. 2011). However, 3D structures of class C sortases reveal a unique feature that is the presence of a "lid", first identified in *S. pneumoniae* pilin-specific sortases SrtC1 and SrtC3 (Manzano et al. 2008). The lid, comprised of a flexible hinge region, a leucine (Leu), and DPW motif hovers over the catalytic triad and surrounding hydrophobic pocket. It has been hypothesized that the lid provides stability to this region as the Asp (D) of the DPW motif has been shown to interact with the reactive residue Arg of the His-Cys-Arg triad. In support of this, deletion of the lid region within the *S. pneumoniae* SrtC-1 led to protein instability (Manzano et al. 2009). Furthermore, mutations of the Asp and Trp residues abrogate pilus polymerization, presumably due to instability of the sortase enzyme (Manzano et al. 2009). In contrast, studies in *A. oris* and *S. agalactiae* showed that mutations of the DWP motif did not affect pilus polymerization (Cozzi et al. 2011; Wu et al. 2012). Intriguingly, Khare and colleagues crystallized the *S. agalactiae* pilin-specific SrtC1 along with an isogenic mutant in which the lid anchor sequence replaced with IPNTG, the sorting signal of the pilin shaft GBS80 (Vengadesan et al. 2011), in place of KDPYS, the lid anchor region of SrtC1. While the mutation did not affect the enzyme integrity or overall structure, the mutant SrtC1 lacked electron density for the introduced 'IPNTG' motif exhibited differences around the active site region, speculating the active site and the lid may play a role in sortase specificity (Khare et al. 2011). Thus, it would be revealing to have co-structures of the pilin-specific sortase SrtC1 and its cognate substrate.

7.5 Cellular Adhesion and Tissue Tropism of *Corynebacterium diphtheriae* Pili

More than three decades ago, *C. renale* expressing pili were observed to aggultinate trypsinized sheep erythrocytes which could be blocked with anti-pili serum (Honda and Yanagawa 1974). Later, it was also shown that *C. renale* pili mediated attachment to mammalian cells (Honda and Yanagawa 1975), thus demonstrating that the pili were important for adhesion. Not much was known about adhesive properties of *C. diphtheriae* pili, except for one study that showed haemagglutination of some *C. diphtheriae* strains associated with the presence of pili (Ermolayev et al. 1987). Until recently, Mandlik and colleagues employed a battery of sortase and pilin mutants to examine the ability of *C. diphtheriae* to adhere to different epithelial cells (Mandlik et al. 2007). The SpaA-type pilus was found to bind specifically to pharyngeal epithelial cells, the major site of corynebacterial infection, whereas the SpaD- and SpaH-type pili displayed certain binding specificity to epithelial laryngeal and lung cells, respectively. The specific binding of SpaA pili to pharyngeal epithelial cells is attributed to the minor pilins SpaB and SpaC. This was supported by drastic reduction in binding to pharyngeal epithelial cells by a mutant that lacks *spaB* and *spaC*, as compared to strain expressing all three pilins (Mandlik et al. 2007). Secondly, it was demonstrated that latex beads coated with SpaB or SpaC protein adhered to pharyngeal epithelial cells, while those conjugated to SpaA did not. More recently, it was shown that a dimer formed between SpaB and SpaC was observed on the bacterial cell surface (Fig. 7.3), leading to the speculation that long-range adhesion can be mediated by long pilus fibers, while monomeric and heteromeric pilins provide a close surface contact with host cells (Chang et al. 2011). It is also possible that the differential binding mediated by various forms of pili and pilins may contribute to efficient delivery of virulence factors such as diphtheria toxin (Mandlik et al. 2008b).

7.6 Concluding Remarks

Klebs first identified *C. diphtheriae* as the causative agent of diphtheria in 1883. Since that time, studies focusing on this organism have advanced our understanding of bacterial pathogenesis by contributing to the mechanistic elucidation of diphtheria, as well as the development of diphtheria vaccines. Genomic studies of *C. diphtheriae* and subsequent investigations of *C. diphtheriae* covalently-linked pili have expanded our knowledge regarding the arsenal of corynebacterial virulence determinants. We now know the SpaA pilus is the major adhesin that targets corynebacteria to pharyngeal epithelial cells, the main site of infection. Work remains to be seen whether SpaA pili contribute to the establishment of the deadly disease. As pili are a common feature of many Gram-positive bacteria and the mode of pilus assembly is conserved, understanding pilus biogenesis, structural biology, and pilus-mediated pathogenesis will facilitate the development of antimicrobial agents and new vaccines in an era of rampant antimicrobial resistance.

Acknowledgments We thank former and current lab members Anjali Mandlik, Anu Swaminathan, Andrew Gasper, Arlene Swierczynski, Arunima Mishra, I-Hsiu Huang, Elizabeth Rogers, Chenggang Wu, and Chungyu Chang for their invaluable contributions to the pilus work, which was supported by grants AI061381 and DE017382 from the NIH to HTT.

References

Barocchi MA, Ries J, Zogaj X, Hemsley C, Albiger B, Kanth A, Dahlberg S, Fernebro J, Moschioni M, Masignani V, Hultenby K, Taddei AR, Beiter K, Wartha F, Von Euler A, Covacci A, Holden DW, Normark S, Rappuoli R, Henriques-Normark B (2006) A pneumococcal pilus influences virulence and host inflammatory responses. Proc Natl Acad Sci U S A 103:2857–2862

Budzik JM, Marraffini LA, Schneewind O (2007) Assembly of pili on the surface of *Bacillus cereus* vegetative cells. Mol Microbiol 66:495–510

Budzik JM, Marraffini LA, Souda P, Whitelegge JP, Faull KF, Schneewind O (2008) Amide bonds assemble pili on the surface of bacilli. Proc Natl Acad Sci U S A 105:10215–10220

Budzik JM, Oh SY, Schneewind O (2009) Sortase D Forms the Covalent Bond That Links BcpB to the Tip of Bacillus cereus Pili. J Biol Chem 284:12989–12997

Cerdeno-Tarraga AM, Efstratiou A, Dover LG, Holden MT, Pallen M, Bentley SD, Besra GS, Churcher C, James KD, De Zoysa A, Chillingworth T, Cronin A, Dowd L, Feltwell T, Hamlin N, Holroyd S, Jagels K, Moule S, Quail MA, Rabbinowitsch E, Rutherford KM, Thomson NR, Unwin L, Whitehead S, Barrell BG, Parkhill J (2003) The complete genome sequence and analysis of *Corynebacterium diphtheriae* NCTC13129. Nucleic Acids Res 31:6516–6523

Chang C, Mandlik A, Das A, Ton-That H (2011) Cell surface display of minor pilin adhesins in the form of a simple heterodimeric assembly in *Corynebacterium diphtheriae*. Mol Microbiol 79:1236–47

Comfort D, Clubb RT (2004) A comparative genome analysis identifies distinct sorting pathways in gram-positive bacteria. Infect Immun 72:2710–2722

Cozzi R, Malito E, Nuccitelli A, D'onofrio M, Martinelli M, Ferlenghi I, Grandi G, Telford JL, Maione D, Rinaudo CD (2011) Structure analysis and site-directed mutagenesis of defined key residues and motives for pilus-related sortase C1 in group B *Streptococcus*. FASEB J 25:1874–1886

Dramsi S, Trieu-Cuot, P, Bierne H (2005) Sorting sortases: a nomenclature proposal for the various sortases of Gram-positive bacteria. Res Microbiol 156:289–297

Dramsi S, Caliot E, Bonne I, Guadagnini S, Prevost MC, Kojadinovic M, Lalioui L, Poyart C, Trieu-Cuot P (2006) Assembly and role of pili in group B streptococci. Mol Microbiol 60:1401–1413

Ermolayev AV, Fish NG, Birger MO, Sanzhakova IE, Lobanova AN (1987) A study on the adhesive properties of *Corynebacterium diphtheriae* and *Corynebacterium. parvum* in the direct haemagglutination reaction. J Hyg Epidemiol Microbiol Immunol 31:313–319

Gaspar AH, Ton-That H (2006) Assembly of distinct pilus structures on the surface of *Corynebacterium diphtheriae*. J Bacteriol 188:1526–1533

Girard AE, Jacius BH (1974) Ultrastructure of *Actinomyces viscosus* and *Actinomyces naeslundii*. Arch Oral Biol 19:71–79

Guttilla IK, Gaspar AH, Swierczynski A, Swaminathan A, Dwivedi P, Das A, Ton-That H (2009) Acyl enzyme intermediates in sortase-catalyzed pilus morphogenesis in Gram-positive bacteria. J Bacteriol 191:5603–5612

Honda E, Yanagawa R (1974) Agglutination of trypsinized sheep erythrocytes by the pili of *Corynebacterium renale*. Infect Immun 10:1426–1432

Honda E, Yanagawa R (1975) Attachment of *Corynebacterium renale* to tissue culture cells by the pili. Am J Vet Res 36:1663–1666

Ilangovan U, Ton-That H, Iwahara J, Schneewind O, Clubb RT (2001) Structure of sortase, the transpeptidase that anchors proteins to the cell wall of *Staphylococcus aureus*. Proc Natl Acad Sci U S A 98:6056–6061

Iwaki M, Komiya T, Yamamoto A, Ishiwa A, Nagata N, Arakawa Y, Takahashi M (2010) Genome organization and pathogenicity of *Corynebacterium diphtheriae* C7(−) and PW8 strains. Infect Immun 78:3791–3800

Kang HJ, Baker EN (2009) Intramolecular isopeptide bonds give thermodynamic and proteolytic stability to the major pilin protein of *Streptococcus pyogenes*. J Biol Chem 284:20729–20737

Kang HJ, Coulibaly F, Clow F, Proft T, Baker EN (2007) Stabilizing isopeptide bonds revealed in gram-positive bacterial pilus structure. Science 318:1625–1628

Kang HJ, Paterson NG, Gaspar AH, Ton-That H, Baker EN (2009) The *Corynebacterium diphtheriae* shaft pilin SpaA is built of tandem Ig-like modules with stabilizing isopeptide and disulfide bonds. Proc Natl Acad Sci U S A 106:16967–16971

Khare B, Fu ZQ, Huang IH, Ton-That H, Narayana SV (2011) The Crystal Structure Analysis of Group B *Streptococcus* Sortase C1: a Model for the "Lid" Movement upon Substrate Binding. J Mol Biol 414:563–577

Klebs E (1883) Über Diphtherie. Verh Cong Inn Med 2:139–154

Kline KA, Kau AL, Chen SL, Lim A, Pinkner JS, Rosch J, Nallapareddy SR, Murray BE, Henriques-Normark B, Beatty W, Caparon MG, Hultgren SJ (2009) Mechanism for sortase localization and the role of sortase localization in efficient pilus assembly in *Enterococcus faecalis*. J Bacteriol 191:3237–3247

Kline KA, Dodson KW, Caparon MG, Hultgren SJ (2010) A tale of two pili: assembly and function of pili in bacteria. Trends Microbiol 18:224–232

Krishnan V, Gaspar AH, Ye N, Mandlik A, Ton-That H, Narayana SV (2007) An IgG-like domain in the minor pilin GBS52 of *Streptococcus agalactiae* mediates lung epithelial cell adhesion. Structure 15:893–903

Krishnan V, Narayana SV (2011) Crystallography of gram-positive bacterial adhesins. Adv Exp Med Biol 715:175–195

Kumazawa N, Yanagawa R (1972) Chemical properties of the pili of *Corynebacterium renale*. Infect Immun 5:27–30

Lauer P, Rinaudo CD, Soriani M, Margarit I, Maione D, Rosini R, Taddei AR, Mora M, Rappuoli R, Grandi G, Telford JL (2005) Genome analysis reveals pili in Group B *Streptococcus*. Science 309:105

Lemieux J, Hava DL, Basset A, Camilli A (2006) RrgA and RrgB are components of a multisubunit pilus encoded by the *Streptococcus pneumoniae rlrA* pathogenicity islet. Infect Immun 74:2453–2456

Maione D, Margarit I, Rinaudo CD, Masignani V, Mora M, Scarselli M, Tettelin H, Brettoni C, Iacobini ET, Rosini R, D'agostino N, Miorin L, Buccato S, Mariani M, Galli G, Nogarotto R, Nardi Dei V, Vegni F, Fraser C, Mancuso G, Teti G, Madoff LC, Paoletti LC, Rappuoli R, Kasper DL, Telford JL, Grandi G (2005) Identification of a universal Group B *Streptococcus* vaccine by multiple genome screen. Science 309:148–150

Mandlik A, Swierczynski A, Das A, Ton-That H (2007) *Corynebacterium diphtheriae* employs specific minor pilins to target human pharyngeal epithelial cells. Mol Microbiol 64:111–124

Mandlik A, Das A, Ton-That H (2008a) The molecular switch that activates the cell wall anchoring step of pilus assembly in gram-positive bacteria. Proc Natl Acad Sci U S A 105:14147–14152

Mandlik A, Swierczynski A, Das A, Ton-That H (2008b) Pili in Gram-positive bacteria: assembly, involvement in colonization and biofilm development. Trends Microbiol 16:33–40

Manzano C, Contreras-Martel C, El Mortaji L, Izore T, Fenel D, Vernet T, Schoehn G, Di Guilmi AM, Dessen A (2008) Sortase-mediated pilus fiber biogenesis in *Streptococcus pneumoniae*. Structure 16:1838–1848

Manzano C, Izore T, Job V, Di Guilmi AM, Dessen A (2009) Sortase activity is controlled by a flexible lid in the pilus biogenesis mechanism of gram-positive pathogens. Biochemistry 48:10549–10557

Mazmanian SK, Liu G, Ton-That H, Schneewind O (1999) *Staphylococcus aureus* sortase, an enzyme that anchors surface proteins to the cell wall. Science 285:760–763

Mazmanian SK, Ton-That H, Schneewind O (2001) Sortase-catalysed anchoring of surface proteins to the cell wall of *Staphylococcus aureus*. Mol Microbiol 40:1049–1057

Mazmanian SK, Ton-That H, Su K, Schneewind O (2002) An iron-regulated sortase anchors a class of surface protein during *Staphylococcus aureus* pathogenesis. Proc Natl Acad Sci U S A 99:2293–2298

Mishra A, Das A, Cisar JO, Ton-That H (2007) Sortase-Catalyzed Assembly of Distinct Heteromeric Fimbriae in *Actinomyces naeslundii*. J Bacteriol 189:3156–3165

Mishra A, Devarajan B, Reardon ME, Dwivedi P, Krishnan V, Cisar JO, Das A, Narayana SV, Ton-That H (2011) Two autonomous structural modules in the fimbrial shaft adhesin FimA mediate Actinomyces interactions with streptococci and host cells during oral biofilm development. Mol Microbiol 81:1205–1220

Mora M, Bensi G, Capo S, Falugi F, Zingaretti C, Manetti AG, Maggi T, Taddei AR, Grandi G, Telford JL (2005) Group A *Streptococcus* produce pilus-like structures containing protective antigens and Lancefield T antigens. Proc Natl Acad Sci U S A 102:15641–15646

Nallapareddy SR, Singh KV, Sillanpaa J, Garsin DA, Hook M, Erlandsen SL, Murray BE (2006) Endocarditis and biofilm-associated pili of *Enterococcus faecalis*. J Clin Invest 116:2799–2807

Nobbs AH, Rosini R, Rinaudo CD, Maione D, Grandi G, Telford JL (2008) Sortase A utilizes an ancillary protein anchor for efficient cell wall anchoring of pili in *Streptococcus agalactiae*. Infect Immun 76:3550–3560

Phan G, Remaut H, Wang T, Allen WJ, Pirker KF, Lebedev A, Henderson NS, Geibel S, Volkan E, Yan J, Kunze MB, Pinkner JS, Ford B, Kay CW, Li H, Hultgren SJ, Thanassi DG, Waksman G (2011) Crystal structure of the FimD usher bound to its cognate FimC-FimH substrate. Nature 474:49–53

Popovic T, Kombarova SY, Reeves MW, Nakao H, Mazurova IK, Wharton M, Wachsmuth IK, Wenger JD (1996) Molecular epidemiology of diphtheria in Russia, 1985–1994. J Infect Dis 174:1064–1072

Proft T, Baker EN (2009) Pili in Gram-negative and Gram-positive bacteria—Structure, assembly and their role in disease. Cell Mol Life Sci 66:613–635

Remaut H, Tang C, Henderson NS, Pinkner JS, Wang T, Hultgren SJ, Thanassi DG, Waksman G, Li H (2008) Fiber formation across the bacterial outer membrane by the chaperone/usher pathway. Cell 133:640–652

Rich RL, Demeler B, Ashby K, Deivanayagam CC, Petrich JW, Patti JM, Narayana SV, Hook M (1998) Domain structure of the *Staphylococcus aureus* collagen adhesin. Biochemistry 37:15423–15433

Rosini R, Rinaudo CD, Soriani M, Lauer P, Mora M, Maione D, Taddei A, Santi I, Ghezzo C, Brettoni C, Buccato S, Margarit I, Grandi G, Telford JL (2006) Identification of novel genomic islands coding for antigenic pilus-like structures in *Streptococcus agalactiae*. Mol Microbiol 61:126–141

Sauer FG, Futterer K, Pinkner JS, Dodson KW, Hultgren SJ, Waksman G (1999) Structural basis of chaperone function and pilus biogenesis. Science 285:1058–1061

Sauer FG, Mulvey MA, Schilling JD, Martinez JJ, Hultgren SJ (2000) Bacterial pili: molecular mechanisms of pathogenesis. Curr Opin Microbiol 3:65–72

Sauer FG, Pinkner JS, Waksman G, Hultgren SJ (2002) Chaperone priming of pilus subunits facilitates a topological transition that drives fiber formation. Cell 111:543–551

Schneewind O, Model P, Fischetti VA (1992) Sorting of protein A to the staphylococcal cell wall. Cell 70:267–281

Sjoquist J, Meloun B, Hjelm H (1972) Protein A isolated from *Staphylococcus aureus* after digestion with lysostaphin. Eur J Biochem 29:572–578

Smith WD, Pointon JA, Abbot E, Kang HJ, Baker EN, Hirst BH, Wilson JA, Banfield MJ, Kehoe MA (2010) Roles of minor pilin subunits Spy0125 and Spy0130 in the serotype M1 *Streptococcus pyogenes* strain SF370. J Bacteriol 192:4651–4659

Soriani M, Telford JL (2010) Relevance of pili in pathogenic streptococci pathogenesis and vaccine development. Future Microbiol 5:735–747

Spirig T, Weiner EM, Clubb RT (2011) Sortase enzymes in Gram-positive bacteria. Mol Microbiol 82:1044–1059

Swaminathan A, Mandlik A, Swierczynski A, Gaspar A, Das A, Ton-That H (2007) Housekeeping sortase facilitates the cell wall anchoring of pilus polymers in *Corynebacterium diphtheriae*. Mol Microbiol 66:961–974

Swierczynski A, Ton-That H (2006) Type III pilus of Corynebacteria: pilus length is determined by the level of its major pilin subunit. J Bacteriol 188:6318–6325

Thanassi DG, Saulino ET, Hultgren SJ (1998) The chaperone/usher pathway: a major terminal branch of the general secretory pathway. Curr Opin Microbiol 1:223–231

Ton-That H, Schneewind O (2003) Assembly of pili on the surface of *Corynebacterium diphtheriae*. Mol Microbiol 50:1429–1438

Ton-That H, Schneewind O (2004) Assembly of pili in Gram-positive bacteria. Trends Microbiol 12:228–234

Ton-That H, Liu G, Mazmanian SK, Faull KF, Schneewind O (1999) Purification and characterization of sortase, the transpeptidase that cleaves surface proteins of *Staphylococcus aureus* at the LPXTG motif. Proc Natl Acad Sci U S A 96:12424–12429

Ton-That H, Marraffini LA, Schneewind O (2004a) Protein sorting to the cell wall envelope of Gram-positive bacteria. Biochim Biophys Acta 1694:269–278

Ton-That H, Marraffini LA, Schneewind O (2004b) Sortases and pilin elements involved in pilus assembly of *Corynebacterium diphtheriae*. Mol Microbiol 53:251–261

Ton-That H, Mazmanian SK, Alksne L, Schneewind O (2002) Anchoring of surface proteins to the cell wall of *Staphylococcus aureus*. Cysteine 184 and histidine 120 of sortase form a thiolate-imidazolium ion pair for catalysis. J Biol Chem 277:7447–7452

Trost E, Blom J, De Castro Soares S, Huang IH, Al-Dilaimi A, Schroder J, Jaenicke S, Dorella FA, Rocha FS, Miyoshi A, Azevedo V, Schneider MP, Silva A, Camello TC, Sabbadini PS, Santos CS, Santos LS, Hirata R Jr, Mattos-Guaraldi AL, Efstratiou A, Schmitt MP, Ton-That H, Tauch A (2012) Pangenomic Study of *Corynebacterium diphtheriae* that provides insights into the genomic diversity of pathogenic isolates from cases of classical diphtheria, endocarditis, and pneumonia. J Bacteriol 194:3199–3215

Vengadesan K, Narayana SV (2011) Structural biology of Gram-positive bacterial adhesins. Protein Sci 20:759–772

Vengadesan K, Ma X, Dwivedi P, Ton-That H, Narayana SV (2011) A Model for Group B *Streptococcus* Pilus Type 1: the Structure of a 35-kDa C-Terminal fragment of the major pilin GBS80. J Mol Biol 407:731–743

Wu C, Mishra A, Reardon ME, Huang IH, Counts SC, Das A, Ton-That H (2012) Structural determinants of Actinomyces sortase SrtC2 required for membrane localization and assembly of type 2 fimbriae for interbacterial coaggregation and oral biofilm formation. J Bacteriol 194:2531–2539

Yanagawa R, Otsuki K, Tokui T (1968) Electron microscopy of fine structure of *Corynebacterium renale* with special reference to pili. Jpn J Vet Res 16:31–37

Yeung MK (1999) Molecular and genetic analyses of *Actinomyces* spp. Crit Rev Oral Biol Med 10:120–138

Zong Y, Bice TW, Ton-That H, Schneewind O, Narayana SV (2004) Crystal structures of *Staphylococcus aureus* sortase A and its substrate complex. J Biol Chem 279:31383–31389

Chapter 8
Toxigenic Corynebacteria: Adhesion, Invasion and Host Response

Lisa Ott and Andreas Burkovski

Abstract *Corynebacterium diphtheriae*, *Corynebacterium ulcerans* and *Corynebacterium pseudotuberculosis* form a distinct group within the genus *Corynebacterium*, the toxigenic corynebacteria. The three species are able to colonize a number of hosts including humans, e.g. in case of diphtheria. However, besides diphtheria toxin action, information about molecular mechanisms of host pathogen interaction is rare. The recent availability of genome sequence information gave new impetus to the characterization of putative virulence factors of *C. ulcerans*, *C. pseudotuberculosis* and *C. diphtheriae* and the recent knowledge about these is summarized here.

Keywords Adhesion · Diphtheria toxin · Inflammatory pathways · Invasion · Phagocytosis

8.1 Introduction

Corynebacteria are morphologically similar, irregular- or clubbed-shaped microorganisms, which belong to the class of Actinobacteria (high G + C Gram-positive bacteria) (Ventura et al. 2007; Zhi et al. 2009). Almost all members of the genus are characterized by complex cell envelope architecture. In these bacteria the cytoplasmic membrane is covered by a peptidoglycan layer, a layer of arabinogalactan polymer, which is covalently linked to the peptidoglycan, and an outer lipid layer composed of mycolic acids (corynomycolates), which is functionally equivalent to the outer membrane of Gram-negatives (Hoffmann et al. 2008; Niederweis et al. 2010; Burkovski 2013).

A. Burkovski (✉) · L. Ott
Lehrstuhl für Mikrobiologie, Friedrich-Alexander-Universität Erlangen-Nürnberg, Staudtstr. 5, 91058, Erlangen, Germany
e-mail: andreas.burkovski@fau.de

A. Burkovski (ed.), *Corynebacterium diphtheriae and Related Toxigenic Species*,
DOI 10.1007/978-94-007-7624-1_8, © Springer Science+Business Media Dordrecht (outside the USA) 2014

Today, the genus *Corynebacterium* comprises 88 species, including soil bacteria with biotechnological importance, commensals found on skin of humans and animals as well as pathogens. Often, members of the latter group are not highly virulent and only immune-compromised patients are infected. Sometimes, virulence is dramatically increased when bacteria are lysogenized by toxin-encoding phages, as in the case of *Corynebacterium diphtheriae* and related toxigenic corynebacteria.

C. diphtheriae is the classical etiological agent of diphtheria and the type species of the genus *Corynebacterium* (Lehmann and Neumann 1896; Barksdale 1970). Its main virulence factor, the diphtheria toxin, is encoded by a corynebacteriophage integrated into the genome of toxigenic strains. As a result of infection with *tox* gene carrying corynebacteriophages, two closely related species, *Corynebacterium ulcerans* and *Corynebacterium pseudotuberculosis*, can produce diphtheria toxin as well (Groman et al. 1984; Buck et al. 1985; Cianciotto and Groman 1996). *C. ulcerans* has been detected as a commensal in domestic and wild animals (Schuhegger et al. 2009; Dixon 2010; Sykes et al. 2010) that may serve as reservoir for zoonotic infections. Human infections usually occur in adults, who consumed raw milk (Bostock et al. 1984; Hart 1984) or had close contact with domestic animals (Wagner et al. 2010); no person to person transmission was reported up to now. *C. pseudotuberculosis* is the etiological agent of caseous lymphadenitis of sheep, goats and other farm and wild animals (Dorella et al. 2006; Baird and Fontaine 2007). Infections due to *C. pseudotuberculosis* are rare in humans; occasionally, the pathogen is isolated from cases of suppurative lymphadenitis in patients with a classical risk exposure of close contact with sheep. Despite the putative presence of a diphtheria toxin in *C. pseudotuberculosis*, diphtheria of respiratory tract or skin is not evoked by this bacterium.

8.2 *Corynebacterium diphtheriae:* Adhesion, Invasion and Host Response

C. diphtheriae causes diphtheria of the upper respiratory tract and cutaneous diphtheria, which outnumbers the classical respiratory diphtheria in tropical areas. Besides these infections, toxigenic and non-toxigenic *C. diphtheriae* strains are associated with other diseases, e.g. endocarditis, osteomyelitis and septic arthritis (Puliti et al. 2006; Hirata et al. 2008), indicating that *C. diphtheriae* is able to colonize not only epithelia but also deeper parts of the body and that the bacteria interact with various types of host cells.

8.2.1 Adhesion

Experimental data regarding host recognition and adhesion by *C. diphtheriae* are limited; however, a few adhesion factors were described, the best investigated being pili structures on the surface of *C. diphtheriae* which are covalently attached to the cell wall peptidoglycan (Navarre and Schneewind 1999). The genome of *C. diphtheriae* NCTC 13129 encodes three distinct pili clusters (*spaABC*, *spaDEF*, *spaGHI*), which are organized together with sortase-encoding genes (*srtA*, *srtB*, *srtC*, *srtD*, *srtE* and *srtF*) (Ton-That and Schneewind 2003). Corynebacterial pili consist of a major pilin subunit SpaA (SpaD and SpaH, respectively), a minor subunit SpaB (SpaE and SpaG, respectively) and the tip protein SpaC (SpaF and SpaI, respectively) (Ton-That and Schneewind 2003). The major pilin proteins carry a cell wall sorting signal at the C-terminus, composed of an LPxTG motif, followed by a hydrophobic membrane-spanning domain and a positively charged tail (Schneewind et al. 1993). This motif is responsible for efficient anchoring of pili to the cell wall by housekeeping sortase SrtF. The additional sortases SrtA-E catalyze the covalent cross-linking of single pilin monomers to form the pilus. The specificity of the pilus subunits for the corresponding sortases seems to be determined by the amino acid sequence LxET, designated as E box and localized between the LPxTG motif and the pilin motif WPK (Ton-That et al. 2004). The E box is conserved in several Gram-positive species that encode sortase and pilin subunit genes with sorting signals and pilin motifs. Minor pilin subunits contain the LPxTG motif but neither the E box nor the pilin motif was detected.

Mandlik and co-workers systematically address the function of the different pili as well as their individual pilin subunits in respect to corynebacterial adherence by the use of different epithelial cell lines and a number of *C. diphtheriae* mutant strains (Mandlik et al. 2007). These analyses revealed that only the SpaA-type pili contribute to corynebacterial adherence to human pharyngeal cells (Detroit 562), while the SpaD and SpaH-type pili support adhesion to laryngeal (HEp-2) and lung cells (A549). Strains expressing only SpaA-type pili were able to adhere to pharyngeal cells properly but showed less adhesion to other cell lines. The deletion of *srtA*, which prevents crosslinking of SpaA-type pili subunits and consequently proper pili formation, impaired binding to pharyngeal cells. Mutant strains lacking all sortases showed decreased adhesion rates to different epithelial cell lines. Interestingly, it was shown that the minor pilin subunits do not only decorate the pilus shaft but are also distributed in clusters all over the bacterial surface. These minor pilins directly anchored to the cell wall are sufficient to mediate specific host cell adhesion. The deletion of either *spaB* or *spaC* resulted in significantly lower adhesion rates and the *spaBC* double mutant revealed marginal binding to Detroit 562 cells. Furthermore, the incubation of bacteria cells with antibodies directed against SpaB, SpaC blocked corynebacterial adhesion to pharyngeal cells; this was not the case after pre-incubation with an antibody directed against SpaA (Mandlik et al. 2007).

In summary, *C. diphtheriae* strain NCTC 13129 carries genes for three pili clusters, with the minor pili subunits being crucial for adhesion and the major subunits determining pili length (Gaspar and Ton-That 2006; Scott and Zahner 2006; Swierczynski and Ton-That 2006). When other isolates were investigated, strain-specific differences in pili formation and adhesion properties were observed; these are summarized in Table 8.1.

Table 8.1 Adhesion rates of different *C. diphtheriae* strains

Strain	Growth medium	Time p. i. [min]	Adhesion rate [%]	Cell line	Reference
ISS3319	HI (Heart Infusion)	90	3.55±0.28	Detroit 562 (pharyngeal carcinoma cells)	(Ott et al. 2010b)
ISS4060			3.99±0.27		
ISS4746			6.54±0.97		
ISS4749			7.34±2.33		
DSM43988			0.69±0.12		
DSM43989			0.34±0.05		
DSM44123			1.84±0.26		
NCTC 13129		60	9.00		(Mandlik et al. 2007)
ISS3319	HI	90	58.39±8.25	HeLa (cervix carcinoma cells)	(Ott et al. 2013)
ISS4060			72.38±13.74		
ISS4746			42.23±2.50		
ISS4749			53.56±5.23		
DSM43988			69.92±5.87		
DSM43989			1.25±0.14		
DSM44123			40.87±1.00		
NCTC 13129		60	25.00	HEp-2 (laryngeal carcinoma cells)	(Mandlik et al. 2007)
241	TSB (Tryptic Soy Broth)	30	61.86		(Moreira et al. 2003)
241		120	32.98		
241	TSB-Fe	30	91.75		
241		120	57.20		
CDC-E8392	TSB	30	18.29		
CDC-E8392		120	59.11		
CDC-E8392	TSB-Fe	30	15.21		
CDC-E8392		120	58.02		
CDC-E8392	TSB	30	23.40		(Hirata et al. 2004)
		60	27.60		
		120	30.10		
		180	38.30		
		360	3.80		
NCTC 13129	HI	60	20.00	A549 (lung carcinoma cells)	(Mandlik et al. 2007)

In order to address these strain-specific differences in adhesion rates, Ott and co-workers investigated the surface structures of several wild type strains by atomic force microscopy. This approach revealed significantly different macromolecular surface structures of the wild type strains tested. Interestingly, not all strains investigated by Ott and co-workers showed pili-like structures on the surface and the pili observed differed in length and number. Some isolates completely lacked pili, some presented short spike-like structures on their surface, while others showed long, hair-like protrusions. Interestingly, adhesion and pili formation were not strictly coupled, as strain ISS4060, which lacks pili-like structures, shows comparable adhesion rates to strain ISS3319, which revealed spike-like pili (Ott et al. 2010b). RNA hybridization experiments carried out in order to investigate mRNA levels of the different *spa* genes (Ott et al. 2010b) revealed distinct differences in the expression pattern of pili subunits for the strains investigated. Transcripts detected belong to the *spaABC* or *spaGHI* cluster, mRNA of genes of the *spaDEF* cluster was not found in these strains. Subsequently carried out PCR experiments indicated corresponding strain-specific differences in the presence of the different genes. A recent pangenomic study of twelve isolates from patients with classical diphtheria, endocarditis and pneumonia provided further evidence for a high degree of genetic variability between different *C. diphtheriae* strains (Trost et al. 2012). A fourth type of pili was observed in this analysis, and despite the fact that all strains contained at least two pili gene cluster, the single clusters were differently distributed among the analyzed isolates. The reasons for strain-specific differences in respect to the pili repertoire remain unclear; probably, these are connected to the binding of different receptors on the eukaryotic surface. This strategy might also be reflected by the existence of adhesion factors besides pili.

One important adhesin in this respect might be the non-fimbrial surface protein 67–72p, named according to its initial identification as two bands with distinct apparent mass in SDS-PAGE (Colombo et al. 2001). Recent studies revealed that the 67–72p protein seems to be encoded by a single gene, DIP0733 (Sabbadini et al. 2012). Both, the 67 and 72 kDa polypeptide bind to erythrocyte receptors, leading to hemagglutination. Hirata and co-workers (Hirata et al. 2004) demonstrated that 67–72p binds not only human erythrocytes but also the human epidermoid laryngeal carcinoma cell line HEp-2 and that binding was effectively blocked by anti-67–72p IgG antibodies. A negative correlation between bacterial adhesion to HEp-2 cells and erythrocytes was found; strains with low hemagglutinating activity showed high adhesion rates to HEp-2 cells and *vice versa*.

Erythrocytes are not the only targets of components of *C. diphtheriae* within the blood; another target is fibrinogen, a major component of the human plasma. The main role of this protein lies in blood clot formation by its conversion into insoluble fibrin, which forms the matrix of the clot. This process is hijacked by many pathogens, e.g. by the majority of virulent *Staphylococcus* strains (Duthie 1954) and by *Yersinia pestis* (Beesley et al. 1967; Rivera et al. 2007), by producing an enzyme termed coagulase. Sabbadini and co-workers demonstrated that *C. diphtheriae* is also able to bind to fibrinogen and convert fibrinogen to fibrin (Sabbadini et al. 2010). Non-toxigenic and toxigenic strains formed bacterial aggregates in the presence of plasma and fibrinogen binding occurred in varying intensities. The fibrinogen polymerisation process might result in a fibrin layer on the bacterial cell wall, and the binding of fibrinogen to the surface of *C. diphtheriae* may be an efficient method to avoid phagocytosis. This was already shown for other Gram-positive pathogens (Schubert et al. 2002; Rennermalm et al. 2004; Pierno et al. 2006). In addition to that, Sabbadini and co-workers (Sabbadini et al. 2010) suggested that the conversion of fibrinogen to fibrin may be connected to pseudomembrane formation, since differences in the abilities to bind and convert fibrinogen may partially explain differences in the extent of pseudomembrane formation during diphtheria.

Besides host factors such as fibrinogen, sugar moieties and sugar-converting enzymes might be involved in bacterial surface decoration and camouflage to mislead the host's immune system. In this respect, sialidases, also called neuramidases, play an important role in the virulence of many pathogenic organisms. Sialidases belong to a class of glycosyl hydrolases that catalyze the removal of terminal sialic acid residues from a variety of glycoconjugates of the host surface (Vimr 1992), which can subsequently used a nutrient or for bacterial surface decoration. In fact, *C. diphtheriae* exposes sialic acids on its surface, but the genes involved in sialic acid cell wall binding have not been identified so far. Sialidase activity was first observed in a crude diphtheria toxin preparation (Blumberg and Warren 1961). Later, exo-sialidase activity was further characterized by Warren and co-workers (Warren and Spearing 1963), who showed that the exo-sialidase is induced in iron-enriched cultures. Moreira and coworkers found that the iron concentration in the culture medium directly affects sialidase production and cell surface carbohydrates of *C. diphtheriae* (Moreira et al. 2003). The work led to the identification of a putative exo-sialidase NanH (DIP0543) from *C. diphtheriae*. Biochemical studies of purified NanH heterologously expressed in *E. coli* revealed the highest cleavage rate for the α-2,6-linked sialic acids of sialyllactose, with comparable activity for the α-2,3-linked sialic acids. Furthermore, *C. diphtheriae* NanH showed trans-sialylation activity using sialyl-α2,3-lactose and sialyl-α2,6-lactose as donors. In addition to that, it could also catalyze the transfer of sialic acids from other sialoconjugates to 4-methylumbelliferyl-a-D-galactopyrano-side (MU-Gal) and the sialidase-mediated transglycosylation reaction using sialic acid-conjugated free oligosaccharides and glycoproteins as donors. *C. diphtheriae* may transport sialic acids hydrolyzed by extracellular sialidases for synthesis of sialic acid derivates. However, it has still to be clarified whether NanH is involved in sialic acids decoration or not (Kim et al. 2010).

Hemagglutination was found to be resistant to heating at 100°C, detergents and trypsin treatment (Colombo et al. 2001), suggesting the additional involvement of non-proteinaceous molecules in this process and probably also in adhesion to other cell types. One candidate factor in respect to adhesion is a 10 kDa lipoglycan of *C. diphtheriae*, designated CdiLAM, which shows similarity to mycobacterial LAM, and is presented on bacterial surface. Moreira and co-workers (Moreira et al. 2008) found that in contrast to LAMs of other actinomycetes, CdiLAM presents an unusual substitution at position 4 of α 1→6 mannan backbone by α-*D*-Ara*f*. Unlike the non-fimbrial adhesin 67–72p, CdiLAM does not bind to human erythrocytes but to human respiratory epithelial cells. Pre-incubation of bacterial cells with anti-CdiLAM IgG antibodies or pre-incubation of cells with purified CdiLAM blocked adhesion effectively, indicating CdiLAM as a specific adhesin to human respiratory cells.

Table 8.2 Adhesion factors of *C. diphtheriae*

Adhesion factor	Function
SpaA type pili	contribute to adherence to Detroit 562 and A549 cells
67–72p protein (DIP0733)	hemagglutination; adherence to HEp-2 cells
coagulase	binding of fibrinogen and conversion to fibrin
NanH	sialidase activity
CdiLAM	binding of HEp-2 cells

8.2.2 Invasion

Besides questions concerning host recognition and adhesion, there are still open questions with respect to the infection process and spread of bacteria. *C. diphtheriae* was generally considered as an extracellular colonizer of the upper respiratory tract; however, failure of penicillin treatment in cases of severe pharyngitis and tonsillitis could only partially be explained by penicillin tolerance (von Hunolstein et al. 2002). Furthermore, the occurrence of diphtheria among vaccinated adults in Rio de Janeiro, as well as an increasing number of cases of endocarditis caused by non-toxigenic strains, indicates the possibility of deeper colonization of the human host (Mattos-Guaraldi et al. 2000; Mattos-Guaraldi et al. 2001). Cases of invasive disease with non-toxigenic strains have been reported for injection drug users in Switzerland (Gubler et al. 1998), Aborigines in Australia (Hogg et al. 1996; Holthouse et al. 1998), homeless people in France (Patey et al. 1997), patients with bone and joint infections in Germany (Funke et al. 1999) and patients with invasive bloodstream infections in North America (Romney et al. 2006). Taken together, these observations lead to the assumption that the bacteria are not only able to adhere to host cells of the respiratory tract but are also able to gain access into deeper tissues.

Hirata and co-workers (Hirata et al. 2002) studied *C. diphtheriae* with respect to its ability to enter and survive within HEp-2 monolayers, using gentamicin protection assays. Gentamicin is an antibiotic which is not able to pass eukaryotic membranes and kills extracellular adherent bacteria, while intracellular bacteria are protected. Strain-specific differences in respect to internalization into HEp-2 cells were found, whereas no correlation between adhesion and invasion rates was observed. Thin-section electron micrograph of *C. diphtheriae* revealed the presence of bacteria inside a membrane-bound vacuole in close proximity to the nuclear membrane. Since it was assumed that actin polymerization is triggered by *C. diphtheriae* to induce uptake in HEp-2 cells (Mattos-Guaraldi et al. 2002), the effect of cytochalasin D, a known inhibitor of eukaryotic microfilament formation, on the adhesion and invasion process was investigated. In fact, internalization of various strains was completely blocked by cytochalasin E, supporting the assumption that cytoskeletal rearrangements are required for bacterial entry. Furthermore, phosphotyrosine-signaling seems to be a relevant mechanism during the invasion process in cultured respiratory cells.

Similar experiments were carried out with the pharyngeal carcinoma cell line Detroit 562 (Bertuccini et al. 2004). Ultrastructural analysis of the internalization process revealed a localized pattern of adherence on colonized Detroit 562 cells, which seemed to be mediated by cellular microvilli, establishing a close contact to the bacterial cell wall. Bertuccini and co-workers observed a ring-like structure, with a ruffle-like appearance, which is part of the host cell surface during the internalization of a single bacteria cell. Cytochalasin D showed a dose-dependent inhibitory effect on the internalization process, pointing to the involvement of F-actin in the bacterial entry. Bacterial binding seems to induce intracellular signal transduction responsible for actin microfilament rearrangements (Rosenshine et al. 1992). In contrast, colchicine had no effect on the number of internalized bacteria, indicating that microtubuli are not involved in bacterial uptake. Staurosporine, an inhibitor of different classes of phosphokinases, positively affected *C. diphtheriae* internalization, indicating that the activation of phosphokinases negatively controlled bacterial uptake. In summary, the possibility to enter the host cell, to hide in epithelial cells even at low numbers and to persist at least for 48 h, gives the bacteria the opportunity to escape from the hosts immune response and antibiotic treatment. The fact that *C. diphtheriae* is able to survive significantly longer within Detroit 562 cells than in HEp-2 cells might indicate better adaption of the bacteria to upper respiratory tract epithelia.

Using an immune-fluorescence microscopy approach, Ott and co-workers (Ott et al. 2010b) observed V-shaped *C. diphtheriae* dimers within the host cell. V-forms are *Corynebacterium*-specific snapping division stages and indicate proliferating bacteria. Consequently, it can be assumed that *C. diphtheriae* is able to replicate within the host cells, and growth and elimination are parallel processes.

The molecular mechanism of invasion is more or less unclear. Ott and co-workers (Ott et al. 2010a) focused their work on the surface-associated protein DIP1281, a member of the NlpC/P60 family (Anantharaman and Aravind 2003), which was annotated as hypothetical invasion-associated protein. NlpC/P60 proteins define a large superfamily of proteins found in bacteria, bacteriophages, RNA viruses and eukaryotes, and several members are conserved among corynebacteria. DIP1281 mutant strains almost completely lacked the ability to adhere to host cells and, in contrast to the wild type, no intracellular bacteria were detectable in gentamicin protection assays. Further characterization of the mutant strains by immuno-fluorescence microscopy, using an antibody specific for the *C. diphtheriae* surface proteome, revealed an increased size of the single bacteria, an altered less club-like shape and formation of chains of bacteria rather than the typical V-like cell division forms or palisades of clustered *C. diphtheriae*. LIVE/DEAD staining showed that all bacteria within the observed chains were still viable and no differences were detectable between wild type and mutants. Since the mutants showed an altered antibody binding in immune-fluorescence microscopy approaches, Ott and co-workers (Ott et al. 2010a) performed a more detailed investigation of the surface proteomes of wild types and mutant strains by two-dimensional gel electrophoresis. The mutants showed a decreased number of spots in the upper molecular weight range and an increased number of spots in the lower molecular weight range. Clear differences between wild type and mutant protein patterns were also found by atomic force microscopy. Wild type cells showed round elevations with a lateral diameter of 10–40 nm at a height of 1–4 nm at their surface, while DIP1281 mutants lost this fine structure and elongated elevation with a width of 50–100 nm was observed. In summary, DIP1281 seems to be crucial for the organization of the outer surface layer of *C. diphtheriae*. Tsuge and co-workers reported similar observations for *C. glutamicum* R, when the DIP1281 homolog cgR_1596 and other members of the NlpC/P60 protein familiy cgR_2070 were mutated (Tsuge et al. 2008). Taken together, DIP1281 and its homologs Ce1659, Cg1735 and Jk0967 in *Corynebacterium efficiens*, *C. glutamicum* and *C. jeikeium*, which were previously annotated as hypothetical invasion-associated proteins, seem to be predominantly involved in the organization of the outer layer of corynebacteria rather than in the separation of the peptidoglycan of dividing bacteria. Mutations induce pleiotropic effects which also influence adhesion and invasion.

Sabbadini and co-workers (Sabbadini et al. 2012) studied the role of surface-exposed non-fimbrial 67–72p protein (DIP0733), previously described as adhesin/hemagglutinin (Colombo et al. 2001; Hirata et al. 2004), in respect to internalization of *C. diphtheriae* in HEp-2 cells. Investigations of the interaction of HEp-2 cells with 67–72p-coated beads by optical and transmission electron microscopy demonstrated the autoaggregative and hydrophobic properties of 67–72p as well as the internalization of 67–72p-adsorbed microspheres by HEp-2 cells. Pretreatment of bacterial suspensions or HEp-2 cells with 67–72p led to an increased number of bacteria in the supernatants of infection assays, while the number of internalized bacteria was decreased (Sabbadini et al. 2012), indicating a shielding of receptors or a jamming of the internalization machinery by 67–72p.

8.2.3 Host Immune Response

The mammalian immune system includes innate and adaptive components, with the innate immunity providing the first line of defense against infectious diseases, while the adaptive immune system is activated only due to exposition to an antigenic challenge. Most bacteria infecting healthy vertebrates are readily cleared within a few days after infection, before activating the adaptive immune response. In this case, microorganisms are recognized by the innate immune system through a limited number of pattern-recognition receptors (PRRs) (Akira et al. 2006). These receptors, which recognize pathogen-associated molecular patterns (PAMPs), are constitutively expressed on all cells of a given type and detect pathogens in any life-cycle stage. The interaction of PRRs with specific PAMPs leads to activation of specific inflammatory signaling pathways, resulting in specific antimicrobial response (Akira et al. 2006). A prominent group of PRRs are Toll-like receptors (TLRs) which are conserved from *Caenorhabditis elegans* to mammals (Medzhitov and Janeway 2002; Hoffmann 2003; Akira and Takeda 2004; Beutler 2004). To date, twelve members of the TLR family have been identified, described as type I integral membrane glycoproteins with extracellular domains containing varying numbers of leucins-rich-repeats (LRRs) and a cytoplasmatic signaling domain, named the Toll/

IL-1 R (TIR) homology domain, due to its homology to interleukin 1 receptor (IL-1 R) (Bowie and O'Neill 2000). Members of the TLR family are able to recognize lipopolysaccharide (LPS) from Gram-negative bacteria, whereas in Gram-positives, which lack LPS, lipoteichoic acids (LTA), mycolic acids and arabinogalactan (in case of mycobacteria, nocardia and corynebacteria) are recognized (Akira et al. 2006). The activation of TRLs leads to activation and recruitment of phagocytic cells, such as macrophages and neutrophils, to the site of infection. Furthermore, TLR signaling through conserved signaling elements leads to transcription of distinct proinflammatory genes and consequently to the production of cytokines, interferons, chemokines, cell surface molecules and chemokine receptors (Sioud 2005).

8.2.4 Cytokine Production During Infection with C. diphtheriae

As mentioned above, besides evoking diphtheria, *C. diphtheriae* is associated with invasive infections, such as endocarditis, septic arthritis, splenic abscesses, bacteraemia and osteomyelitis, which are usually caused by non-toxigenic strains (Alexander 1984; Tiley et al. 1993; Poilane et al. 1995; Funke et al. 1999; Belko et al. 2000; Mattos-Guaraldi et al. 2001). The clinical disease pattern of human diphtheroid arthritis can occur as an acute, purulent process or as a chronic arthritis with plasmacytic or histiocytic inflammation. Puliti and coworkers introduced mice as a model system for septic arthritis (Puliti et al. 2006). Depending on the number of bacteria injected, mice died or developed clinical signs of arthritis, most frequently in ankles and wrists. By quantitative monitoring of *C. diphtheriae,* Puliti and coworkers (Puliti et al. 2006) also demonstrated that *C. diphtheriae* bacteria were cleared from the bloodstream within five days after infection, but the bacteria were able to persist in kidneys and spleens for more than two weeks after infection. Similar to cases reported of osteoarthritis in humans, histopathological examinations of mice inoculated with *C. diphtheriae* revealed a mild to moderate arthritis including the presence of inflammatory cells. Cytokine levels of serum and joint samples of infected mice were determined by ELISA. The investigated non-toxigenic strains induced systemic and local IL-6, IL-1β and TNFα secretion, which might contribute to articular damage in cases of septic arthritis based on the fact that some strains evoked severe arthritis symptoms and high levels of IL-6 and IL-1β, while infection with other strains led to low IL-6 and IL-1β synthesis together with a low incidence and severity of arthritis. Interestingly, induction of high and early IFN-γ production was only observed in mice infected with a distinct isolate, indicating that IFN-γ production counterbalances the proinflammatory response.

Induction of host response was investigated *in vitro* at a molecular level by Ott and co-workers, using luciferase-expressing reporter cell lines to monitor the induction of various host signaling pathways. Strain- and cell line-specific differences were found in respect to induction of N-FAT, NFκB, STAT5RE and AP-1 (unpublished observations). Furthermore, Ott and co-workers demonstrated that induction of the NFκB transduction pathway in epithelial cells only occurred in response to incubation with living bacteria and required invasion of the bacteria into the epithelial cells (Ott et al. 2013).

8.2.5 Phagocytosis and Induction of Apoptosis

Antibodies are present in plasma and extracellular liquids and protect against pathogens or toxigenic products by binding of antigens to inhibit their interaction with the host cells and thus avoiding their damage, a process termed neutralization. In case of bacteria proliferating in the extracellular space, antibodies recruit phagocytes that internalize bacteria for degradation. This mechanism is specific for bacteria that are able to escape direct recognition of phagocytes. In this case, the bacteria are labeled with antibodies for ingestion by phagocytes, a process called opsonization. Additionally, antibodies activate the complement system, a system of plasma proteins, which is enabling phagocytes to ingest and degrade bacterial cells that are otherwise undetectable for the cells.

Non-opsonic phagocytosis, mediated by phagocyte receptors that recognize structures on microbial surfaces, have become of greater interest as a potential host defense mechanism against intracellular pathogens (Ofek et al. 1995). Dos Santos and co-workers (dos Santos et al. 2010) investigated the ability of homologous non-toxigenic and toxigenic *C. diphtheriae* strains to be phagocytozed by the human macrophage cell line U937 in the absence of immune serum (opsonins). The fact, human macrophages were able to ingest *C. diphtheriae* bacilli by an antibody-independent mechanism. The highest number of cell-associated bacteria of the non-toxigenic strain was observed 3 h post infection and non-toxigenic *C. diphtheriae* able to survive within U937 cells for a time period of 24 h. The observation that the internalized non-toxigenic bacteria were able to survive longer periods within the macrophages might indicate that human macrophages may be not effective in killing *C. diphtheriae* in the absence of opsonins (dos Santos et al. 2010). The *tox*$^+$ strain showed significant higher numbers of associated bacteria after 1 to 2 h post infection, but was not able to survive within the cells over a time period of 24 h. Non-opsonic phagocytosis of both toxigenic and non-toxigenic strains induced cytopathogenicity in U937 cells. More than 42 % of U937 cells were killed during infection with the *tox*$^+$ strain (3h post infection), while roughly 18 % were killed during infection with the non-toxigenic strain. Changes of nuclear morphology upon *C. diphtheriae* infection were indicating that *C. diphtheriae* induces apoptosis and necrosis in human macrophages, independent of the presence of the diphtheria toxin.

Sabbadini and co-workers (Sabbadini et al. 2012) made similar observations concerning apoptosis induction during the infection of non-phagocytic cell line HEp-2 with *C. diphtheriae*. Blebs and "budding" cells were found within 3 to 6 h of treatment with 67–72p. The assessment of apoptosis by double staining propidium iodide (PI)/Annexin V (AV) revealed significant numbers of early (PI^-/AV^+) and late (PI^+ /AV^+) apoptotic HEp-2 cells treated for 6 h with 67–72p. Using the trypan blue exclusion method Sabbadini and co-workers (Sabbadini et al. 2012) showed

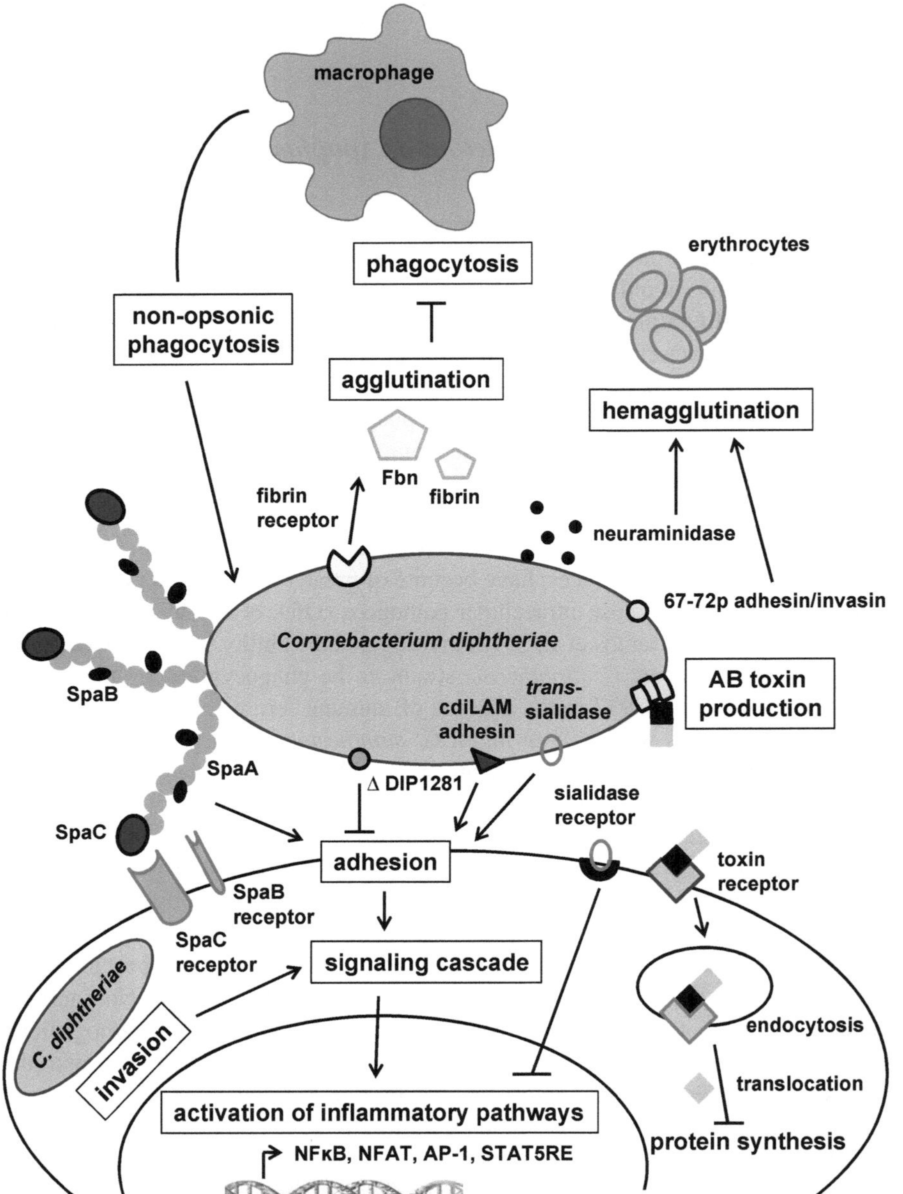

Fig. 8.1 Infection properties of *C. diphtheriae.* The most prominent virulence factor of *C. diphtheriae* is the diphtheria toxin which is encoded by a β-corynephage integrated in the bacterial chromosome. This AB toxin is released by the bacterial cell and binds the toxin receptor (HB-EGF) on the host's surface. The toxin gets internalized by endocytosis which leads to translocation of the A-fragment and inhibition of the protein biosynthesis by binding of the elongation factor 2. Best investigated adhesion factors of *C. diphtheriae* might be the pili structures. Strain NCTC 13129 exhibits three different pilus types, the SpaABC type pilus is depicted here exemplarily. SpaA forms the pilus shaft, SpaB is linked to SpaA via transpeptidation, and SpaC forms the tip

that the plasma membrane of HEp-2 cells treated with 67–72p suffered permeability changes, vacuolization and DNA fragmentation after 12 h. A high occurrence of apoptotic bodies was observed after 24 h, as well as a detachment of HEp-2 cells post treatment with 67–72p. DNA condensation and aggregation was observed by DAPI staining of 67–72p treated cells. Furthermore, the incubation of HEp-2 cells with 67–72p caused a reduction in cell volume which is a distinct marker of apoptosis, termed apoptotic volume decrease (Bortner and Cidlowski 2002).

The ability of *C. diphtheriae* to kill non-phagocytic cells might be beneficial for the survival and the invasion of deeper tissue of this organism and thus to disperse in the whole body. In summary, the interaction of the Gram-positive pathogen *C. diphtheriae* with its host is much more complex than initially expected. Up to now, several different interaction mechanisms were elucidated (summarized in Fig. 8.1), but there is still room for research, especially regarding interactions at the molecular level. Based on the already described results and ongoing research, *C. diphtheriae* might work as a model for other pathogenic but molecular biologically less addressable corynebacteria such as *C. jeikeium*.

8.3 Pathogenicity of *Corynebacterium ulcerans* and *Corynebacterium pseudotuberculosis*

Besides *C. diphtheriae*, the most prominent member of toxigenic corynebacteria, two further species might carry toxin-encoding corynebacteriophages and might have impact on human health, *C. ulcerans* and *C. pseudotuberculosis*.

---

of the pilus. SpaB and SpaC are able to bind receptors on the surface of the host cell thus enabling colonization of the cell. Besides pili formation the non-fimbrial protein 67–72p (DIP0733) might be an important adhesin. Via this protein *C. diphtheriae* is able to bind erythrocytes, a process called hemagglutination which facilitates *C. diphtheriae* to disseminate throughout the body via the blood. Furthermore, the 67–72p protein seems to be involved in the internalization process and might contribute to apoptosis and necrosis of the host cell. Neuraminidase-secreting strains were also found to induce agglutinating human erythrocytes effectively. In the case of the *C. diphtheriae* sialidase NanH, it is still unclear whether it is involved in sialic acids decoration or not. Besides proteins as adhesion factors, *C. diphtheriae* exhibits a 10 kDa lipoglycan on its surface, the CdiLAM, which binds HEp-2 cells. Additionally, *C. diphtheriae* is able to bind fibrinogen and to convert it to fibrin, resulting in a fibrin layer on the bacterial surface that may be an efficient method to avoid phagocytosis. In addition to adhesion, *C. diphtheriae* is able to internalize into non-phagocytic cells in a strain-specific manner. The invasion process seems to involve actin polymerization along with a phosphotyrosine signaling event in cultured respiratory cells. Investigations of DIP1281 (annotated as invasion-associated protein) revealed the involvement in the organization of the outer protein layer which resulted in an adhesion- and invasion-negative phenotype of the corresponding mutant strains. Furthermore, infection of epithelial cells leads to inflammatory host response, especially the activation of the NFκB signal transduction pathway. NFκB induction is strictly dependent on intracellular bacteria, since bacterial adhesion was not sufficient to activate inflammatory pathways. *C. diphtheriae* taken up by human macrophages via non-opsonic phagocytosis are able to persist inside the macrophages and to induce apoptosis and necrosis

C. ulcerans was first isolated in 1926 from a patient with throat lesions and symptoms of a diphtheria-like illness (Gilbert and Stewart 1926). Infections of humans have been classically associated with consumption of raw milk products or contact with cattle, but the bacterium was increasingly found in domestic animals, such as dogs and cats, suggesting that clinical cases in humans are more likely devolved from domestic pets (Taylor et al. 2002; De Zoysa et al. 2005; Lartigue et al. 2005). In the last decade the majority of diphtheria cases reported in Europe were caused by *C. ulcerans* (Wagner et al. 2010), indicating that this organism plays an increasing role as a zoonotic pathogen.

C. pseudotuberculosis biovar *ovis* and *equi* cause caseous lymphadentitis in sheep and goats (Batey 1986) and ulcerative lymphangitis in horses (Brumbaugh and Ekman 1981). Interestingly, *C. pseudotuberculosis* is a facultative intracellular pathogen that is able to survive and grow in macrophages, thus escaping the immune response of the host (McKean et al. 2005; Dorella et al. 2006).

8.3.1 Gene Regions Encoding Adhesive Pili Subunits in **C. ulcerans** *and* **C. pseudotuberculosis**

As in case of *C. diphtheriae*, also *C. ulcerans* and *C. pseudotuberculosis* genome sequences encode different putative pili. In *C. ulcerans* one gene cluster is identically organized compared to *spaDEF* from *C. diphtheriae* NCTC 13129, while the second pili cluster, *spaBC*, lacks the gene encoding major pili subunit SpaA (Trost et al. 2011). It is very likely that homodimeric/heterodimeric SpaB/SpaC proteins in *C. ulcerans*, anchored covalently to the cell wall, can provide close contact between the bacterial surface and host cell tissues even without the shaft protein as already shown for the minor pilin SpaB in *C. diphtheriae* NCTC 13129 (Mandlik et al. 2007; Chang et al. 2011). The genes in *C. pseudotuberculosis* are organized in two clusters (*spaABC* and *spaDEF*) together with their corresponding sortase-encoding genes (*srtB* and *srtC*) (Trost et al. 2010). The pili of *C. pseudotuberculosis* FRC41 are composed of major pilin subunits (SpaA and SpaD), minor pilin subunits (SpaB and SpaE) and tip proteins (SpaC and SpaF). Interestingly, there are two hypothetical proteins, SpaX (encoded in cluster *spaABC*) and SpaY (encoded in cluster *spaDEF*), with unknown function (Trost et al. 2010).

8.3.2 Phospholipase D Activity in **C. ulcerans** *and* **C. pseudotuberculosis**

C. ulcerans and *C. pseudotuberculosis* carry the gene for phospholipase D (PLD), the so called "ovis toxin", a sphingomylinase (phosphatidylcholine phosphohydrolase, EC 3.1.4.4). Phospholipase D activity has been found in no other species of the genus *Corynebacterium* which makes PLD a distinctive taxonomic marker (Barksdale et al. 1981).

Although the role of phospholipases in disease and pathogenesis must often be interpreted with caution, phospholipases may be directly toxic, with their interaction with the cell's plasma membrane being the main factor. Additionally, a metabolic function was discussed as well (Ostroff et al. 1989; Shortridge et al. 1992). An illustrative example for cell damage is the formation of membrane vesicles in erythrocytes infected with *C. pseudotuberculosis* (Brogden et al. 1990), as consequence of sphingomyelin depletion of the outer leaf and invagination of the inner sheet of the membrane (Low et al. 1974). Detrimental effects of *C. pseudotuberculosis* PLD were also shown for ovine neutrophils (Jozwiak et al. 1993). Furthermore, Muckle and co-workers (Muckle and Gyles 1983) found that *C. pseudotuberculosis* PLD increases vascular permeability which might be advantageous for the bacteria to spread from the initial site of infection to regional lymph nodes (McNamara et al. 1994).

8.3.3 Further Putative Virulence Factors of **C. ulcerans** *and* **C. pseudotuberculosis**

Trost and co-workers identified a gene encoded secreted protease CP40, also designated Cpp, in *C. ulcerans* (Trost et al. 2011) as well as in *C. pseudotuberculosis* (Trost et al. 2010). Since the extracellular CP40 enzyme has proteolytic activity, it was assumed to play a role in virulence of *C. pseudotuberculosis*, but an enzymatic activity of CP40 was not detectable in the supernatant (Wilson et al. 1995). Nevertheless, vaccination of sheep with this antigen protects against caseous lymphadenitis, supporting the idea that the protein plays a major role in this disease (Walker et al. 1994). The amino acid sequences of CP40 from *C. ulcerans* 809 and *C. pseudotuberculosis* FRC41 showed peculiar similarities to the α-domain of the extracellular endoglycosidase EndoE from *Enterococcus faecalis* which is involved in the degradation of N-linked glycans from ribonuclease B and the hydrolysis if the conserved glycans on IgG (Collin and Fischetti 2004). Hence, both *C. ulcerans* strains 809 and BR-AD22 might be able to interact with the mammalian host cell by modulation of the host glycoproteins.

In addition to CP40, the extracellular neuraminidase NanH poses a putative virulence factor in *C. ulcerans* and *C. pseudotuberculosis* (Trost et al. 2010; Trost et al. 2011). Neuraminidases and sialidases are glycoside hydrolase enzymes that cleave the glycosidic linkages of neuraminic acids and can contribute to the recognition of sialic acids on host cell surfaces (Mattos-Guaraldi et al. 2000; Vimr et al. 2004). NanH of *C. diphtheriae* KCTC3075 was identified as an active extracellular sialidase that could transfer a sialic acid from sialylconjugates to asialoglycans via transglycosylation (Kim et al. 2010). Furthermore, Trost and co-workers could detect a GlxR-binding site upstream of *nanH* in *C. pseudotuberculosis* FRC41, indicating that this virulence factor is under control of the cAMP-sensing transcription regulator GlxR (Trost et al. 2010).

Table 8.3 Prophage like elements in *C. ulcerans* genomes

Prophage	Size	G + C content	Integration site	Attachment site
ΦCULC809I	41.4 kb	53 %	CULC809_01141	not detected
ΦCULC22I	42 kb	53 %	CULC22_01157	not detected
ΦCULC22II	44.9 kb	55 %	between CULC22_01663 and CULC22_01724	TTAGATAC
ΦCULC22III	14 kb	57 %	tRNALys	TTCAAGTCCCTGATGGCGCAC
ΦCULC22IV	41 kb	54 %	tRNALys	TTGAGCTGGAGATGGGACTT-GAACCC

8.3.4 Further Putative Virulence Factors of C. ulcerans: *Prophage-like Sequences and Toxins*

Since the important corynebacterial virulence factor diphtheria toxin is encoded by corynephages integrated in the genome of toxigenic strains, the genome sequences of the *C. ulcerans* 809 and BR-AD22 were screened for prophage-like elements. These analyses revealed one putative prophage in the genome of *C. ulcerans* 809 and four prophage-like regions in the genome sequence of BR-AD22 (Trost et al. 2011) (Table 8.3, modified after (Trost et al. 2011).

Further characterization of the prophage-like regions revealed that ΦCULC809I from *C. ulcerans* 809 and ΦCULC22I from *C. ulcerans* BR-AD22 are closely related genetic elements. The two prophages share 36 genes which encode proteins comprising approximately 98 % sequence identity. The previously suggested presence of a diphtheria toxin-encoding β-corynephage in strain 809 (Mattos-Guaraldi et al. 2008) was not confirmed. However, sequence analysis of a tyrosine recombinase of *C. ulcerans* and the integrase of the β-corynephage integrated in the genome of *C. diphtheriae* NCTC 13129 showed 92 % sequence identity, supporting the assumption that a β-corynephage-like element had been integrated in *C. ulcerans* 809 and BR-AD22 genomes in former times.

While a diphtheria toxin-encoding gene is absent in the genome, *C. ulcerans* 809 carries a gene coding for a Shiga toxin-like ribosome-binding protein, Rbp. The deduced amino acid sequence of this protein shares only 24 % identity with the A chains of the Shiga-like toxins SLT-1 and SLT-2 of *E. coli*, but comprises all highly conserved amino acid residues needed for the catalytic N-glycosidase activity (Trost et al. 2011). SLT-1 usually consists of a catalytic A domain that is non-covalently linked with a pentamer of chains B which are essential for binding to the specific glycolipid receptor and the translocation of the toxin into the endoplasmatic reticulum (ER) of the host cell. This leads to the retranslocation of the catalytic chain A into the cytosol, followed by the abortion of the protein biosynthesis by depurination of a single adenosine residue in the 28S rRNA of the eukaryotic ribosome (O'Loughlin and Robins-Browne 2001). Interestingly, the amino acid sequence of Rbp from *C. ulcerans* 809 lacks the ER-targeting sequence at the C-terminal end of the protein. Instead the secretion of the putative toxin into the cytosol is supported by a typical signal sequence at the N-terminus of the protein (Trost et al. 2011).

8.3.5 Inflammatory Response to C. ulcerans *Infection in Mice*

When mice were intravenously infected with different doses of *C. ulcerans* strain 809, a human respiratory tract isolate, BR-AD22, an isolate from an asymptomatic dog, and CDC-KC279, an animal clinical isolate (Dias et al. 2011), all three strains had a lethal effect on mice, but at different levels. Strain 809 revealed the highest lethality rate, followed by CDC-KC279 and BR-AD22 which led to the lowest mortality. When *C. ulcerans* was recovered from blood, kidney, liver, spleen, heart, lung, joints and brain, viable bacteria of strain 809 and CDC-KC279 were detectable until day 3 post infection and of BR-AD22 until day 2 post infection. Viable bacteria were also found in kidneys, livers and spleens, independent of the strains tested, whereas colonization of lung, heart and brain was not observed. The three strains were able to persist in joints until day 20 post infection, but only mice infected with strain 809 and strain CDC-KC279 showed clinical symptoms of arthritis. In contrast, *C. ulcerans* BR-AD22 and the toxigenic *C. diphtheriae* strain ATCC 27012 were not able to induce signs of arthritis in mice. To investigate the arthritis induction during *C. ulcerans* and *C. diphtheriae* infection in more detail, Dias and co-workers (Dias et al. 2011) measured the systemic production levels of IL-6 and TNF-α at different time points of infection with different *C. ulcerans* and *C. diphtheriae*. The data indicated that TNF-α and IL-6 levels were always higher than the negative control, with the exception of TNF-α mice group infected with the toxigenic *C. diphtheriae* ATCC27012, which is in accordance with the observation that this strain is not able to induce clinical symptoms of arthritis. *C. ulcerans* BR-AD22, which was also not able to induce clinical signs of arthritis, reached the peak of TNF-α production on day 1 post infection while 809 and CDC-KC279 reached its peaks on day 3 post infection. The results of Dias and co-workers show a strong correlation between the arthritis index and *in vivo* pro-inflammatory cytokine production, demonstrating that *C. ulcerans* strains have the capacity to induce arthritis in conventional Swiss Webster mice and furthermore demonstrating that some *C. ulcerans* strains are more virulent than *C. diphtheriae*, independent of the presence of the diphtheria toxin.

8.3.6 C. pseudotuberculosis Regulons Involved in Iron Metabolism, Oxidative Stress Response and Detoxification of Nitric Oxide

Iron is an essential element of proteins containing heme, iron-sulfur clusters or mono- and binuclear iron species. Since many of these genes are play important roles in pathogen-host interaction, iron uptake and metabolism are closely connected to virulence.

It was already known that *C. pseudotuberculosis* encodes an iron uptake system, *fagCBA-fagD* that is regulated by iron *in vitro* (Billington et al. 2002). Additionally, the complete genome of *C. pseudotuberculosis* FRC41 was screened for other iron uptake systems and the corresponding transcriptional regulators. Trost and co-workers (Trost et al. 2010) were able to assign a number of genes to the DtxR regulon of *C. pseudotuberculosis* FRC41 that are involved in the utilization of various host compounds as iron source. In addition to that, *htaA*-like genes (*htaD* and *htaF*) that are associated with genes encoding membrane proteins (*htaE* and *htaG*) were detected, which might be responsible for the acquisition of iron from the host. The *hmuO* gene of the DtxR regulon, a heme oxygenase, is able to release iron from the protoporphyrin ring of heme and enables the acquisition of iron from heme and hemoglobin (Schmitt 1997b, a). Trost and co-workers (Trost et al. 2010) identified two response regulators that are similar to ChrA and HrrA within the *C. pseudotuberculosis* FRC41 genome, supporting the theory that a complex hierarchical control of *hmuO* gene expression might be established in this organism.

The genome of *C. pseudotuberculosis* FRC41 contains two DtxR-regulated gene clusters that might be involved in siderophore biosynthesis, independent of non-ribosomal peptide synthetases (Trost et al. 2010). The first one is the *ciu* locus which consists of the *ciuABCD* gene cluster, encoding an ABC-type transporter system, together with *ciuE*, encoding a siderophore biosynthesis-related protein, and *ciuF*, encoding a putative efflux protein. The *ciu* gene region is homologous to the *C. diphtheriae* NCTC 13129, which additionally carries the *ciuG* gene, encoding a protein of unknown function. The second gene cluster consists of four genes, which might be involved in siderophore biosynthesis pathway (*ogs, ocd, odc* and *tsb*), one gene encoding an efflux protein (*mdtK*) and four genes encoding an ABC-type transporter (*stsABCD*) (Trost et al. 2010).

In addition to that, genes encoding Dps-like proteins were detected in *C. pseudotuberculosis* (Trost et al. 2010). Dps proteins mainly protect DNA from redox stress; they also can act as iron-binding and storage proteins, but prefer H_2O_2 as oxidant instead of O_2, indicating that its primary function is in protecting DNA against the combined action of ferrous iron and H_2O_2 and rather in iron storage (Andrews et al. 2003). The protein that shares a number of physiological properties with Dps-like proteins in *C. pseudotuberculosis* FRC14 is the multifunctional histone-like protein and transcription regulator Lsr2. Lsr2 proteins protect the DNA against reactive oxygen intermediates, as it was shown for mycobacteria, and thus Lsr2 may also protect corynebacterial DNA (Colangeli et al. 2009). RipA-binding sites were detected in front of *dps-fpg1* in *C. pseudotuberculosis* (Trost et al. 2010). RipA, a DNA-binding transcription regulator is under direct control of DtxR in *C. glutamicum.* When under conditions of iron-restriction, RipA binds the DNA and acts as a repressor for genes encoding iron proteins in *C. glutamicum* (Wennerhold et al. 2005). Another gene (*fpg2*), encoding a formamidopyrimidine-DNA glycosylase, was detected by Trost and co-workers in the genome of *C. pseudotuberculosis*. These enzymes function mainly in the repair of DNA lesions caused by oxidative damage, but *fpg2* in *C. pseudotuberculosis* FRC41 lacks the RipA-binding site.

Nevertheless, the interaction of the DtxR regulon and RipA shows a complex regulatory network of iron metabolism and oxidative stress response in corynebacteria, with regard to virulence of *C. pseudotuberculosis*.

Besides DNA protection, many pathogenic bacteria developed mechanisms for the detoxification of reactive oxygen and nitrogen species produced from macrophages as part of their antimicrobial response (Nathan and Shiloh 2000; Zahrt and Deretic 2002). Examples for such detoxification proteins in important pathogens are the SodC protein of *Mycobacterium tuberculosis*, which is involved in the resistance against oxidative burst species produced by macrophages (Dussurget et al. 2001; Piddington et al. 2001), and the Cu, Zn-dependent superoxide dismutases, which have been reported in *Neisseria meningitidis* and *Hemophilus ducreyi* (Wilks et al. 1998; San Mateo et al. 1999).

In the *C. pseudotuberculosis* FRC41 genome four genes, which are probably involved in the detoxification process of reactive oxygen species, are found (Trost et al. 2010): *ahpCD* (alcyl hydroperoxide reductase), *sodA* (manganese-dependent superoxide dismutase) and *sodC* (copper, zinc-dependent superoxide dismutase). However, the function in virulence of these genes in *C. pseudotuberculosis* FRC41 has to be proven.

A further potential virulence factor of *C. pseudotuberculosis* FRC41detected by Trost and co-workers might be the *nor* gene, coding for nitric oxide reductase, which is generally involved in the detoxification of nitric oxide and therefore important for pathogens to persist within macrophages (Luthra et al. 2008). Interestingly, McKean and co-workers (McKean et al. 2005) did not find induction of the expression of the *nor* gene upon the infection of macrophages by animal pathogenic *C. pseudotuberculosis*. Thus, the regulation pattern of *nor* transcription and its part in protection against nitric oxide, is still unclear.

8.4 Conclusions

The development of new molecular tools allowed major progress in the analysis of virulence factors of toxigenic corynebacteria. Especially the availability of genome sequence information gave new insights in the presence of putative virulence factors of *C. ulcerans*, *C. pseudotuberculosis* and *C. diphtheriae* (summarized in Table 8.4). However, there are still a number of open questions concerning strain-specific differences interaction with epithelial cells and macrophages as well as induction of host pathways, which might have implication for disease development and will hopefully elucidated in future.

Table 8.4 Putative virulence factors of *C. Diphtheriae* NCTC 13129, *C. Ulcerans* 809, BR-AD22 and *C. Pseudotuberculosis* FRC41

ID *C. diphtheriae* NCTC 13129	ID *C. ulcerans* 809	ID *C. ulcerans* BR-AD22	ID *C. pseudotuberculosis* FRC41	Gene name	Annotated function of deduced protein
DIP0222	–	–	–	*tox*	Diphtheria toxin precursor
DIP2013	–	–	cpfrc_01874	*spaA*	Surface-anchored protein (pilus subunit)
DIP2011	–	CULC22_02131	cpfrc_01872	*spaB*	Surface-anchored protein (pilus subunit)
DIP2010	CULC809_01980	CULC22_02130	cpfrc_01870	*spaC*	Surface-anchored protein (pilus subunit)
DIP0235	CULC809_01979	CULC22_02106	cpfrc_01904	*spaD*	Surface-anchored protein (pilus subunit)
DIP0237	CULC809_01952	CULC22_02104	cpfrc_01902	*spaE*	Surface-anchored protein (pilus subunit)
DIP0238	CULC809_01950	CULC22_02103	cpfrc_01901	*spaF*	Surface-anchored protein (pilus subunit)
DIP2227	CULC809_01949	–	–	*spaG*	Surface-anchored protein (pilus subunit)
DIP2226	–	–	-	*spaH*	Surface-anchored protein (pilus subunit)
DIP2223	–	–	–	*spaI*	Surface-anchored protein (pilus subunit)
DIP0733	CULC809_00602	CULC22_00609	cpfrc_00553	-	67–72p hemagglutinin
DIP0543	CULC809_00434	CULC22_00437	cpfrc_00386	*nanH*	Sialidase precursor (neuraminidase H)
DIP1281	CULC809_01133	CULC22_01148	cpfrc_01079	*rpfI*	Resuscitation-promoting factor interacting protein (D, L-endopeptidase)
–	CULC809_00040	CULC22_00038	cpfrc_00029	*pld*	Phospholipase D
–	CULC809_01974	CULC22_02125	cpfrc_01895	*cpp*[b]	Corynebacterial protease CP40 (Endoglycosidase Endo E)
–	CULC809_00177	–	–	*rbp*	(Shiga toxin-like) ribosome-binding protein
–	Integration site CULC809_01141	–	–	–	Prophage ΦCULC809I
–	–	Integration site CULC22_01157	–	–	Prophage ΦCULC22I
–	–	Integration site between CULC22_01663 and CULC22_01724	–	–	Prophage ΦCULC22II

Table 8.4 (continued)

ID *C. diphtheriae* NCTC 13129	ID *C. ulcerans* 809	ID *C. ulcerans* BR-AD22	ID *C. pseudotuberculosis* FRC41	Gene name	Annotated function of deduced protein
–	–	Integration site tRNALys	–	–	Prophage ΦCULC22III
–	–	Integration site tRNALys	–	–	Prophage ΦCULC22IV
DIP1621	CULC809_01521	CULC22_01537	–	*cwlH*	Cell wall-associated hydrolase
DIP1414	–	CULC22_01326	cpfrc_01219	*dtxR*	Diphtheria toxin repressor
–	CULC809_02066	CULC22_02219	cpfrc_01988	*htaD*	Cell-surface hemin receptor
–	CULC809_02125	CULC22_02281	cpfrc_01477	*htaF*	Cell-surface hemin receptor
–	CULC809_02065	CULC22_02218	cpfrc_01987	*htaE*	Membrane protein
–	–	–	cpfrc_01476	*htaG*	Membrane protein
DIP1669	CULC809_01565	CULC22_01581	cpfrc_01487	*hmuO*	Heme oxygenase
DIP2327	CULC809_02136	CULC22_02292	cpfrc_02058	*chrA*	Two-component system transcriptional regulatory protein
[a]	CULC809_02085	CULC22_02242	cpfrc_02010	*hrr*	Two-component system transcriptional regulatory protein
[a]	CULC809_01040	CULC22_01055	cpfrc_00987	*ciuA*	Iron ABC transport system substrate-binding protein
[a]	CULC809_01041	CULC22_01056	cpfrc_00988	*ciuB*	Iron ABC transport system permease protein
[a]	CULC809_01042	CULC22_01057	cpfrc_00989	*ciuC*	Iron ABC transport system permease protein
[a]	CULC809_01545	CULC22_01058	cpfrc_00990	*ciuD*	Iron ABC transport system ATP-binding protein
[a]	CULC809_01044	CULC22_01059	cpfrc_00991	*ciuE*	Siderophore biosynthesis-related protein
[a]	CULC809_01045	CULC22_01060	cpfrc_00992	*ciuF*	Putative efflux protein
[a]	–	–	–	*ciuG*	Conserved membrane protein
–	CULC809_00665	CULC22_00672	cpfrc_00624	*ogs*	Oxygenase

Table 8.4 (continued)

ID *C. diphtheriae* NCTC 13129	ID *C. ulcerans* 809	ID *C. ulcerans* BR-AD22	ID *C. pseudotuberculosis* FRC41	Gene name	Annotated function of deduced protein
–	CULC809_00664	CULC22_00671	cpfrc_00623	*ocd*	Ornithine cyclodeaminase
–	CULC809_00663	CULC22_00670	cpfrc_00622	*odc*	Ornithine decarboxylase
–	–	CULC22_00669	cpfrc_00621	*tsb*	Cysteine synthase
–	–	–	cpfrc_00620	*mdtk*	Putative membrane protein
–	CULC809_00666	CULC22_00673	cpfrc_00625	*stsA*	ABC transporter solute-binding protein
–	CULC809_00667	CULC22_00674	cpfrc_00626	*stsB*	ABC transporter permease protein
–	CULC809_00668	CULC22_00675	cpfrc_00627	*stsC*	ABC transporter permease protein
–	CULC809_00669	CULC22_00676	cpfrc_00628	*stsD*	ABC transporter ATP-binding protein
DIP2266	CULC809_02084	CULC22_02241	cpfrc_02009	*lsr2*	Protein Lsr2
DIP1543	CULC809_02127	CULC22_02283	cpfrc_02047	*fpg1*	Formamidopyrimidine DNA glycosylase
–	CULC809_01439	CULC22_01453	cpfrc_01339	*fpg2*	Formamidopyrimidine DNA glycosylase
[a]	CULC809_01318	CULC22_01332	cpfrc_01226	*ahpC*	Alkyl hydroperoxide reductase AhpD
DIP1419	CULC809_01317	CULC22_01331	cpfrc_01225	*ahpD*	Alkyl hydroperoxide reductase AhpD
DIP2261	CULC809_02075	CULC22_02231	cpfrc_01998	*sodA*	Superoxide dismutase
[a]	CULC809_00652	CULC22_00659	cpfrc_00610	*sodC*	Superoxide dismutase [Cu–Zn]
–	–	–	cpfrc_00128	*nor*	Nitric oxide reductase

[a] not detected in strain NCTC 13129, but in several other *C. diphtheriae* strains

References

Akira S, Takeda K (2004) Functions of toll-like receptors: lessons from KO mice. C R Biol 327(6):581–589

Akira S, Uematsu S, Takeuchi O (2006) Pathogen recognition and innate immunity. Cell 124(4):783–801

Alexander D (1984) Splenic abscess caused by *Corynebacterium diphtheriae*. Clin Pediatr (Phila) 23 (10):591–592.

Anantharaman V, Aravind L (2003) Evolutionary history, structural features and biochemical diversity of the NlpC/P60 superfamily of enzymes. Genome Biol 4(2):R11

Andrews SC, Robinson AK, Rodriguez-Quinones F (2003) Bacterial iron homeostasis. FEMS Microbiol Rev 27(2–3):215–237

Baird GJ, Fontaine MC (2007) *Corynebacterium pseudotuberculosis* and its role in ovine caseous lymphadenitis. J Comp Pathol 137(4):179–210

Barksdale L (1970) *Corynebacterium diphtheriae* and its relatives. Bacteriol Rev 34(4):378–422

Barksdale L, Linder R, Sulea IT, Pollice M (1981) Phospholipase D activity of *Corynebacterium pseudotuberculosis* (*Corynebacterium ovis*) and *Corynebacterium ulcerans*, a distinctive marker within the genus *Corynebacterium*. J Clin Microbiol 13(2):335–343

Batey RG (1986) Pathogenesis of caseous lymphadenitis in sheep and goats. Aust Vet J 63(9):269–272

Beesley ED, Brubaker RR, Janssen WA, Surgalla MJ (1967) Pesticins. 3. Expression of coagulase and mechanism of fibrinolysis. J Bacteriol 94(1):19–26

Belko J, Wessel DL, Malley R (2000) Endocarditis caused by *Corynebacterium diphtheriae*: case report and review of the literature. Pediatr Infect Dis J 19(2):159–163

Bertuccini L, Baldassarri L, von Hunolstein C (2004) Internalization of non-toxigenic *Corynebacterium diphtheriae* by cultured human respiratory epithelial cells. Microb Pathog 37(3):111–108

Beutler B (2004) Toll-like receptors and their place in immunology. Where does the immune response to infection begin? Nat Rev Immunol 4(7):498

Billington SJ, Esmay PA, Songer JG, Jost BH (2002) Identification and role in virulence of putative iron acquisition genes from *Corynebacterium pseudotuberculosis*. FEMS Microbiol Lett 208(1):41–45

Blumberg BS, Warren L (1961) The effect of sialidase on transferrins and other serum proteins. Biochim Biophys Acta 50:90–101

Bortner CD, Cidlowski JA (2002) Apoptotic volume decrease and the incredible shrinking cell. Cell Death Differ 9(12):1307–1310

Bostock AD, Gilbert FR, Lewis D, Smith DC (1984) *Corynebacterium ulcerans* infection associated with untreated milk. J Infect 9(3):286–288

Bowie A, O'Neill LA (2000) The interleukin-1 receptor/Toll-like receptor superfamily: signal generators for pro-inflammatory interleukins and microbial products. J Leukoc Biol 67(4):508–514

Brogden KA, Engen RL, Songer JG, Gallagher J (1990) Changes in ovine erythrocyte morphology due to sphingomyelin degradation by *Corynebacterium pseudotuberculosis* phospholipase D. Microb Pathog 8(2):157–162

Brumbaugh GW, Ekman TL (1981) *Corynebacterium pseudotuberculosis* bacteremia in two horses. J Am Vet Med Assoc 178(3):300–301

Buck GA, Cross RE, Wong TP, Loera J, Groman N (1985) DNA relationships among some tox-bearing corynebacteriophages. Infect Immun 49(3):679–684

Burkovski A (2013) Cell Envelope of Corynebacteria: Structure and Influence on Pathogenicity. ISRN Microbiology Article ID 935736, doi:10.1155/2013/935736.

Chang C, Mandlik A, Das A, Ton-That H (2011) Cell surface display of minor pilin adhesins in the form of a simple heterodimeric assembly in *Corynebacterium diphtheriae*. Mol Microbiol 79(5):1236–1247

Cianciotto NP, Groman NB (1996) Extended host range of a beta-related corynebacteriophage. FEMS Microbiol Lett 140(2–3):221–225

Colangeli R, Haq A, Arcus VL, Summers E, Magliozzo RS, McBride A, Mitra AK, Radjainia M, Khajo A, Jacobs WR Jr, Salgame P, Alland D (2009) The multifunctional histone-like protein Lsr2 protects mycobacteria against reactive oxygen intermediates. Proc Natl Acad Sci U S A 106(11):4414–4418

Collin M, Fischetti VA (2004) A novel secreted endoglycosidase from *Enterococcus faecalis* with activity on human immunoglobulin G and ribonuclease B. J Biol Chem 279(21):22558–22570

Colombo AV, Hirata R Jr, de Souza CM, Monteiro-Leal LH, Previato JO, Formiga LC, Andrade AF, Mattos-Guaraldi AL (2001) *Corynebacterium diphtheriae* surface proteins as adhesins to human erythrocytes. FEMS Microbiol Lett 197(2):235–239

De Zoysa A, Hawkey PM, Engler K, George R, Mann G, Reilly W, Taylor D, Efstratiou A (2005) Characterization of toxigenic *Corynebacterium ulcerans* strains isolated from humans and domestic cats in the United Kingdom. J Clin Microbiol 43(9):4377–4381

Dias AA, Silva FC Jr, Santos LS, Ribeiro-Carvalho MM, Sabbadini PS, Santos CS, Filardy AA, Myioshi A, Azevedo VA, Hirata R Jr, Villas-Boas MH, Mattos-Guaraldi AL (2011) Strain-dependent arthritogenic potential of the zoonotic pathogen *Corynebacterium ulcerans*. Vet Microbiol 153(3–4):323–331

Dixon B (2010) Sick as a dog. Lancet Infect Dis 10(2):73

Dorella FA, Pacheco LG, Oliveira SC, Miyoshi A, Azevedo V (2006) *Corynebacterium pseudotuberculosis*: microbiology, biochemical properties, pathogenesis and molecular studies of virulence. Vet Res 37(2):201–218

dos Santos CS, dos Santos LS, de Souza MC, dos Santos DF, de Souza deODias AA, Sabbadini PS, Pereira GA, Cabral MC, Hirata Junior R, de Mattos-Guaraldi AL (2010) Non-opsonic phagocytosis of homologous non-toxigenic and toxigenic *Corynebacterium diphtheriae* strains by human U-937 macrophages. Microbiol Immunol 54(1):1–10

Dussurget O, Stewart G, Neyrolles O, Pescher P, Young D, Marchal G (2001) Role of *Mycobacterium tuberculosis* copper-zinc superoxide dismutase. Infect Immun 69(1):529–533

Duthie ES (1954) Evidence for two forms of staphylococcal coagulase. J Gen Microbiol 10(3):427–436

Funke G, Altwegg M, Frommelt L, von Graevenitz A (1999) Emergence of related nontoxigenic *Corynebacterium diphtheriae* biotype *mitis* strains in Western Europe. Emerg Infect Dis 5(3):477–480

Gaspar AH, Ton-That H (2006) Assembly of distinct pilus structures on the surface of *Corynebacterium diphtheriae*. J Bacteriol 188(4):1526–1533

Gilbert R, Stewart FC (1926) *Corynebacterium ulcerans*: a pathogenic microorganism resembling *C. diphtheriae*. J Lab Clin Med 12:756–761.

Groman N, Schiller J, Russell J (1984) *Corynebacterium ulcerans* and *Corynebacterium pseudotuberculosis* responses to DNA probes derived from corynephage beta and *Corynebacterium diphtheriae*. Infect Immun 45(2):511–517

Gubler J, Huber-Schneider C, Gruner E, Altwegg M (1998) An outbreak of nontoxigenic *Corynebacterium diphtheriae* infection: single bacterial clone causing invasive infection among Swiss drug users. Clin Infect Dis 27(5):1295–1298

Hart RJ (1984) *Corynebacterium ulcerans* in humans and cattle in North Devon. J Hyg (Lond) 92 (2):161–164.

Hirata R Jr, Souza SM, Rocha-de-Souza CM, Andrade AF, Monteiro-Leal LH, Formiga LC, Mattos-Guaraldi AL (2004) Patterns of adherence to HEp-2 cells and actin polymerisation by toxigenic *Corynebacterium diphtheriae* strains. Microb Pathog 36(3):125–130

Hirata R, Napoleao F, Monteiro-Leal LH, Andrade AF, Nagao PE, Formiga LC, Fonseca LS, Mattos-Guaraldi AL (2002) Intracellular viability of toxigenic *Corynebacterium diphtheriae* strains in HEp-2 cells. FEMS Microbiol Lett 215(1):115–119

Hirata R, Pereira GA, Filardy AA, Gomes DL, Damasco PV, Rosa AC, Nagao PE, Pimenta FP, Mattos-Guaraldi AL (2008) Potential pathogenic role of aggregative-adhering *Corynebacterium diphtheriae* of different clonal groups in endocarditis. Braz J Med Biol Res 41:986–991

Hoffmann C, Leis A, Niederweis M, Plitzko JM, Engelhardt H (2008) Disclosure of the mycobacterial outer membrane: cryo-electron tomography and vitreous sections reveal the lipid bilayer structure. Proc Natl Acad Sci U S A 105(10):3963–3967

Hoffmann JA (2003) The immune response of Drosophila. Nature 426(6962):33–38

Hogg GG, Strachan JE, Huayi L, Beaton SA, Robinson PM, Taylor K (1996) Non-toxigenic *Corynebacterium diphtheriae* biovar *gravis*: evidence for an invasive clone in a south-eastern Australian community. Med J Aust 164(2):72–75

Holthouse DJ, Power B, Kermode A, Golledge C (1998) Non-toxigenic *Corynebacterium diphtheriae*: two cases and review of the literature. J Infect 37(1):62–66

Jozwiak S, Gornicki J, Michalowicz R, Gastol P (1993) [A case of renal clear cell sarcoma in a child with tuberous sclerosis]. Wiad Lek 46(21–22):846–848

Kim S, DB Oh, Kwon O, Kang HA (2010) Identification and functional characterization of the NanH extracellular sialidase from *Corynebacterium diphtheriae.* J Biochem 147(4):523–533

Lartigue MF, Monnet X, Le Fleche A, Grimont PA, Benet JJ, Durrbach A, Fabre M, Nordmann P (2005) *Corynebacterium ulcerans* in an immunocompromised patient with diphtheria and her dog. J Clin Microbiol 43(2):999–1001

Lehmann KB, Neumann R (1896) Atlas und Grundriss der Bakteriologie und Lehrbuch der speziellen bateriologischen Diagnostik. Lehmann, Munich.

Low DKR, Freer JH, Arbuthnott JP, Möllby R, Wadström T (1974) Consequences of sphingomyelin degradation in erythrocyte ghost membranes by staphylococcal β-toxin (sphingomyelinase C). Toxicon 12(3):279–282, IN215–IN218, 283–285

Luthra A, Malik SS, Ramachandran R (2008) Cloning, purification and comparative structural analysis of two hypothetical proteins from *Mycobacterium tuberculosis* found in the human granuloma during persistence and highly up-regulated under carbon-starvation conditions. Protein Expr Purif 62(1):64–74

Mandlik A, Swierczynski A, Das A, Ton-That H (2007) *Corynebacterium diphtheriae* employs specific minor pilins to target human pharyngeal epithelial cells. Mol Microbiol 64(1):111–124

Mattos-Guaraldi AL, Duarte Formiga LC, Pereira GA (2000) Cell surface components and adhesion in *Corynebacterium diphtheriae*. Microbes Infect 2(12):1507–1512

Mattos-Guaraldi AL, Formiga LC, Camello TC, Pereira GA, Hirata R Jr, Halpern M (2001) *Corynebacterium diphtheriae* threats in cancer patients. Rev Argent Microbiol 33(2):96–100

Mattos-Guaraldi AL, Formiga LCD, Andrade AFB, Hirata RJ (2002) Patterns of adherence to HEp-2 cells and ability to induce actin polymerization by toxigenic Corynebacterium diphtheriae strains. Seventh International Meeting of the European Laboratory Working Group on Diphtheria—ELWGD, Vienna

Mattos-Guaraldi AL, Sampaio JL, Santos CS, Pimenta FP, Pereira GA, Pacheco LG, Miyoshi A, Azevedo V, Moreira LO, Gutierrez FL, Costa JL, Costa-Filho R, Damasco PV, Camello TC, Hirata Jr R (2008) First detection of *Corynebacterium ulcerans* producing a diphtheria-like toxin in a case of human with pulmonary infection in the Rio de Janeiro metropolitan area, Brazil. Mem Inst Oswaldo Cruz 103(4):396–400

McKean S, Davies J, Moore R (2005) Identification of macrophage induced genes of *Corynebacterium pseudotuberculosis* by differential fluorescence induction. Microbes Infect 7(13):1352–1363

McNamara PJ, Bradley GA, Songer JG (1994) Targeted mutagenesis of the phospholipase D gene results in decreased virulence of *Corynebacterium pseudotuberculosis.* Mol Microbiol 12(6):921–930

Medzhitov R, Janeway CA Jr (2002) Decoding the patterns of self and nonself by the innate immune system. Science 296(5566):298–300

Moreira LO, Andrade AF, Vale MD, Souza SM, Hirata R Jr, Asad LM, Asad NR, Monteiro-Leal LH, Previato JO, Mattos-Guaraldi AL (2003) Effects of iron limitation on adherence and cell surface carbohydrates of *Corynebacterium diphtheriae* strains. Appl Environ Microbiol 69(10):5907–5913

Moreira LO, Mattos-Guaraldi AL, Andrade AF (2008) Novel lipoarabinomannan-like lipoglycan (CdiLAM) contributes to the adherence of *Corynebacterium diphtheriae* to epithelial cells. Arch Microbiol 190(5):521–530

Muckle CA, Gyles CL (1983) Relation of lipid content and exotoxin production to virulence of *Corynebacterium pseudotuberculosis* in mice. Am J Vet Res 44(6):1149–1153

Nathan C, Shiloh MU (2000) Reactive oxygen and nitrogen intermediates in the relationship between mammalian hosts and microbial pathogens. Proc Natl Acad Sci U S A 97(16):8841–8848

Navarre WW, Schneewind O (1999) Surface proteins of gram-positive bacteria and mechanisms of their targeting to the cell wall envelope. Microbiol Mol Biol Rev 63(1):174–229

Niederweis M, Danilchanka O, Huff J, Hoffmann C, Engelhardt H (2010) Mycobacterial outer membranes: in search of proteins. Trends Microbiol 18(3):109–116

O'Loughlin EV, Robins-Browne RM (2001) Effect of Shiga toxin and Shiga-like toxins on eukaryotic cells. Microbes Infect 3(6):493–507

Ofek I, Goldhar J, Keisari Y, Sharon N (1995) Nonopsonic phagocytosis of microorganisms. Annu Rev Microbiol 49:239–276

Ostroff RM, Wretlind B, Vasil ML (1989) Mutations in the hemolytic-phospholipase C operon result in decreased virulence of *Pseudomonas aeruginosa* PAO1 grown under phosphate-limiting conditions. Infect Immun 57(5):1369–1373

Ott L, Höller M, Gerlach RG, Hensel M, Rheinlaender J, Schäffer TE, Burkovski A (2010a) *Corynebacterium diphtheriae* invasion-associated protein (DIP1281) is involved in cell surface organization, adhesion and internalization in epithelial cells. BMC Microbiol 10:2

Ott L., Höller M, Rheinlaender J, Schäffer TE, Hensel M, Burkovski A (2010b) Strain-specific differences in pili formation and the interaction of *Corynebacterium diphtheriae* with host cells. BMC Microbiol 10:257

Ott L, Scholz B, Holler M, Hasselt K, Ensser A, Burkovski A (2013) Induction of the NFk-B signal transduction pathway in response to *Corynebacterium diphtheriae* infection. Microbiology 159(Pt 1):126–135

Patey O, Bimet F, Riegel P, Halioua B, Emond JP, Estrangin E, Dellion S, Alonso JM, Kiredjian M, Dublanchet A, Lafaix C (1997) Clinical and molecular study of *Corynebacterium diphtheriae* systemic infections in France. Coryne Study Group. J Clin Microbiol 35(2):441–445

Piddington DL, Fang FC, Laessig T, Cooper AM, Orme IM, Buchmeier NA (2001) Cu, Zn superoxide dismutase of *Mycobacterium tuberculosis* contributes to survival in activated macrophages that are generating an oxidative burst. Infect Immun 69(8):4980–4987

Pierno M, Maravigna L, Piazza R, Visai L, Speziale P (2006) FbsA-driven fibrinogen polymerization: a bacterial "deceiving strategy". Phys Rev Lett 96(2):028108

Poilane I, Fawaz F, Nathanson M, Cruaud P, Martin T, Collignon A, Gaudelus J (1995) *Corynebacterium diphtheriae* osteomyelitis in an immunocompetent child: a case report. Eur J Pediatr 154(5):381–383

Puliti M, von Hunolstein C, Marangi M, Bistoni F, Tissi L (2006) Experimental model of infection with non-toxigenic strains of *Corynebacterium diphtheriae* and development of septic arthritis. J Med Microbiol 55(Pt 2):229–235

Rennermalm A, Nilsson M, Flock JI (2004) The fibrinogen binding protein of *Staphylococcus epidermidis* is a target for opsonic antibodies. Infect Immun 72(5):3081–3083

Rivera J, Vannakambadi G, Hook M, Speziale P (2007) Fibrinogen-binding proteins of Gram-positive bacteria. Thromb Haemost 98(3):503–511

Romney MG, Roscoe DL, Bernard K, Lai S, Efstratiou A, Clarke AM (2006) Emergence of an invasive clone of nontoxigenic *Corynebacterium diphtheriae* in the urban poor population of Vancouver, Canada. J Clin Microbiol 44(5):1625–1629

Rosenshine I, Duronio V, Finlay BB (1992) Tyrosine protein kinase inhibitors block invasin-promoted bacterial uptake by epithelial cells. Infect Immun 60(6):2211–2217

Sabbadini PS, Genovez MR, Silva CF, Adelino TL, Santos CS, Pereira GA, Nagao PE, Dias AA, Mattos-Guaraldi AL, Hirata Junior R (2010) Fibrinogen binds to nontoxigenic and toxigenic *Corynebacterium diphtheriae* strains. Mem Inst Oswaldo Cruz 105(5):706–711

Sabbadini PS, Assis MC, Trost E, Gomes DL, Moreira LO, Dos Santos CS, Pereira GA, Nagao PE, Azevedo VA, Hirata Junior R, Dos Santos AL, Tauch A, Mattos-Guaraldi AL (2012) *Corynebacterium diphtheriae* 67–72p hemagglutinin, characterized as the protein DIP0733, contributes to invasion and induction of apoptosis in HEp-2 cells. Microb Pathog 52(3):165–176

San Mateo LR, Toffer KL, Orndorff PE, Kawula TH (1999) Neutropenia restores virulence to an attenuated Cu, Zn superoxide dismutase-deficient *Haemophilus ducreyi* strain in the swine model of chancroid. Infect Immun 67(10):5345–5351

Schmitt MP (1997a) Transcription of the *Corynebacterium diphtheriae hmuO* gene is regulated by iron and heme. Infect Immun 65(11):4634–4641

Schmitt MP (1997b) Utilization of host iron sources by *Corynebacterium diphtheriae*: identification of a gene whose product is homologous to eukaryotic heme oxygenases and is required for acquisition of iron from heme and hemoglobin. J Bacteriol 179(3):838–845

Schneewind O, Mihaylova-Petkov D, Model P (1993) Cell wall sorting signals in surface proteins of gram-positive bacteria. EMBO J 12(12):4803–4811

Schubert A, Zakikhany K, Schreiner M, Frank R, Spellerberg B, Eikmanns BJ, Reinscheid DJ (2002) A fibrinogen receptor from group B *Streptococcus* interacts with fibrinogen by repetitive units with novel ligand binding sites. Mol Microbiol 46(2):557–569

Schuhegger R, Schoerner C, Dlugaiczyk J, Lichtenfeld I, Trouillier A, Zeller-Peronnet V, Busch U, Berger A, Kugler R, Hormansdorfer S, Sing A (2009) Pigs as source for toxigenic *Corynebacterium ulcerans*. Emerg Infect Dis 15(8):1314–1315

Scott JR, Zahner D (2006) Pili with strong attachments: Gram-positive bacteria do it differently. Mol Microbiol 62(2):320–330

Shortridge VD, Lazdunski A, Vasil ML (1992) Osmoprotectants and phosphate regulate expression of phospholipase C in *Pseudomonas aeruginosa*. Mol Microbiol 6(7):863–871

Sioud M (2005) Induction of inflammatory cytokines and interferon responses by double-stranded and single-stranded siRNAs is sequence-dependent and requires endosomal localization. J Mol Biol 348(5):1079–1090

Swierczynski A, Ton-That H (2006) Type III pilus of corynebacteria: Pilus length is determined by the level of its major pilin subunit. J Bacteriol 188(17):6318–6325

Sykes JE, Mapes S, Lindsay LL, Samitz E, Byrne BA (2010) *Corynebacterium ulcerans* bronchopneumonia in a dog. J Vet Intern Med 24(4):973–976

Taylor DJ, Efstratiou A, Reilly WJ (2002) Diphtheria toxin production by *Corynebacterium ulcerans* from cats. Vet Rec 150(11):355

Tiley SM, Kociuba KR, Heron LG, Munro R (1993) Infective endocarditis due to nontoxigenic *Corynebacterium diphtheriae*: report of seven cases and review. Clin Infect Dis 16(2):271–275

Ton-That H, Schneewind O (2003) Assembly of pili on the surface of *Corynebacterium diphtheriae*. Mol Microbiol 50(4):1429–1438

Ton-That H, Marraffini LA, Schneewind O (2004) Sortases and pilin elements involved in pilus assembly of *Corynebacterium diphtheriae*. Mol Microbiol 53(1):251–261

Trost E, Ott L, Schneider J, Schröder J, Jaenicke S, Goesmann A, Husemann P, Stoye J, Dorella FA, Rocha FS, Soares Sde C, D'Afonseca V, Miyoshi A, Ruiz J, Silva A, Azevedo V, Burkovski A, Guiso N, Join-Lambert OF, Kayal S, Tauch A (2010) The complete genome sequence of *Corynebacterium pseudotuberculosis* FRC41 isolated from a 12-year-old girl with necrotizing lymphadenitis reveals insights into gene-regulatory networks contributing to virulence. BMC Genomics 11:728

Trost E, Al-Dilaimi A, Papavasiliou P, Schneider J, Viehoever P, Burkovski A, Soares SC, Almeida SS, Dorella FA, Miyoshi A, Azevedo V, Schneider MP, Silva A, Santos CS, Santos LS, Sabbadini P, Dias AA, Hirata R Jr, Mattos-Guaraldi AL, Tauch A (2011) Comparative analysis of two complete *Corynebacterium ulcerans* genomes and detection of candidate virulence factors. BMC Genomics 12(1):383

Trost E, Blom J, Soares Sde C, Huang IH, Al-Dilaimi A, Schroder J, Jaenicke S, Dorella FA, Rocha FS, Miyoshi A, Azevedo V, Schneider MP, Silva A, Camello TC, Sabbadini PS, Santos CS, Santos LS, Hirata R Jr, Mattos-Guaraldi AL, Efstratiou A, Schmitt MP, Ton-That H, Tauch A (2012) Pangenomic study of *Corynebacterium diphtheriae* that provides insights into the genomic diversity of pathogenic isolates from cases of classical diphtheria, endocarditis, and pneumonia. J Bacteriol 194(12):3199–3215

Tsuge Y, Ogino H, Teramoto H, Inui M, Yukawa H (2008) Deletion of *cgR_1596* and *cgR_2070*, encoding NlpC/P60 proteins, causes a defect in cell separation in *Corynebacterium glutamicum* R. J Bacteriol 190(24):8204–8214

Ventura M, Canchaya C, Tauch A, Chandra G, Fitzgerald GF, Chater KF, van Sinderen D (2007) Genomics of Actinobacteria: tracing the evolutionary history of an ancient phylum. Microbiol Mol Biol Rev 71(3):495–548

Vimr ER (1992) Microbial sialidases: does bigger always mean better? Trends Microbiol 2:271–277

Vimr ER, Kalivoda KA, Deszo EL, Steenbergen SM (2004) Diversity of microbial sialic acid metabolism. Microbiol Mol Biol Rev 68(1):132–153

von Hunolstein C , Scopetti F, Efstratiou A, Engler K (2002) Penicillin tolerance amongst non-toxigenic *Corynebacterium diphtheriae* isolated from cases of pharyngitis. J Antimicrob Chemother 50(1):125–128

Wagner KS, White JM, Crowcroft NS, De Martin S, Mann G, Efstratiou A (2010) Diphtheria in the United Kingdom, 1986–2008: the increasing role of *Corynebacterium ulcerans*. Epidemiol Infect 138(11):1519–1530

Walker J, Jackson HJ, Eggleton DG, Meeusen EN, Wilson MJ, Brandon MR (1994) Identification of a novel antigen from *Corynebacterium pseudotuberculosis* that protects sheep against caseous lymphadenitis. Infect Immun 62(6):2562–2567

Warren L, Spearing CW (1963) Sialidase (Neuraminidase) of *Corynebacterium diphtheriae*. J Bacteriol 86:950–955

Wennerhold J, Krug A, Bott M (2005) The AraC-type regulator RipA represses aconitase and other iron proteins from *Corynebacterium* under iron limitation and is itself repressed by DtxR. J Biol Chem 280(49):40500–40508

Wilks KE, Dunn KL, Farrant JL, Reddin KM, Gorringe AR, Langford PR, Kroll JS (1998) Periplasmic superoxide dismutase in meningococcal pathogenicity. Infect Immun 66(1):213–217

Wilson MJ, Brandon MR, Walker J (1995) Molecular and biochemical characterization of a protective 40-kilodalton antigen from *Corynebacterium pseudotuberculosis*. Infect Immun 63(1):206–211

Zahrt TC, Deretic V (2002) Reactive nitrogen and oxygen intermediates and bacterial defenses: unusual adaptations in *Mycobacterium tuberculosis*. Antioxid Redox Signal 4(1):141–159

Zhi XY, Li WJ, Stackebrandt E (2009) An update of the structure and 16S rRNA gene sequence-based definition of higher ranks of the class Actinobacteria, with the proposal of two new suborders and four new families and emended descriptions of the existing higher taxa. Int J Syst Evol Microbiol 59(Pt 3):589–608

Chapter 9
Detection Methods for Laboratory Diagnosis of Diphtheria

Anja Berger, Michael Hogardt, Regina Konrad and Andreas Sing

Abstract Although diphtheria has to be diagnosed primarily on clinical symptoms, the rapid and reliable detection and identification of the potentially toxigenic *Corynebacterium* species, *C. diphtheriae, C. ulcerans* and *C. pseudotuberculosis*, is essential for the definite diagnosis and management of diphtheria with respect to both the individual patient and the public health measures to be undertaken. Laboratory confirmation of suspected diphtheria has to aim for the isolation of the etiologic pathogen (including species identification and antibiotic susceptibility testing) as well as for differentiation of toxigenic from non-toxigenic strains (by using *tox* gene detection and toxigenicity testing). The recent introduction of the Matrix-assisted laser desorption/ionization time-of-flight mass spectrometry (MALDI-TOF MS) into the microbiological algorithm of laboratory diagnosis of diphtheria allows the specific and escalating identification of the three potentially toxigenic *Corynebacterium* species from growing colonies. Finally, molecular typing techniques may be applied to explore the clonal relatedness of clinical isolates and their potential routes of transmission. The most relevant laboratory procedures fulfilling these requirements (microbiological culture, conventional biochemical tests, molecular methods for species identification and toxigenicity testing) will be presented here.

Keywords Immunochromatographic strip test (ICS) · Elek Ouchterlony test · Cytotoxicity testing · Matrix-assisted-laser-desorption/ionization-time-of-flight-mass-spectrometry (MALDI-TOF) · Non-toxigenic *tox*-gene bearing strains (NTTB)

A. Sing (✉) · A. Berger · M. Hogardt · R. Konrad
Bayerisches Landesamt für Gesundheit und Lebensmittelsicherheit (LGL),
Veterinärstraße 2, 85764, Oberschleißheim, Germany
e-mail: Andreas.sing@lgl.bayern.de

A. Berger
e-mail: anja.berger@lgl.bayern.de

M. Hogardt
e-mail: michael.hogardt@lgl.bayern.de

R. Konrad
e-mail: regina.konrad@lgl.bayern.de

A. Burkovski (ed.), *Corynebacterium diphtheriae and Related Toxigenic Species*,
DOI 10.1007/978-94-007-7624-1_9,

9.1 Introduction

Diphtheria is a potentially fatal disease caused by toxigenic strains of the three *Corynebacterium* spp. *Corynebacterium diphtheriae, Corynebacterium ulcerans* and *Corynebacterium pseudotuberculosis* harbouring lysogenic beta-corynephages bearing the *tox* gene. Laboratory confirmation of suspected diphtheria has to aim for (i) isolation of the respective pathogen—for species identification, antibiotic susceptibility testing and where required strain typing purposes[1]—and (ii) for toxigenicity testing, since diphtheria toxin is responsible for the characteristic systemic symptoms of diphtheria.

In many cases of classical respiratory diphtheria, medical and public health actions are mainly based on clinical diagnosis prompting immediate activity, i.e. isolation of the index patient, antitoxin administration, sampling of diagnostic material for microbiological diagnosis, supportive antibiotic treatment and contact or source tracing. Therefore, in unambiguous cases clinical diagnosis usually precedes microbiological confirmation thus allowing straightforward management of the patient. However, rapid diphtheria diagnosis might be hampered for several reasons. For instance, in low incidence countries clinical awareness might be low and proper diagnosis delayed. Moreover, in patients presenting with milder symptoms, e.g. pharyngitis without "pathognonomic" pseudomembranes or with cutaneous diphtheria, establishment of the definite diagnosis might only be achieved by the detection of a toxigenic *Corynebacterium* species.

Basically, the main role of the laboratory is to provide simple, rapid and reliable methods to assist clinicians in confirming a clinical diagnosis (Efstratiou and George 1996). In these cases, microbiological diagnosis is complementary to the clinical diagnosis and basically supports the clinical decision making process by proving the suspected diagnosis *ex post*. However, the rapid identification of the three potentially toxigenic *Corynebacterium* species is not only essential for diagnosis and treatment of diphtheria and diphtheria-like diseases with respect to the single patient, but is also important for public health reasons, since a suspected diphtheria case prompts a variety of public health actions including contact tracing and WHO notification. Notably, due to the rarity of diphtheria in most industrialized countries experience in microbiological diagnosis of diphtheria has diminished even in reference laboratories as international external quality assays organized by the European Diphtheria Surveillance Network and the ECDC have shown (Engler et al. 2001a; Neal and Efstratiou 2009; European Centre for Disease Prevention and Control 2011); capability of toxigenicity testing is mostly restricted to reference laboratories, either due to the shortage of expertise or lack of required reagents, e.g.—with respect to the Elek test—of antitoxin (Wagner et al. 2009).

[1] Typing methods are beyond the scope of this article and will therefore only briefly be addressed here.

9.2 Microbiology

Corynebacteria are partially acid-fast Gram-positive rods. *C. diphtheriae*, the classical diphtheria agent, and the two zoonotic species *C. ulcerans* and *C. pseudotuberculosis* are facultative anaerobic, non-motile, non-sporulating, unencapsulated, pleomorphic bacteria.

9.3 Laboratory Safety Issues

C. diphtheriae, *C. ulcerans* and *C. pseudotuberculosis* are Biosafety level (BSL)-2 organisms. The infective dose for causing human *C. diphtheriae* colonization and/or infection is unknown. The tenacity of corynebacteria in general is relatively high; *C. diphtheriae* can survive in a dry, light-protected environment for several months.

Laboratory-acquired infections have been reported in the last 20 years at least three times in the UK (Wagner et al. 2010) and once in Germany (Thilo et al. 1997), some of them due to handling of toxigenic *C. diphtheriae* reference strains while performing an Elek test. In contrast to other bacterial aerosol-transmitted diseases, however, laboratory diphtheria infections seem to be considerably less frequent even in an era of markedly higher diphtheria prevalence than today: between 1930 and 1950, 15 diphtheria cases among laboratory staff were observed in the United States, while 224 cases of brucellosis, 153 of tuberculosis or 65 of tularaemia were reported, respectively (Pike et al. 1965). Also, in a review of published and unpublished cases of laboratory infections until 1974, laboratory-acquired diphtheria only ranked as fourth of primarily aerosol-transmitted laboratory infections worldwide (24 cases in the United States, 9 in other countries, with a case fatality rate of 0) (Pike 1976).

To minimise the risk of laboratory-acquired infection, all cultures suspected to contain potentially toxigenic corynebacteria (including reference strains) should be handled in a safety cabinet. Aerosol production should be avoided to reduce the risk of droplet transmission. Laboratory workers working with potentially toxigenic *Corynebacterium* spp. should have adequate immune protection with documented vaccination status (basic immunisation plus recommended booster doses) according to the respective national immunization recommendations. Laboratory staff possibly exposed to diphtheria in the course of work should have a minimum protective level of 0.01 International Units per mL (IU/mL). Laboratory workers regularly handling toxigenic strains are recommended to maintain a higher level of 0.1 IU/mL (Efstratiou and George 1999).

9.4 Collection and Transport of Specimens for Laboratory Diagnosis

Since microbiological culture is essential for confirming diphtheria, a clinical specimen should be taken as soon as possible when any form of diphtheria is suspected, ideally before starting antibiotic treatment. Suitable specimens are swabs from the nose, the pharynx, from the pseudomembrane, if present, or from wounds. Visible pseudomembranes should be removed under laryngoscopic control (cave: risk of blocking the respiratory tract). Both sections of the pseudomembrane and swabs taken from the area under the pseudomembrane where bacteria are concentrated should be obtained. Multiple site sampling should be considered in any suspected diphtheria case to increase the recovery rate. Especially in patients with known or suspected cutaneous diphtheria swabs should be obtained not only from the skin ulcer, but also from the nasopharynx to detect a possible simultaneous colonisation of the respiratory tract. Nasopharyngeal swabs should also be taken from close contacts of an index patient.

In two outbreak situations in Manchester (Butterworth et al. 1974) and London (Taylor et al. 1962) the recovery rate from throat swabs was considerably higher than from nose swabs (82 vs. 36 % of 47 and 88 vs. 36 % of 140 carriers, respectively). Since a significant percentage of carriers in both studies, however, was positive only in one specimen (throat only 64 % in both studies, nose only 12–18 %, both specimens positive 18–24 %), swabbing of both anatomical regions is warranted. Similarly, a higher PCR-recovery rate from throat than from nose swabs was reported in a study from the Republic of Georgia in the 1990s (Kobaidze et al. 2000).

Notably, colonisation or infection with both toxigenic and non-toxigenic *C. diphtheriae* strains of the same biotype in a single individuum have been described for at least seven persons (Chang et al. 1978; Mouton 1960; Simmons et al. 1980). Therefore, both clinicians and microbiologists should be aware that differentiation and characterisation of a single bacterial colony might not be sufficient for obtaining a definite laboratory-assisted diagnosis of diphtheria in a clinically suspicious patient. Therefore, pooling of several colonies for toxigenicity testing and use of a single bacterial colony for further characterisation was proposed (Jellard 1980).

After specimen collection, the swabs should be sent to the microbiological laboratory quickly, as rapid inoculation on special culture media is important. Transport at room temperature is sufficient. If an immediate transport and inoculation of plates is not possible, a transport medium (e.g. Amies- or a silica-gel medium) should be used (Efstratiou and Maple 1994; Funke and Bernard 2007). Specimens must be assigned to UN 3373 "Biological substance, Category B" and transported in compliance with 49 CFR, Part 173.199 or IATA Packing Instruction 650.

It is important that the clinician informs the laboratory of a diphtheria suspicion, since special diagnostic procedures will be needed (e.g. selective media, toxigenicity testing).

9.5 Microscopic Appearance

Corynebacteria including the three potentially toxigenic species are pleomorphic, slightly curved Gram-positive rods with tapered or clubbed ends. They often stain weakly and unevenly giving them a beaded appearance. Coccobacillary forms are often seen in older cultures. The bacteria may occur as single cells or in pairs, often in a "V"-form or in other angular arrangements resembling palisades or Chinese letters. Although a specific Gram stain morphology for each *C. diphtheriae* biotype has been described (e.g. *mitis* with long, pleomorphic, rigid club-shaped rods; *intermedius* with highly pleomorphic rods, very variable in size; *gravis* with usually short, coccoid or pyriform rods) (Noble and Dixon 1990), Gram stain-based species differentiation is obviously not reliable due to morphological heterogenicities and the dependence on the culture medium on which the bacteria were grown (Funke et al. 1997a). Upon Loeffler methylene blue staining reddish purple metachromatic granules or bars might be visible (Synder 2004).

In conclusion, the diagnosis of diphtheria based upon direct microscopy of a clinical specimen is unreliable. Both in respiratory and skin specimens various non-pathogenic corynebacteria or other Gram-positive coryne-like bacteria can be present as part of the respective resident flora. Especially *C. pseudodiphtheriticum* is able to present a false positive Loeffler stain.

9.6 Primary Cultivation

Diagnostic material should be inoculated as soon as possible both on standard blood agar (e.g. Columbia agar) and on a suitable selective tellurite-containing agar (e.g. Hoyle's agar, Clauberg II agar or freshly prepared Tinsdale agar) in parallel (Efstratiou and Maple 1994). Any infusion blood agar base supplemented with potassium tellurite and blood produce satisfying results; the classical morphological features of colonies used for presumptive corynebacterial species or *C. diphtheriae* biovar identification are best obtained when using media containing sheep rather than horse blood (Efstratiou and George 1996). The tellurite-containing media are selective for tellurite-reductase positive bacteria (e.g. the three potentially toxigenic *Corynebacterium* species, but also *C. jeikeium* and *C. striatum*) which grow as brownish-black colonies due to tellurite reduction.

The non-selective blood agar culture helps to identify also tellurite-sensitive *C. diphtheriae* strains and other clinically relevant pathogens (e.g. ß-haemolytic streptococci). Since corynebacteria are usually resistant to fosfomycin, either fosfomycin disks or supplementation of blood agar with fosfomycin might be used to select for *C. diphtheriae* by simultaneous growth inhibition of *viridans* group streptococci and *Neisseria* species which are found 1 to 2 logs more often in throat cultures than corynebacteria. Using fosfomycin disks Wirsing von Koenig et al. increased the isolation rate of corynebacteria from throat cultures by a factor of 5; all 150 toxigenic *C. diphtheriae* strains tested fosfomycin-resistant (Wirsing von Koenig et al. 1988) as were 18 different *Corynebacterium* spp. reference strains including *C. diphtheriae* in a study using fosfomycin-supplemented blood agar (von Graevenitz et al. 1998).

Tinsdale agar (Tinsdale 1947) or other cystine tellurite blood agar preparations additionally exploit the cysteinase-activity of the three potentially toxigenic *Corynebacterium* species which appear as black colonies (due to tellurite reduction) surrounded by a brownish halo (due to their cystinase activity). The halo is caused by the production of H_2S from cystine, interacting with the tellurite salt. Since corynebacteria other than *C. diphtheriae*, *C. ulcerans* and *C. pseudotuberculosis* do not show cystinase activity, cystine-containing media may serve as screening and/or confirmation tool for suspicious colonies grown on non-selective blood agar or selective tellurite-containing media (Efstratiou and George 1999). However, due to their relatively short stability and the lower detection rates obtained when using them these highly selective media are not recommended for primary cultivation purposes (Jellard 1971; Efstratiou and Maple 1994). Moreover, toxigenic *C. diphtheriae* colonies subcultured from modified Tinsdale medium were found to be inhibited in Elek test activity when compared to colonies isolated from Hoyle's medium (Jellard 1971). Additionally, at least in one study halo production by *C. diphtheriae* biovar *mitis* and *intermedius* was found to be inconsistent on Tinsdale medium (i.e. nearly two-thirds of the tested strains were halo-non producers), while a modified cystine-serum-tellurite agar correctly identified all tested *C. diphtheriae* biovar *gravis*, *mitis* and *intermedius* isolates (Segal et al. 1973).

Typical colony morphology features for the three potentially toxigenic *Corynebacterium* spp. seen on different agars are summarized in Table 9.1. It should be noted, however, that variations in colony morphology occur (Coyle et al. 1993); therefore, species or biotype identification cannot be based on colony morphology alone. Interestingly, the morphology of *C. diphtheriae* strains seems to be stable even after extensive subcultivation and also throughout an epidemic, although colony types could not been linked to distinct genotypes (Krech 1994).

All media are usually incubated at 36 +/− 1 °C for 18–48 h ideally in 5–10 % CO_2 atmosphere if not indicated otherwise by the manufacturer.

For screening for *C. ulcerans* strains, a charcoal-tellurite blood agar containing activated charcoal (0.05 %), potassium-tellurite (0.03 %) and sheep blood (10 %) has been proposed, since *C. ulcerans* growth on Tinsdale agar might be hindered by a higher potassium-tellurite concentration (Katsukawa et al. 2009); selective isolation of *C. ulcerans* was much better using this novel agar when compared to blood agar (Katsukawa et al. 2012).

Table 9.1 Typical colony morphology features for the three potentially toxigenic *Corynebacterium* spp. on different agars. (adapted from Health Protection Agency 2008; Efstratiou and George 1996; Dorella et al. 2006; Riegel et al. 1995)

Species and/or biovar	Blood agar	Hoyle's tellurite agar	Tinsdale agar
C. diphtheriae biovar *gravis*	Non-hemolytic	Dull, grey/black, opaque, 1.5–2 mm in diameter, matt surface, friable, moveable or tending to break into small segments when touched with a loop	Black with brownish-black halo
C. diphtheriae biovar *mitis*	Small zone of β-hemolysis	Grey/black, opaque, 1.5–2 mm in diameter, entire edge and glossy smooth surface, variable in size	Black with brownish-black halo
C. diphtheriae biovar *intermedius*	Small zone of β-hemolysis	Grey/black, opaque, small, 0.5-1 mm in diameter, shiny surface, discrete, translucent, partly with black center	Black with brownish-black halo
C. diphtheriae biovar *belfanti*	Small zone of β-hemolysis	Grey/black, opaque, 1.5–2 mm in diameter, entire edge and glossy smooth surface, variable in size	Black with brownish-black halo
C. ulcerans	Small zone of β-hemolysis, grey/white, dry, waxy consistency, circular, slightly convex with an entire margin	Grey/black, opaque, very dry	Black with brownish-black halo
C. pseudotuberculosis	Small zone of β-hemolysis, cream to orange coloured, concentrally ringed	Grey/black, opaque, very dry	Black with brownish-black halo
C. striatum	Non-hemolytic, white, moist, smooth	Grey/black	Black without halo
C. jeikeium	Non-hemolytic, grey/white, low convex	Grey/black	Black without halo

9.7 Species Identification of Potentially Toxigenic *Corynebacterium* spp.

9.7.1 Presumptive Identification of Potentially Toxigenic Corynebacterium *spp.*

Colonies suggestive to represent corynebacteria (e.g. due to microscopic staining features, colony morphology or fosfomycin-resistance) might initially be screened

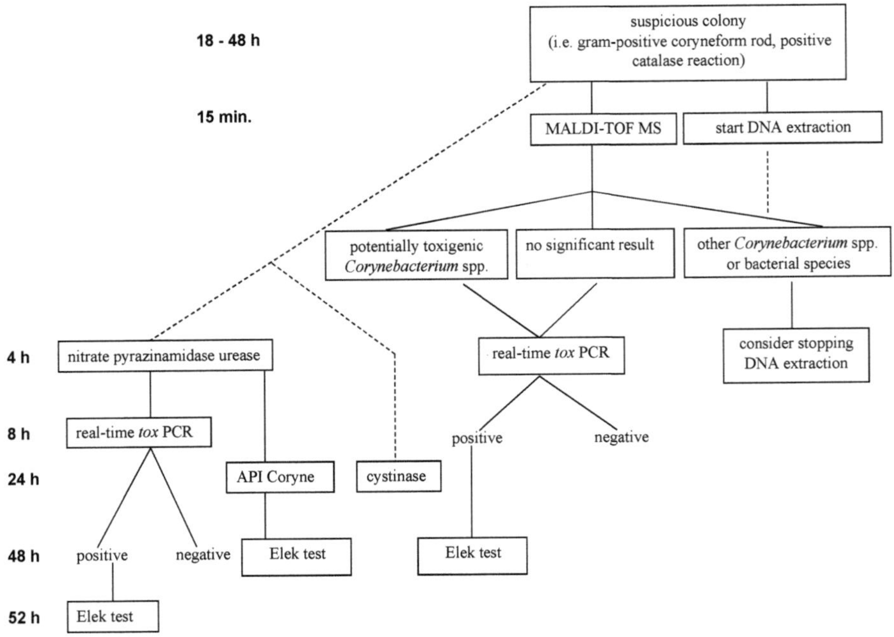

Fig. 9.1 Algorithm for rapid identification of toxigenic *Corynebacterium* species

as part of a step-wise diagnostic work-flow prior to further biochemical differentiation (Efstratiou and Maple 1994; see also Fig. 9.1). As minimal requirement for distinguishing potentially toxigenic *Corynebacterium* spp. from other bacterial genera often encountered in throat or wound specimens or from other coryneform bacteria, the following screening tests have been recommended, especially for resource-poor laboratories (Colman et al. 1992; Efstratiou and Maple 1994; Efstratiou and George 1999): catalase production (positive in all *Corynebacterium* spp.), pyrazinamidase production (negative in *C. diphtheriae*, *C. ulcerans* and *C. pseudotuberculosis*, but positive in other coryneform bacteria most often encountered in throat or wound specimens, e.g. *C. amycolatum*, *C. striatum*, *C. imitans*, but variable in *C. pseudodiphtheriticum*), cystinase production (positive only in the three potentially toxigenic *Corynebacterium* spp., but not in other corynebacteria), and motility (corynebacteria are non-motile) (von Graevenitz and Funke 1996; Efstratiou and George 1999). Procedures for a presumptive identification of *C. diphtheriae* (and to some degree of its biovars), *C. ulcerans* and *C. pseudotuberculosis* additionally should include screening for urease hydrolysis (negative for *C. diphtheriae*, positive for *C. ulcerans* and *C. pseudotuberculosis*) and nitrate reduction (positive for *C. diphtheriae* except the biovar *belfanti*, negative for *C. ulcerans* and *C. pseudotuberculosis*).

Based on cystinase-positivity and pyrazinamidase-negativity screening alone, Colman et al. were able to correctly identify all 61 potentially toxigenic *Corynebacterium* spp. (56 *C. diphtheriae,* 2 *C. ulcerans,* 3 *C. pseudotuberculosis*) strains of the English and Welsh Diphtheria Reference Laboratory strain collection tested (Colman et al. 1992); only 2 out of 3 *C. pseudodiphtheriticum* strains yielded a negative pyrazinamidase test result as did the three potentially toxigenic *Corynebacterium* spp. strains (in contrast to them, however, they were cystinase-negative). Including the 38 strains of non-toxigenic *Corynebacterium* strains of at least 7 different species, this study yielded a sensitivity and specificity for the cystinase test both of 100 % and for the pyrazinamidase test of 96.8 % and of 94.7 %, respectively.

As alternative screening test for differentiating *C. diphtheriae* from *Corynebacterium*-like colonies the DNAse test was proposed (Pimenta et al. 2008b). After incubation of colonies inoculated on DNAse-test media for 24 to 48 h, all 91 tested *C. diphtheriae* isolates (toxigenic and non-toxigenic) showed DNAse activity, while 93.9 % of 564 nondiphtherial *Corynebacterium* strains (including 3 *C. pseudotuberculosis* isolates) yielded negative results. DNAse positive results were mainly due to *C. minutissimum* (about 5 % of isolates) and *C. glucuronolyticum* (about 23 % of strains); the only *C. ulcerans* isolate tested was DNAse positive (Pimenta et al. 2008b). The authors suggest the DNAse test as an additional alternative screening procedure supplementing the pyrazinamidase test, especially with respect to increasing numbers of pyrazinamidase-negative species, namely *C. pseudodiphtheriticum*, which may be encountered more frequently in certain geographic regions. However, due to the long time needed until a test result is obtained, this method might be useful mainly in low resource settings where biochemical differentiation kits are not easily available.

In laboratories with access to Matrix-Assisted-Laser-Desorption/Ionization-Time-Of-Flight-Mass-Spectrometry (MALDI-TOF), all screening procedures mentioned above might turn out to be obsolete and will probably not be used in the near future.

9.7.2 Basic Species Identification Methods

Besides the screening assays mentioned above (tests for catalase, urease, pyrazinamidase, cystinase and nitrate reduction), the following basic microbiological techniques are required for the presumptive species identification of the most relevant humanpathogenic *Corynebacterium* spp. encountered in respiratory or skin and soft tissue samples, especially in resource-poor settings without the availability of commercial biochemical identification kits or MALDI-TOF: fermentation tests for glucose, maltose, sucrose, glycogen (Efstratiou and George 1999; Thompson et al. 1983), xylose and mannitol, esculin hydrolysis, fermentative or oxidative metabolism and the CAMP reaction with a β-hemolysin-producing strain of *Staphylococcus aureus* (ATCC 25923) (Funke et al. 1997a). Using nitrate reduction, urease activity, glucose, maltose, sucrose and glycogen fermentation alone, Thompson et al. were able to correctly identify all *C. diphtheriae* (86) and *C. ulcerans* (7) isolates from two Canadian Diphtheria reference strain collections within 1 to 4 h, while 2 of the 4 tested *C. pseudotuberculosis* strains yielded false-negative results both in the nitrate and the urease assay (Thompson et al. 1983). Table 9.2 summarizes the most important basic biochemical identification tests for clinically relevant coryne-bacteria.

9.7.3 Biochemical Differentiation Methods

Several biochemical differentiation systems for corynebacteria are commercially available (e.g. API Coryne, Vitek 2, BioMérieux; Phoenix and BBL Crystal, BD Diagnostics; RapID CB, Remel; Biolog). Their reliability depends not only on the selected identification reactions and the laboratory test procedure performance done by the technicians, but also on the quality and quantity of the strains evaluated for and included in the database used for species identification. In general, they seem to be very reliable in identifying *C. diphtheriae* (but not necessarily for biovar identification). The biochemical differentiation between *C. ulcerans* and *C. pseudotuberculosis*, however, might be difficult and therefore additional tests could be needed.

For historical reasons, the API Coryne system is the best studied biochemical differentiation scheme. The API Coryne kit is a microcupule system consisting of 20 different tests (11 enzymatic reactions, 8 carbohydrate fermentation tests and a fermentation control; catalase reaction is additionally tested by adding 3% H_2O_2) comprising 33 taxa (Janda 1999). Freney et al. found that 99.0% of 198 *Corynebacterium* spp. isolates of 19 different species were correctly identified to the species level, with about a third of the total strains requiring additional tests to arrive at a final identification (Freney et al. 1991). However, only 11 strains of potentially toxigenic *Corynebacterium* spp. were included (7 *C. diphtheriae*; 3 *C. ulcerans* and 1 *C. pseudotuberculosis*). All of these were correctly identified to the species level without further extra tests needed; however, *C. diphtheriae* biovar identification was not correct in one strain prompting the study authors to express concerns regarding the ability of this early API Coryne 1.0 version for biovar determination.

In another evaluation of the API Coryne system yielding an overall correct identification of 93.2% for 104 *Corynebacterium* spp. strains tested (with only about 7% needing additional biochemical testing), Gavin et al. correctly identified all 18 *C. diphtheriae* strains of their collection; however, *C. diphtheriae* biovar *intermedius* strains were misidentified as biovar *mitis*. For the correct identification of two out of five *C. pseudotuberculosis* strains and the only *C. ulcerans* strain in their study a trehalose test was needed (Gavin et al. 1992). Similarly, in a Spanish study (overall correct identification for 95 *Corynebacterium* spp. isolates: 93.7%) all tested *C. diphtheriae* (2), *C. ulcerans* (6) and *C. pseudotuberculosis* (4) strains were correctly identified using the API Coryne system (Soto et al. 1994).

The updated API Coryne 2.0 system covering the same 20 biochemical reactions as its precursor version, but including both more taxa (54 instead of previously 33) and additional differential tests in the database was evaluated in two studies. In a large multi-center study yielding an overall correct identification of 93.8% for 242 *Corynebacterium* spp. strains (with 58.2% of strains needing additional

Table 9.2 Basic biochemical identification of clinically significant corynebacteria. (adapted from Funke et al. 1997a; Efstratiou and George 1999)

Species	CYS	PYZ	Nitrate	Urease	Glucose	Maltose	Sucrose	Mannitol	Xylose	Glycogen	CAMP	Metabolism	Additional reactions
C. diphtheriae biotype *gravis*	+	–	+	–	+	+	–	–	–	+	–	F	
C. diphtheriae biotype *mitis*	+	–	+	–	+	+	–	–	–	–	–	F	
C. diphtheriae biotype *intermedius*	+	–	+	–	+	+	–	–	–	–	–	F	
C. diphtheriae biotype *belfanti*	+	–	–	–	+	+	–	–	–	–	–	F	
C. ulcerans	+	–	–	+	+	+	–	–	–	+	REV	F	
C. pseudo-tuberculosis	+	–	–	+	+	+	v	–	–	–	REV	F	
C. amycolatum	–	+	v	v	+	v	v	–	–	–	–	F	
C. imitans	–	(+)	–	–	+	+	(+)	–	–	–	+	F	Tyrosin -; O/129 resistant
C. pseudo-diphtheriticum	–	v	+	+	–	–	–	–	–	–	–	O	
C. striatum	–	+	+	–	+	–	v	–	–	–	v	F	Tyrosin +
C. argentoratense	–	+	–	–	+	–	–	–	–		–	F	Chymotryp-sin+
C. accolens	–	v	+	–	+	–	v	v	–		–	F	
C. propinquum	–	v	+	–	–	–	–	–	–		–	O	

(+) weak reaction, *v* variable reaction, *CYS* cystinase production, *PYZ* pyrazinamidase activity, *CAMP* test with β-hemolysin-producing strain of *Staphylococcus aureus*, *F* fermentation, *O* oxidation biochemical testing), Funke et al. correctly identified all tested *C. diphtheriae* (24), *C. ulcerans* (6) and *C. pseudotuberculosis* (3) strains (Funke et al. 1997b). With respect to biovar identification, all 6 *gravis* biovar strains were correctly identified, while the other 18 strains had to be analyzed by additional tests (i.e. examination for colony size to differentiate between the nonlipophilic biovars *mitis* and *belfanti* from the small-colony-forming biovar *intermedius*). The only other study evaluating the API Coryne 2.0 correctly identified 92.3 % of 78 *Corynebacterium* strains of 10 different species, but did not include potentially toxigenic *Corynebacterium* spp. (Almuzara et al. 2006).

In a comparison of API Coryne, *rpoB* sequencing and MALDI-TOF for 116 clinical isolates of 18 different *Corynebacterium* spp., API Coryne yielded concordant species-level results with the two other methods in only half of the tested species (9/18); however, all isolates representing the three potentially toxigenic *Corynebacterium* species (78 *C. diphtheriae*, 8 *C. ulcerans* and 4 *C. pseudotuberculosis*) were concordantly identified by all three methods (Konrad et al. 2010).

Another commercially available detection kit is the cuvette-based RapID CB Plus system which includes 18 single-test wells (14 enzyme and 4 carbohydrate tests). Only seven reactions (the four "basic standard" reactions nitrate reduction, urease, glucose and sucrose fermentation as well as alpha-glucosidase, N-acetyl-beta-glucosaminidase and ribose fermentation) are shared with the API Coryne system (Hudspeth et al. 1998; Janda 1999). The database comprises 57 taxa including 23 *Corynebacterium* spp. The test is performed within 4 h and therefore more rapid than the API Coryne system (24–48 h). As for the API Coryne system, a relative heavy inoculum size is needed for obtaining robust results. An evaluation of the RapID CB Plus system correctly identified 95 % of 42 *Corynebacterium* spp. strains tested including the ATCC reference strains of *C. diphtheriae* and *C. pseudotuberculosis* (Hudspeth et al. 1998). However, clinical isolates of the three potentially toxigenic *Corynebacterium* spp. were not analyzed in this study. In another evaluation of the RapID CB Plus system Funke et al. identified only 85.1 % of 215 *Corynebacterium* spp. strains to the species level (Funke et al. 1998), including all 3 *C. ulcerans* and 2 *C. pseudotuberculosis* strains tested. However, two out of 13 *C. diphtheriae* strains were only identified to the genus level which seems not to be acceptable for biochemical diagnosis of the most important diphtheria agent.

The Anaerobe and *Corynebacterium* identification card for Vitek 2 (bioMérieux) was compared with 16S rRNA gene sequencing in a multicenter study reaching an overall correct identification rate of 98 % in 51 clinical isolates of seven different *Corynebacterium* spp. (Rennie et al. 2008). All three *C. diphtheriae* (two isolates, however, with a low level of discrimination meaning that alternative species identification proposals were also given by the system) and two *C. ulcerans* isolates were correctly identified.

The accuracy of the BBL Crystal system has so far been evaluated only in a study comparing 3 phenotypic (API Coryne, RapID, BBL Crystal) and 3 genotypic (broad-range and coryneform-optimized 16S *rRNA* sequencing, *rpoB* sequencing) methods for species determination of coryneform bacteria (Adderson et al. 2008). When tested for 12 reference strains of 12 different species, BBL Crystal correctly identified 58.3% of strains to the species level (API Coryne: 75.0%, RapID CB Plus: 41.7%) including the only strain of a potentially toxigenic species, i.e. *C. pseudotuberculosis*, which was also correctly identified by API Coryne, but not by RapID CB Plus. With respect to 38 clinical isolates of different coryneform species, the highest concordance for species-level differentiation among the three biochemical test systems was found between API Coryne and BBL Crystal. In the absence of a "gold standard", species-level identification rates were 81.6, 71.1 and 55.3% for BBL Crystal, API Coryne and RapID CB Plus, respectively (Adderson et al. 2008). Interestingly, all three methods showed identical identification results (twice *C. striatum*, once *C. aquaticum*) for only three out of 38 clinical isolates suggesting relatively low identification rates for coryneform bacteria. However, in the context of laboratory-assisted diphtheria diagnosis, at least no *C. diphtheriae* and *C. ulcerans* isolates were included in this evaluation study.

In an evaluation of the Biolog system (release 3.50), a biochemical identification system based on 95 different carbon sources, about a third of 80 strains belonging to 12 different *Corynebacterium* species were correctly identified to the species level after 4 h and only less than half after 24 h; disturbingly, even after 24 h incubation up to a quarter of the respective isolates was not identified to the species level (*C. diphtheriae*: 1 out of 7; *C. pseudotuberculosis*: 1 out of 4) (Lindemann et al. 1995). Confusingly, all 3 tested *C. ulcerans* isolates were correctly identified to the species level after 4 h, but none after 24 h. Therefore, the 3.50 version cannot be regarded as a reliable tool for identifying potentially toxigenic *Corynebacterium* spp. Studies evaluating available updates of Biolog's database have not yet been published.

In conclusion, although commercially available biochemical differentiation kits are used for laboratory-assisted diphtheria diagnosis in routine microbiology laboratories throughout the world, most published evaluation studies have included only a small number of isolates of the three potentially toxigenic *Corynebacterium* spp. In general, however, most biochemical differentiation kits—and particularly API Coryne, which is the biochemical test most often used by 34 diphtheria reference laboratories of the International Diphtheria Surveillance Network (DIPNET) (Neal and Efstratiou 2009)- seem to be acceptable tools for identification of most of the clinically relevant coryneform species. In a steadily evolving taxonomic field as "corynebacteriology" with many new species described in the last few years, the underlying database is crucial for the correct identification of isolates, esp. with respect to *Corynebacterium* spp. other than the potentially toxigenic species. Regarding diphtheria-causing agents, problems may arise mainly in differentiating the four *C. diphtheriae* biotypes (esp. between *mitis, intermedius* and *belfanti*) and distinguishing *C. ulcerans* from *C. pseudotuberculosis*. The latter is due to the fact that many of the reactions included in the commercial systems are either present or absent in both species (or at least in a certain proportion of strains of the respective species) and thus cannot act as an exclusive marker for one of them. Therefore, additional biochemical tests and observation of colony morphology are needed to get a final biochemical diagnosis.

For instance, to differentiate *C. ulcerans* from *C. pseudotuberculosis*, glycogen (included in API Coryne, but not in RapID CB-Plus), starch and trehalose fermentation capacities of *C. ulcerans* can be used (Groman et al. 1984; Funke et al. 1997a). Moreover, in contrast to *C. pseudotuberculosis, C. ulcerans* was found to be susceptible to the vibriostatic agent O129, to be able to fermentate maltotriose and to show a positive *p*-nitrophenyl phosphate hydrolysis reaction (Groman et al. 1984).

Highly sophisticated methods (e.g. cell wall or cellular fatty acids analysis) are mainly used for chemotaxonomic investigations and not applicable in the routine microbiological laboratory (Funke et al. 1997a; Funke and Bernard 2007).

9.7.4 Molecular Species Differentiation Methods

To date, the use of molecular detection or differentiation methods directly from clinical human specimens has not been evaluated. Similarly, valid single gene-based nucleic acid amplification tests (NAAT) for species determination or differentiation of the 3 potentially toxigenic *Corynebacterium* species from either each other or from other coryneform bacteria have so far not been described. In contrast, sequence-based NAATs targeting either the 16S rRNA or *rpoB* genes for determination of *Corynebacterium* species including the 3 potentially toxigenic *Corynebacterium* species have been developed and evaluated.

An amplified rDNA-restriction analysis (ARDRA) using the enzymes *Alu*I, *Cfo*I and *Rsa*I which was found to correctly identify—among 26 other coryneform bacterial species—4 *C. diphtheriae* and 2 *C. pseudotuberculosis* strains within 8 h due to their unique restriction pattern (Vaneechoutte et al. 1995) is probably not applicable in a routine laboratory and might be obsolete due to the increasing availability of sequencing facilities.

An early commercially available 16S rDNA sequencing kit, the MicroSeq 500 Perkin-Elmer system, which is designed to sequence the first 527 bp of the 16S rRNA gene correctly identified about two third of 42 *Corynebacterium* isolates from 12 different species to the species level when compared to phenotypic methods including biochemical methods (conventional, API Coryne, Biolog) and cellular fatty acid profile analysis serving as "gold standard" (Tang et al. 2000). All four *C. diphtheriae* isolates tested were correctly identified, 1 of 7 *C. pseudodiphtheriticum* isolates was misidentified as *C. pseudotuberculosis* using the MicroSeq database available at that time.

While initial phylogenetic studies of the *Corynebacterium* genus and other coryneform genera were based on rRNA sequences (Ruimy et al. 1995; Pascual et al. 1995) and subsequently 16S rDNA sequencing was used as species differentiation tool for coryneform species, more recent data showed that *rpoB* gene sequencing is more discriminatory than 16S rDNA sequencing for differentiation between *Corynebacterium* species—both for type strains as well as for clinical isolates (Khamis et al. 2004, 2005). This higher discriminatory power is especially important when differentiating between the closely related species *C. ulcerans* and *C. pseudotuberculosis*. While 16S rRNA sequence differences between type strains of the three potentially toxigenic *Corynebacterium* species are less than 2 % (*C. diphtheriae* to both *C. ulcerans* and *C. pseudotuberculosis* 98.5 %) and the respective type strains of *C. ulcerans* and *C. pseudotuberculosis* are basically not distinguishable from each other by 16S rDNA sequencing (similarity 99.7 %), *rpoB* gene sequencing reliably allows differentiating *C. ulcerans* from *C. pseudotuberculosis* (similarity 93.6 % for the complete gene). *rpoB* sequence differences between *C. diphtheriae* and either *C. ulcerans* (86 %) or *C. pseudotuberculosis* (84.9 %) are clearly greater than 10 % (Khamis et al. 2004). Even partial sequencing of 434 to 452 bases yields similar sequence differences and can therefore be considered as a useful tool for differentiating the three potentially toxigenic *Corynebacterium* species. Similarly, when comparing 16S rRNA and partial *rpoB* gene sequencing on 168 corynebacterial isolates in a routine microbiological laboratory, the same authors found a significantly higher proportion of strains positively identified by *rpoB* than by 16S rDNA sequencing (91 vs. 81 %) (Khamis et al. 2005); the only isolate belonging to the three potentially toxigenic *Corynebacterium* species was correctly identified by both sequencing methods as *C. diphtheriae*.

In an evaluation of three phenotypic (API Coryne, RapID, BBL Crystal) and three genotypic (broad-range and coryneform-optimized 16S rDNA sequencing, *rpoB* gene sequencing) methods for species determination of coryneform bacteria, only *rpoB* gene sequencing was able to correctly identify the only reference strain of a potentially toxigenic *Corynebacterium* species (*C. pseudotuberculosis*) tested by genotypic methods (Adderson et al. 2008). When additionally evaluating these methods for 38 clinical isolates of different species, concordance between phenotypic and genotypic systems and between different methods among these two groups was quite poor, with only one out of 50 isolates showing the same result in all six methods tested in this study. In the context of laboratory-assisted diphtheria diagnosis, however, it should be noted that no clinical isolates of *C. diphtheriae* or *C. ulcerans* were tested.

MALDI-TOF MS is a novel technology for species identification based on the protein composition of microbial cells. Due to its most prominent advantages, i.e. speed of less than 15 min until obtaining an identification result from a single bacterial colony, low running costs and continuous expansion of a quality-controlled database of reference spectra currently comprising more than 3,200 medically relevant microorganisms (including over 150 mass spectral entries of *Corynebacterium* spp. (Alatoom et al. 2012)), MALDI-TOF MS might be useful not only for final species differentiation of clinically relevant coryneform bacteria, but also as a screening tool for coryneform bacterial colonies guiding the laboratory decision making process which colonies might be chosen for toxigenicity testing (Konrad et al. 2010). So far, two studies have evaluated MALDI-TOF MS for differentiation of *Corynebacterium* spp. In an evaluation study of MALDI-TOF MS as a method supporting laboratory-assisted diphtheria diagnosis, 115 out of 116 clinical *Corynebacterium* isolates (99.1 %) of 18 different coryneform species showed identical results in both MALDI-TOF MS and *rpoB* sequencing; the only strain not reaching a MALDI-TOF log(score) of > 2.0 needed for correct species identification according to the manufacturer's recommendations (Bruker Daltonics) was determined as *C. tuberculostearicum* by *rpoB* gene sequencing and—albeit with a lower log(score) of 1.8—also by MALDI-TOF (Konrad et al. 2010). All 90 isolates of potentially toxigenic *Corynebacterium* species (78 *C. diphtheriae*, 8 *C. ulcerans* and 4 *C. pseudotuberculosis*) were concordantly identified by both methods. A similarly high concordance between MALDI-TOF and *rpoB* sequencing was found in a study by Alatoom et al. focusing on non-diphtheriae *Corynebacterium* spp.: all 45 (100 %) clinical isolates of 13 different species (including one strain of *C. ulcerans* as single representative of a potentially toxigenic species) and 71 of 75 (94.7 %) clinical isolates of 15 different species were concurrently identified to the species level by both methods, when MALDI-TOF log scores of >2 and >1.7 are taken, respectively; the only inconclusive results were obtained with isolates identified by MALDI-TOF MS as *C. minutissimum* and by *rpoB* gene sequencing as *C. aurimucosum* (albeit only with a 95–96 % sequence similarity to the respective database entries) (Alatoom et al. 2012).

In conclusion, MALDI-TOF MS might not only be used as a species determination tool, but—due to its rapidity (MALDI-TOF MS is much faster than most other screening laboratory methods mentioned above) and robustness at least for potentially toxigenic *Corynebacterium* species- also as a screening method in the laboratory-assisted diphtheria diagnosis. A proposed algorithm for rapid identification of toxigenic *Corynebacterium* spp. is shown in Fig. 9.1; this approach might help to decide whether suspicious colonies should be analysed for toxigenicity.

For *C. pseudotuberculosis* diagnosis both from pure cultures and from clinical veterinary samples a multiplex PCR targeting *C. pseudotuberculosis*-specific 16S rRNA, *rpoB* and *pld* sequences was developed (Pacheco et al. 2007). The primer sequences for *pld* gene encoding phospholipase D, a major virulence factor exclusively found within the genus *Corynebacterium* both in *C. ulcerans* and *C. pseudotuberculosis* (Barksdale et al. 1981; Dorella et al. 2006), were chosen to be specific for both species (PLD-F and PLD-R1) or for *C. pseudotuberculosis* alone (PLD-F and PLD-R2) (see Table 9.3). *C. ulcerans* and *C. pseudotuberculosis pld* genes share 80 % DNA sequence identity (McNamara et al. 1995). All 41 clinical *C. pseudotuberculosis* isolates were correctly identified by the multiplex PCR, while the 4 *C. ulcerans* isolates tested negative in the respective *pld* PCR reaction. 53 out of 56 pus samples from animals with caseous lymphadenitis (CLA) were also multiplex PCR positive suggesting the usefulness for this assay in the work-up of CLA-afflicted animals. So far, this is the only study evaluating a molecular species differentiation method directly from clinical samples both in medical and veterinary microbiology. Similarly, Seto et al. investigated seven *C. ulcerans* strains from humans and animals (killer whales, lion) in Japan by multiple methods such as biotyping, sequencing of diphtheria-like toxin (DLT) and PLD, ribotyping and PFGE analysis and concluded that due to the PLD production and high level of similarity between *pld* genes among the isolates, PLD may be useful as a marker of *C. ulcerans* (Seto et al. 2008).

In contrast, an alternative 16S rDNA-based PCR method for CLA diagnosis in sheep and goats allows no differentiation between *C. ulcerans* and *C. pseudotuberculosis* (Cetinkaya et al. 2002).

Table 9.3 PCR primers for different *tox* and *dtxr* detection systems or systems for *C. pseudotuberculosis* species determination

Target	Name	Primer	binding position	fragment size (bp)	subunit	Organism	Source
tox	Tox 1 (rev)	ATCCACTTTTAGTGCGAGAACCTTC-GTCA	366–338[b]	249	A	*C. diphtheriae* and *C. ulcerans*	Pallen et al. 1994; Nakao Popovic 1997
	Tox 2 (fw)	GAAAACTTTTCTTCGTACCACGGGAC-TAA	118–146[b]				
tox	Dipht 6 F	ATACTTCCTGGTATCGGTAGC	991–1011[b]	297	B	*C. diphtheriae* and *C. ulcerans*	Nakao Popovic 1997
	Dipht 6 R	CGAATCTTCAACAGTGTTCCA	1287–1267[b]				
tox	DT1	CGGGGATGGTGCTTCGCG	456–473[b]	910	A + B	*C. diphtheriae* and *C. ulcerans*	Hauser et al. 1993
	DT2	CGCGATTGGAAGCGGGGT	1348–1365[b]				
tox	DTOX-1	GTTCGGTGATGGTGCTTCGC	453–472[b]	201	A	*C. diphtheriae*[a]	Lucchini et al. 1992
	DTOX-2	CGCCTGACACGATTTCCTGC	653–634[b]				
Tox	DT1	CGGGGATGGTGCTTCGCG	456–473[c]	1035	A + B	*C. ulcerans*	Sing et al. 2003
	1467R	CGGCAAAAGGTTGTAGCATC	1490–1471[c]				
Tox	DT1	CGGGGATGGTGCTTCGCG	456–473[c]	1150	A + B	*C. ulcerans*	Sing et al. 2003
	1586R	GTCTATGGAGCTCAACGGAG	1605–1586[c]				
Tox	Forward	GGCGTGGTCAAAGTGACGTA	310–329[b]	118	A	*C. diphtheriae*	Mothershed et al. 2002
	Reverse	CTTGCTCCATCAACGGTTCA	427–408[b]				
	Probe	CCAGGACTGACGAAGGTTCTCGCACT	331–356[b]				
Tox	Forward	CGCCCTAAATCTCCTGTTTATGTT	1489–1512[b]	129	B	*C. diphtheriae*	Mothershed et al. 2002
	Reverse	GTACCCAAGAACGCCTATGGAA	1617–1596[b]				
	Probe	TTCACAGAAGCAGCTCGGAGA-AAATTCATTC	1547–1577[b]				
tox	RTDT_fw	TTATCAAAAGGTTCGGTGATGGTG	443–460[b]	128	A	*C. diphtheriae* and *C. ulcerans*	Schuhegger et al. 2008b
	RTDT_rev2	AATCTCAAGTTCTACGCTTAAC	570–549[b]				
	RTDT_So	CGCGTGTAGTGCTCAGCCTTCCCT	470–493[b]				

Table 9.3 (continued)

Target	Name	Primer	binding position	fragment size (bp)	subunit	Organism	Source
tox	CD-toxF	GAAAACTTTTCTTCGTACCACGGGAC-TAA	118–146[b]	249	A	*C. diphtheriae* and *C. ulcerans*	Sing et al. 2011
	CD-toxR	ATCCACTTTTAGTGCGAGAACCTTC-GTCA	366–338[b]				
	CD-HP-3	AATAAATACGACGCTGCGGGATAC	247–270[b]				
	CD-HP-4	CTGTAGATAATGAAAACCCGCTC	272–294[b]				
dtxR	dtxR1F	GGGACTACAACGCAACAAGAA	51–71[d]	258		*C. diphtheriae*	Pimenta et al. 2008a
	dtxR1R	CAACGGTTTGGCTAACTGTA	308–289[d]				
16S rRNA	16S-F	ACCGCACTTTAGTGTGTGTG	150–169[e]	816		*C. pseudotuberculosis*	Cetinkaya et al. 2002; Pacheco et al. 2007
	16-S-R	TCTCTACGCCGATCTTGTAT	965–946[e]				
rpoB	C2700F	CGTATGAACATCGGCCAGGT	2712–2731[f]	446		*C. pseudotuberculosis*	Khamis et al. 2004; Pacheco et al. 2007
	C3130R	TCCATTTCGCCGAAGCGCTG	3157–3138[f]				
pld	PLD-F	ATAAGCGTAAGCAGGGAGCA	506–526[g]	203		*C. pseudotuberculosis*	Pacheco et al. 2007
	PLD-R2	ATCAGCGGTGATTGTCTTCCAGG	708–686[g]				
pld	PLD-F	ATAAGCGTAAGCAGGGAGCA	506–526[g]	203		*C. pseudotuberculosis* and *C. ulcerans*	Pacheco et al. 2007
	PLD-R1	ATCAGCGGTGATTGTCTTCC	708–689[g]				

[a] Tested only for *C. diphtheriae* isolates, but the primer should also amplify *C. ulcerans* and *C. pseudotuberculosis tox* genes as predicted by sequence analysis
[b] ab602357
[c] fj858272
[d] m80337
[e] gq118342
[f] ay492239
[g] l16586

9.8 Toxigenicity Testing

The most important test in the microbiological diagnosis of diphtheria is the detection of strains which produce active diphtheria toxin (DT). With increasing availability of nucleic acid amplification tests (NAAT) in many routine laboratories, two important features have to be borne in mind which were only detected by the application of molecular techniques: (i) the existence of *tox*-bearing *C. diphtheriae* (Efstratiou et al. 1998; Mokrousov 2009)—and also *C. ulcerans* (own unpublished data)—strains which are not able to produce active DT (so-called NTTB = **n**on-**t**oxigenic *tox*-gene **b**earing strains); and (ii) the heterogeneity of *tox* sequences both between different species (so far described for *C. diphtheriae* and *C. ulcerans* (Sing et al. 2003, 2005)) and between different strains of the same species (mainly within *C. ulcerans* (Schuhegger et al. 2008a)). Notably, for most of the non-NAAT-based toxigenicity assays the medium used for culturing bacteria to be tested might be of critical importance due to the dependence of DT production from iron; usually, Elek broth or comparable media are used.

9.8.1 In Vivo Toxigenicity Testing

In vivo DT detection is the traditional "gold standard" for toxigenicity testing, but will probably become completely obsolete due to its long duration of 2 to 5 days (depending on the test), its inconvenient dependence on laboratory animal facilities and the availability of non-*in vivo* alternatives (Efstratiou et al. 1998). Both the dermatonecrosis assay (Fraser and Weld 1926; Fraser 1937) and the subcutaneous test for virulence depend on the application of filtered culture supernatants either intradermally or subcutaneously into guinea pigs, respectively (Efstratiou et al. 1998). The latter is regarded as the most reliable test for detecting DT production by the WHO Manual, but not recommended for inexperienced laboratory personnel (Efstratiou and Maple 1994). Readout for a positive assay are either the presence of specific dermonecrotic lesions (after 48 h) or death including a *post mortem* pathological examination (after 2 to 5 days) in the animals, respectively, and the absence of these outcomes in negative control animals protected by diphtheria antitoxin (either by intravenous or intraperitoneal administration of antitoxin or by challenging animals in parallel with culture supernatants which were preincubated with antitoxin). Similar assays in rabbits are also described (Fraser 1931).

Reports on *in vivo* assays for DT determination in *C. ulcerans* are very scanty. Notably, for a true positive *in vivo* test a negative readout result is required in antitoxin-treated animals. Amino acid sequences, however, differ between *C. diphtheriae* and *C. ulcerans* DT variants; moreover, the protective effect of *C. diphtheriae*-derived antitoxin against *C. ulcerans* DT is not yet elucidated. Moreover, *C. ulcerans* strains seem to produce lesser DT amounts than *C. diphtheriae* at least *in vitro* (Sing et al. 2003; Seto et al. 2008). So far, only four *C. ulcerans* isolates with sufficient clinical information available have been tested for toxigenicity *in vivo*: only one isolate clearly showed DT-specific toxicity (as demonstrated by antitoxin protection in the control animal) (Meers 1979), one Elek-test positive isolate showed no DT-specific toxicity (Bostock et al. 1984), and the observed *in vivo* toxigenicity could only be partially neutralized in two *tox*-positive isolates (one of them Elek-and Vero cell test-positive, the other probably a true NTTB due to Elek- and Vero cell test-negativity (Komiya et al. 2010)). Additionally, 13 *C. ulcerans* isolates from several European strain collections tested positive for DT in both rabbit skin and guinea-pig inoculation assays, as did the only 2 so far *in vivo* toxigenicity tested *C. pseudotuberculosis* isolates—both obtained from Egyptian buffaloes—known so far in the literature (Maximescu et al. 1974).

9.8.2 In Vitro Cytotoxicity Testing

In vitro cytotoxicity assays allow the detection of active DT. Based on findings from the late 1950s that purified DT acts on a variety of mammalian cell lines (Lennox and Kaplan 1957; Strauss and Hendee 1959) several tissue culture methods for the detection of DT produced by live *C. diphtheriae* were developed subsequently to the first published proof of principle (Andre et al. 1960). In most assays, *C. diphtheriae* culture filtrates are added to cell monolayers in the presence or absence (as negative control to rule out unspecific cytotoxic effects) of antitoxin and incubated for mostly 24 h to 5 days. Directly visible or colorimetrically measurable (after addition of a cell staining dye) cytotoxic effects are usually the readout. By diluting the culture supernatants cytotoxicity titres can be measured allowing quantification of DT production by a given inoculum of the respective *C. diphtheriae* strain. Optimization strategies for diverse tissue culture systems mainly aimed at the cell types to be used, the broth media needed for growing *Corynebacterium* colonies (including serum supplementation) and the amount of antitoxin to be added (Schubert et al. 1968; Laird and Groman 1973; Murphy et al. 1978). Detection limits for DT range between 10 and 100 pg/ml, depending on the system used (Murphy et al. 1978; Hoy and Sesardic 1994; Efstratiou et al. 1998). In comparison to *in vivo* tests, the detection level for active DT is estimated to be about 10 to 100-fold higher in *in vitro* cytotoxicity assays (Lennox and Kaplan 1957). Cell types commonly used for cytotoxigenicity testing are Vero cells (African green monkey kidney cells; probably the cell line most often used in this setting (Neal and Efstratiou 2009)), human heteroploid HeLa cells, and to a lesser degree Chinese hamster ovary (CHO) cells and rabbit kidney cells. Notably, cytotoxicity assays lack standardization and have to be fine-tuned and evaluated in each individual laboratory. As with immunoprecipitation tests, the clarity and accuracy are critically dependent upon the quality of the bacterial growth medium, both the concentration and quality of the antitoxin, and the use of appropriate DT-negative control strains.

Optimized *in vitro* cytotoxicity systems show 100% concordance with *in vivo* assays using toxigenic and non-toxigenic *C. diphtheriae* strains (22 toxigenic, 28 non-toxigenic strains in Efstratiou et al. 1998; 3 toxigenic, 4 non-toxigenic strains in Murphy et al. 1978; 138 toxigenic, 191 non-toxigenic strains in Schubert et al. 1968; 32 toxigenic, 16 non-toxigenic strains in Laird and Groman 1973).

Besides, a colony overlay test has been described in which agar-grown colonies are inoculated on agar medium overlaying a HeLa cell monolayer (Laird and Groman 1973). After 24 h incubation the agar overlay is removed and cytotoxic effects can be analyzed. The authors found complete concordance with *in vivo* results for all tested 32 toxigenic and 16 non-toxigenic strains, respectively.

With respect to *C. ulcerans* cytotoxogenicity testing, the same caveats are applicable which have been outlined for *in vivo* toxicity testing, since reliable positive results depend on the absence of a cytotoxic effect in cells treated with *C. diphtheriae*-derived antitoxin (Schuhegger et al. 2008b; Dias et al. 2011). Antitoxin specifically raised against *C. ulcerans* DT has so far not been produced; in the light of the world-wide observed difficulties to obtain—even for therapeutic purposes—commercially available and quality-controlled *C. diphtheriae* antitoxin preparations (Wagner et al. 2009), the chance to get *C. ulcerans*-specific antitoxin for diagnostic needs seems to be very unlikely. In the majority of published cases on *in vitro C. ulcerans* cytotoxogenicity testing (including two studies on comparing different methods for toxigenicity testing in *Corynebacteria* spp. or *C. ulcerans* prevalence in dogs, respectively), Elek- or *in vivo* test positive *C. ulcerans* isolates also yielded positive *in vitro* cytotoxogenicity results (56 isolates: Efstratiou et al. 1998: all 4 isolates; Hatanaka et al. 2003: 1 isolate; Katsukawa et al. 2009: 1 isolate; Katsukawa et al. 2012: all 42 isolates; Komiya et al. 2010: 1 isolate; Mattos-Guaraldi et al. 2008: 1 isolate; Seto et al. 2008: 6 isolates; Sing et al. 2003: 1 isolate); however, negative *in vitro* results were also found in some Elek-, immunochromatographic strip test (ICS)- or *tox*-PCR-positive isolates in which either no cytotoxic effect at all (Sing et al. 2003: 1 isolate) or no DT-specific cytotoxic effect could be observed (complete lack of or only partial inhibition by *C. diphtheriae*-based antitoxin) (5 isolates: Komiya et al. 2010: 1 isolate, probably true NTTB; Mattos-Guaraldi et al. 2008: 1 isolate; Schubert et al. 1968:2 isolates; Sing et al. 2003: 1 isolate).

9.8.3 Immunological Toxicity Testing

The Elek-Ouchterlony-Immunodiffusion or Immunoprecipitation assay Based on the observation of at that time so-called Liesegang rings surrounding toxigenic *C. diphtheriae* colonies on DT antitoxin containing medium (Petrie and Steabben 1943 using 4 different *C. diphtheriae* strains for their preliminary report), Ouchterlony and Elek simultaneously and independently published an *in vitro* method for detection of DT producing *Corynebacterium* spp. by visualizing toxin-antitoxin immunoprecipitates on semi-solid media (Ouchterlony 1948; Elek 1948). The method was evaluated on 545 and 20 toxigenic and non-toxigenic *C. diphtheriae* isolates, respectively, showing 100% concordance with *in vivo* results. Basically, streaks from single corynebacterial colonies were plated on agar in a 90 degree angle towards either a trench cut into the agar containing immune serum (Ouchterlony

1948) or a filter paper strip dipped into highly concentrated antitoxin (Elek 1948)[2]. Subsequently, DT produced by toxigenic corynebacteria and antitoxin diffuses[3] into the medium and finally interacts with each other. After 24 to 48 h of incubation, DT and antitoxin precipitates[4] become visible to the naked eye within about 1 cm distance from the respective antitoxin source at an angle of about 45 ° to the line of bacterial growth; the precipitin lines were described as fine "moustache-like streaks" (Ouchterlony 1948) or "arrow-heads" virtually marking toxigenic strains (Elek 1948). The position of the precipitin lines and consequently the readability of the test depend on the initial concentration and the diffusion velocity of the two reacting substances, e.g. DT and antitoxin. Notably, unspecific precipitin lines, probably due to the interaction of other bacterial substances with corresponding serum components directed against them, may develop after prolonged incubation periods; this is especially the case, when undiluted serum or high amounts of serum (> 500 U/ml antitoxin) are used (Young and Smith 1976; Efstratiou and Maple 1994). Therefore, issues influencing both the sensitivity and the specificity have been experimentally addressed in several studies aiming to optimize the original protocols, e.g. type of medium and its components, distance between colony streaks and antitoxin source (trench, strip or disk), density of the bacterial inoculum, agar thickness, antitoxin quality and concentration, time needed to obtain a definite and reliable result, etc. (Bickham and Jones 1972; Carter and Wilson 1949; Colman et al. 1992; Elek 1949; Engler et al. 1997; Freeman 1950; Frobisher et al. 1951; Reinhardt et al. 1998; Young and Smith 1976). A modified Elek test based on a methodology developed in Russia and Ukraine and optimized by Engler et al. allowing to obtain specific results after 16 h is probably the most widely used variant of this test (Engler et al. 1997); an optimum distance between inoculum and antitoxin disc [10 IU] test of 9 mm and the use of Elek base medium supplemented with newborn bovine serum as well as of a quality-checked antitoxin were found to be the main issues for getting reliable test results (Engler et al. 1997). While rabbit, calf and adult bovine serum produce also reasonable results—with newborn bovine serum being the best standard recommended by the WHO Manual (Efstratiou and Maple 1994; European Centre for Disease Prevention and Control 2011)—, the use of equine-derived serum should be discouraged due to the high percentage of false-positive cross-reactions (Neal and Efstratiou 2009). However, the scarce availability of quality-checked antitoxin for diagnostic purposes is of major concern (Wagner et al. 2009). The detection limit for the Elek test and its modifications is difficult to estimate: scanty available data suggest a higher detection limit than that for the ICS (0.5 ng/ml) (Sing et al. 2003; Engler et al. 2002); however, even the isolate with the lowest cytotoxic capacity

[2] The test was historically named Elek-Ouchterlony test. Due to the easier performance of Elek's method using antitoxin-drenched filter strips this assay was continuously improved by different laboratories. Although Ouchterlony evaluated his assay on much more isolates than Elek in the respective seminal papers, the test is now known as Elek test by most authors. Curiously, many authors consider Elek as an acronym wrongly writing the test's name with capital letters as ELEK.

[3] Therefore, the alternative name immunodiffusion assay is coined.

[4] Therefore, the alternative name immunoprecipitation assay is coined.

yielded a positive Elek test result in an evaluation study of different pheno- and genotypic DT detection methods (Efstratiou et al. 1998).

Notably, since immunoprecipitation assays test for the presence of antitoxin irrespective of its biologically specific function, toxigenic *C. ulcerans* strains are also reliably detected by the Elek test due to the high amino acid similarity between *C. diphtheriae* and *C. ulcerans* DT of about 95 % (Sing et al. 2003). So far, 108 *C. ulcerans* isolates with an additional positive phenotypic toxigenicity test result (Bostock et al. 1984: 1 isolate; Efstratiou et al. 1998: all 4 isolates; Engler et al. 1997: 10 isolates; Engler et al. 2002: 39 isolates; Hatanaka et al. 2003: 1 isolate; Katsukawa et al. 2009: 1 isolate; Katsukawa et al. 2012: all 42 isolates; Komiya et al. 2010: 1 isolate; Mattos-Guaraldi et al. 2008: 2 isolates; Seto et al. 2008: 6 isolates; Sing et al. 2003: 1 of 2 strains) and 70 *C. ulcerans* without an additional phenotypic toxicity assay result (Berger et al. 2011a: 1 isolate; Berger et al. 2011b: 1 isolate; Bonmarin et al. 2009: 7 isolates; Cassiday et al. 2008: 11 isolates; De Zoysa et al. 2005: 45 isolates; Schuhegger et al. 2009: 1 isolate; Sing et al. 2005: 1 isolate; Tiwari et al. 2008: 2 isolates; Wellinghausen et al. 2002: 1 isolate) were reported to be Elek test positive.

Similarly, only a few *C. pseudotuberculosis* isolates were reported to yield positive results in the Elek test (with an additional positive phenotypic toxigenicity test result: Engler et al. 1997: 1 isolate; Maximescu et al. 1974: 2 isolates; without an additional phenotypic toxicity assay result reported: Bonmarin et al. 2009: 2 isolates).

In conclusion, once available in many routine microbiological laboratories worldwide the Elek test—as a consequence of the decrease in diphtheria incidence—is now performed only in reference centers at least in non-endemic countries both due to the needed specialized expertise for obtaining reliable test results and the increasing lack of antitoxin for diagnostic purposes (Neal and Efstratiou 2009; Wagner et al. 2009). However, even in reference centers around Europe the knowledge especially in toxigenicity testing is not everywhere available as seen by various external quality assurance exercises conducted under the auspices of the Diphtheria Surveillance Network (DIPNET) and the European Centre for Disease Prevention and Control (ECDC) (Neal and Efstratiou 2009; European Centre for Disease Prevention and Control 2011).

The Immunochromatographic Strip Test (ICS) Based on reagents incorporated into a rapid enzyme immunoassay (EIA; Engler and Efstratiou 2000), an immunochromatographic strip (ICS) test was developed for DT detection both from bacterial cultures and from clinical specimens directly inoculated into broth cultures by using an equine polyclonal antibody (as capture antibody) and a colloidal gold-labeled monoclonal DT fragment A-specific antibody (as detection antibody) (Engler et al. 2002). About 3.5 h incubation time is needed until a toxicity test result can be obtained from bacterial colonies which have to be suspended in serum-supplemented Elek broth (SSEB); 10 min after adding an ICS test strip the test can be visually read. For toxicity testing directly from clinical specimens, swabs are inoculated into SSEB and incubated for 16 h. Subsequently, an ICS test strip is added to the broth tube and read after 10 min. Concordance between ICS, Elek and EIA was 100% for 915 pure clinical *Corynebacterium* isolates (851 *C. diphtheriae* and 64 *C. ulcerans*) and 99% for clinical specimens directly inoculated into broth cultures (848 of 850 specimens) resulting in a sensitivity and specificity for the ICS of 98 and 99%, respectively. In a subgroup of obviously DT low-producers also one *C. pseudotuberculosis* isolate was concordantly positive in ICS, Elek test and EIA. Due to the high amino acid similarity between *C. diphtheriae* and *C. ulcerans* DT, also other authors used ICS for the detection of *C. ulcerans* DT (Mattos-Guaraldi et al. 2008; Sing et al. 2003: 2 isolates each). The limit of detection was approximately 0.5 ng/ml, which is comparable to the amplified EIA (Engler and Efstratiou 2000) and makes the ICS 10- to 20-fold more sensitive than other EIA or agglutination assays. Unfortunately, this easy-to-use, low-tech, fast, robust, reliable, highly sensitive and specific test which was evaluated in field trials in an epidemic situation in the 1990s is not yet produced anymore (Neal and Efstratiou 2009).

Other immunological assays Several immunological approaches including agglutination (Jalgaonkar and Saoji 1993; Maximescu and Fîciu 1980; Toma et al. 1997), counterimmunoelectrophoresis (Thompson and Ellner 1978), enzyme immunoassays (Engler and Efstratiou 2000; Hallas et al. 1990; Krech and Wittelsbürger 1987; Nielsen et al. 1987) or dot immunobinding methods (Pietrzak et al. 1990) have been chosen to detect DT from *Corynebacterium* isolates or clinical samples. However, most of these assays are technically demanding, lack sensitivity, are relatively expensive, were only published in non-English journals (mainly in Russian and Romanian) or have been evaluated only with a very small panel of isolates. Consequently, they were not developed to further stages.

9.8.4 Genotypic Toxigenicity Testing

A key advance in the diagnosis of diphtheria was the use of PCR to detect the *tox* gene. However, since NAAT only amplify the DT encoding gene, but do not give information regarding the real production of DT, both physicians and microbiologists need to be aware that an Elek test must be performed on all *tox*-positive isolates to test for DT production.

In both *C. diphtheriae* (Efstratiou et al. 1998) and *C. ulcerans* (Cassiday et al. 2008; Komiya et al. 2010; Seto et al. 2008; Sing et al. 2003) *tox*-positive non-toxigenic (i.e. Elek-test negative) strains (NTTB) have been described.

The PCR systems probably most widely used are those developed by Pallen et al. 1994 (from the WHO Collaborating Centre for Reference and Research on Diphtheria at the Health Protection Agency, UK), by Hauser et al. 1993 (from the Institut Pasteur, France) and an extension of the British PCR system adding a second primer set targeting a fragment within the B subunit encoding part of the *tox* gene by Nakao and Popovic 1997 (used at the Centers for Disease Control and Prevention, USA). Details for these commonly used primer sets are summarized in Table 9.3.

In general, the three PCR systems were found to yield 100% concordant results for *C. diphtheriae* isolates with *in vivo*, ADP-ribosylation activity or Elek testing in different settings (evaluation of the Pallen et al. 1994 method[5]: 17 toxigenic, 67 non-toxigenic isolates, possibly one NTTB strain; Aravena-Román 1995 8 toxigenic, 56 non-toxigenic isolates; Efstratiou et al. 1998: 22 toxigenic, 22 non-toxigenic isolates, 6 NTTB; Mikhailovich et al. 1995: 170 toxigenic, 80 non-toxigenic isolates; data for Hauser et al. 1993: 24 toxigenic, 8 non-toxigenic isolates; Sing et al. 2011: 8 toxigenic and 1 non-toxigenic strains, 2 NTTB; Sulakvelidze et al. 1999: 78 toxigenic isolates; the PCR system by Nakao and Popovic 1997 was only tested on direct clinical samples; evaluation on a strain collection was done by Mothershed et al. 2002: 25 toxigenic and 9 non-toxigenic strains).

The limit of detection was calculated to range between 50 to 1,500 CFU (Nakao and Popovic 1997; Mothershed et al. 2002). Besides these widely used *tox* PCR detection methods described above, a further PCR assay was evaluated on 5 toxigenic and 11 non-toxigenic *C. diphtheriae* strains reaching 100% concordance with *in vivo* or Elek results (Lucchini et al. 1992).

So far, only the PCR method by Nakao and Popovic was evaluated for *tox* amplification directly from clinical samples. 10 out of 13 throat swabs from clinically diagnosed diphtheria patients were positive, whereas only 5 positive culture results were obtained (Nakao and Popovic 1997). When comparing the *toxA* amplifying PCR primer set (Pallen et al. 1994) with their own *toxB* primer set, the latter was found to be less sensitive (Nakao and Popovic 1997). Similar results were obtained when amplifying *toxA* and *toxB* directly from 100 clinical paired throat and nose swab samples from diphtheria patients after prolonged storage of swabs (7–14 months with a detection rate of 53 and 15% for the two PCR primer sets, respectively) (Kobaidze et al. 2000). Consequently, the authors recommend considering specimens with only one *tox* subunit-positive result as an only preliminary result to be followed by culture when working-up suspected cases.

Despite approximately 5% discrepancies in basepair composition between *C. ulcerans* and *C. diphtheriae tox* sequences, the three most widely used *tox* PCR systems originally designed for *C. diphtheriae tox* were frequently used to successfully amplify and detect *C. ulcerans tox*—e.g. the *toxA*-PCR by Pallen et al. 1994 (which is identical to the *toxA*-PCR subset of the Nakao and Popovic 1997 method) (Efstratiou et al. 1998: 4 isolates; Hatanaka et al. 2003: 1 isolate; Katsukawa et al. 2009: 1 isolate; Katsukawa et al. 2012: 42 isolates; Komiya et al. 2010: 2o isolates; Seto et al. 2008: 6 isolates; Wellinghausen et al. 2002: 1 isolate), the Nakao and Popovic 1997 extension (*toxA* and *toxB*; Tiwari et al. 2008: 2 isolates) as well as the *toxAB*-PCR by Hauser et al. 1993 (Schuhegger et al. 2009: 1 isolate; Sing et al. 2003: 2 isolates; Sing et al. 2005: 1 isolate). So far, only one false negative *C. ulcerans tox*-PCR result was reported from an Elek-positive isolate grown on blood agar; when grown on tellurite-agar the same isolate was *tox*-PCR positive (Pallen et al. 1994).

[5] Notably, the first published PCR system (Pallen 1991) was later found to give negative results in other laboratories (Hauser et al. 1993); therefore, a new first primer was designed by the same author to overcome these difficulties (Pallen et al. 1994).

However, a dependence of PCR outcome on the medium used for culture was not reported in any of the other studies reported here.

For the specific detection of *C. ulcerans tox* alone- without amplifying *C. diphtheriae tox*—two different primer sets were developed based on the first published *tox* gene sequences (Sing et al. 2003) and used in the work-up of four different strains (Schuhegger et al. 2009; Sing et al. 2003, 2005).

To enhance rapidity, sensitivity and specificity of NAA methods for *tox* detection, three different real-time PCR methods have been developed and evaluated (Mothershed et al. 2002; Schuhegger et al. 2008b; Sing et al. 2011; for primer information see also Table 9.3). The TaqMan-based real-time PCR method designed at the CDC aims to amplify both the A and B subunits of the *tox* gene (Mothershed et al. 2002). Evaluated on 23 toxigenic and 9 non-toxigenic *C. diphtheriae* as well as on 44 non-*C. diphtheriae* strains its sensitivity and specificity were 100%, respectively. The limit of detection was 2 CFU thus being 750-fold more sensitive than the conventional PCR used at the CDC (Nakao and Popovic 1997). The higher sensitivity of the real-time PCR is also obvious when used directly on 36 specimens collected from patients with clinical diphtheria: in 34 of them (94%) one or both *tox* subunits were amplified, while conventional PCR results yielded only 9 positive results (25%). When using this real-time method in the work-up of two patients with diphtheria-like illness caused by toxigenic *C. ulcerans*, however, atypical amplification of the *tox* A subunit and no amplification of the *tox* B subunit, respectively, was realized, while conventional PCR was positive for both subunits in both isolates (Tiwari et al. 2008). Prompted by this disturbing observation, a subsequent study on one NTTB and 9 toxigenic *C. ulcerans* isolates which were conventional PCR-positive for both *tox* A and B subunits showed that only four of these were detected by real-time PCR (Cassiday et al. 2008). In order to avoid missing toxigenic corynebacteria a TaqMan-based real-time PCR method for identification of both *C. diphtheriae* and *C. ulcerans tox* was designed at the German National Consiliary Laboratory for Diphtheria (Schuhegger et al. 2008b). In an evaluation study, all 19 toxigenic (9 *C. diphtheriae* [2 of them NTTB] and 10 *C. ulcerans* [1 of them NTTB]) isolates were correctly identified, whereas all 70 *tox*-negative isolates of seven different *Corynebacterium* spp. tested negative. The limit of detection equals 100 fg/reaction corresponding to a 25-fold higher sensitivity than the conventional PCR (Hauser et al. 1993). So far, this real-time PCR was also used in the diagnosis of *tox*-positive *C. diphtheriae* (Hall et al. 2010) and *C. ulcerans* isolates (Berger et al. 2011a, b).

Additionally, a LightCycler-based real-time *tox* PCR was designed in order to have also a detection system for the other of the two most widely applied real-time PCR formats (Sing et al. 2011). Evaluated with 11 conventional *tox*-PCR positive *C. diphtheriae* (9 toxigenic, 2 NTTB; among them 3 NCTC control strains) and 12 *C. ulcerans* isolates (10 toxigenic, 2 NTTB) as well as 70 *tox*-negative isolates, both a sensitivity and specificity of 100% were obtained, respectively. The method is 25-fold more sensitive than the conventional PCR (limit of detection: 100 fg/reaction). Interestingly, the LightCycler melting curve analysis allows clear differentiation between *C. ulcerans* and at least all thus-far-published human-derived *C. diphtheriae tox* genes.

Recently, a novel group of feline-derived *C. diphtheriae* strains possibly representing a new subspecies of *C. diphtheriae* was described (Hall et al. 2010). These strains harbor a variant *tox* sequence with a higher sequence similarity to *C. ulcerans tox* than to *tox* sequences of human-derived *C. diphtheriae* strains. Due to a 1-nt deletion at position 55 and a cytosine-to-thymine substitution at position 74 these novel feline *C. diphtheriae* strains cannot functionally express *tox* and are therefore non-toxigenic. Although this *C. diphtheriae tox* variant would be detected by the LightCycler real-time PCR as it was by the German, but not by the CDC TaqMan-based real-time PCR (Hall et al. 2010), it should be mentioned that the LightCycler melting curve analysis might misclassify this *C. diphtheriae tox* gene as *C. ulcerans tox* (Sing et al. 2011).

Besides PCR systems for *tox* detection, also a PCR method for the *tox*-regulating *dtxR* gene was described for differentiating *C. diphtheriae* from non-*C. diphtheriae* or *Corynebacterium*-like colonies (Pimenta et al. 2008a). In this evaluation study, all 91 *C. diphtheriae* isolates (54 non-toxigenic and 37 toxigenic) were *dtxR*-PCR-positive, while all 111 non-*C. diphtheriae* isolates of 14 different species (including one *C. ulcerans* and three *C. pseudotuberculosis* isolates, all non-toxigenic) yielded negative results. Importantly, when comparing the *dtxR* primer binding sites of *C. diphtheriae* with *C. ulcerans* and *C. pseudotuberculosis* published *dtxR* gene sequences, an amplification of the *dtxR* gene fragment of the two possibly toxigenic non-*C. diphtheriae* species is excluded.

In conclusion, if the equipment is available real-time PCR methods like those of Schuhegger et al. 2008b or Sing et al. 2011 are recommended as reliable and fast screening tests to identify both *tox*-positive *C. diphtheriae* and *C. ulcerans*.

9.9 Antimicrobial Resistance Testing

Antimicrobial resistance against either penicillin or erythromycin, the two antibiotics of choice in eliminating diphtheria-causing corynebacteria, has been described both in *C. diphtheriae*—albeit very rarely (Coyle et al. 1979; Dixon 1984; Engler et al. 2001b; Kneen et al. 1998; Mina et al. 2011; Pereira et al. 2008; von Hunolstein et al. 2002) and *C. ulcerans* (Schuhegger et al. 2009; Tiwari et al. 2008). Therefore, antibiotic sensitivity testing is highly recommended. The CLSI prepared standardized MIC determination procedures for corynebacteria including *C. diphtheriae* via broth microdilution (document M45; http://www.clsi.org/). However, so far neither CLSI nor EUCAST (http://www.eucast.org/) have published standardized procedures or interpretation criteria for the more easily to perform disk diffusion method or the E-test. Moreover, there are only a few studies comparing different antibiotic susceptibility testing methods for corynebacteria (Martinez-Martinez et al. 1995; Weiss et al. 1996; Zapardiel et al. 1994). Therefore, in the absence of standardized criteria and guidelines for agar diffusion tests many authors use CLSI guidelines for other Gram-positive bacteria (Mina et al. 2011; Pereira et al. 2008; Sing et al. 2003). Usually, blood supplemented Mueller-Hinton agar is used for antimicrobial resistance testing by agar diffusion assays

9.10 Typing Methods for *C. diphtheriae*

Typing techniques are needed not only to identify a common source during an local outbreak situation by exploring the strain identity of different isolates, but also for monitoring clonal relatedness and population structure of endemically or epidemically spreading strains. Distinguishing between epidemic, endemic or imported cases can help to timely implement and reconsider adequate preventive measures by public health authorities. Depending on the main scientific purpose and the resources available in a given laboratory setting the respective typing method has to be chosen. A detailed summary of typing methods for *C. diphtheriae* is beyond the scope of this article. So far, several typing techniques have been developed. Traditional techniques were based upon serologic, phage and biotyping methods. However, since the methods provide limited resolution, molecular typing techniques, including amplified fragment length polymorphisms (AFLP) (De Zoysa and Efstratiou 2000), random amplified polymorphic DNA (RAPD) (De Zoysa and Efstratiou 1999; Nakao and Popovic 1998), multilocus enzyme electrophoresis (MEE) (Popovic et al. 1996), spoligotyping (Mokrousov et al. 2005) and pulsed-field gel electrophoresis (PFGE) (De Zoysa et al. 1995), have been developed and show significant intraspecies genetic diversity. In a comparison of the different typing techniques for *C. diphtheriae* it was shown that the most discriminative method was ribotyping (De Zoysa et al. 2008). Recently, a novel multilocus sequence typing (MLST) scheme was proposed which might overcome the difficulties seen with the highly user-dependent ribotyping (Grimont et al. 2004) and consequently help to guarantee a higher degree of reproducibility among different laboratories (Bolt et al. 2010). The MLST scheme includes the 7 housekeeping genes *atpA* (ATP synthase alpha chain), *dnaE* (DNA polymerase III alpha subunit), *dnaK* (chaperone protein DnaK), *fusA* (elongation factor G), *leuA* (2-isopropylmalate synthase), *odhA* (2-oxoglutarate dehydrogenase) and *rpoB* (RNA polymerase beta chain) The mean avarage of the proportion of variable sites at each locus was 7 %. Bolt and colleagues found 73 sequence types (ST) among 150 *C. diphtheriae* isolates (Bolt et al. 2010). For analysis and determination of the ST an internet platform is available (http://pubmlst.org/cdiphtheriae/). So far 233 STs for *C. diphtheriae* have been defined, for *C. ulcerans* such a database has yet to be established. Both ribotyping (De Zoysa et al. 2005; Seto et al. 2008; Schuhegger et al. 2009; Komiya et al. 2010; Berger et al. 2011a) and MLST (Schuhegger et al. 2009; Berger et al. 2011a) have been shown to be useful for characterising *C. ulcerans* strains in a limited number of isolates.

References

Adderson EE, Boudreaux JW, Cummings JR, Pounds S, Wilson DA, Procop GW, Hayden RT (2008) Identification of clinical coryneform bacterial isolates: comparison of biochemical methods and sequence analysis of 16S rRNA and *rpoB* genes. J Clin Microbiol 46(3):921–927

Alatoom AA, Cazanave CJ, Cunningham SA, Ihde SM, Patel R (2012) Identification of non-*diphtheriae Corynebacterium* by use of matrix-assisted laser desorption ionization-time of flight mass spectrometry. J Clin Microbiol 50(1):160–163

Almuzara MN, De Mier C, Rodríguez CR, Famiglietti AMR, Vay CA (2006) Evaluación del sistema API Coryne, versión 2.0, para la identificación de bacilos gram-positivos difteroides de importancia clínica. Rev Argent Microbiol 48(4):197–201

Andre J, Audebaud G, Chambon L (1960) Diagnostic de la diphtérie sur cultures de tissu. Ann Inst Pasteur 99(1):179–187

Aravena-Román M, Bowman R, O'Neill G (1995) Polymerase chain reaction for the detection of toxigenic *Corynebacterium diphtheria*. Pathology 27(1):71–73

Barksdale L, Linder R, Sulea IT, Pollice M (1981) Phospholipase D activity of *Corynebacterium pseudotuberculosis* (*Corynebacterium ovis*) and *Corynebacterium ulcerans*, a distinctive marker within the genus *Corynebacterium*. J Clin Microbiol 13(2):335–343

Berger A, Huber I, Merbecks SS, Ehrhard I, Konrad R, Hörmansdorfer S, Hogardt M, Sing A (2011a) Toxigenic *Corynebacterium ulcerans* in woman and cat. Emerg Infect Dis 17(9):1767–1769

Berger A, Schmidt-Wieland T, Huber I, Sing A (2011b) Human diphtheria infection from a cat bite. Ann Intern Med 155(9):646–648

Bickham ST, Jones WL (1972) Problems in the use of the in vitro toxigenicity test for *Corynebacterium diphtheria*. Am J Clin Pathol 57(2):244–246

Bolt F, Cassiday P, Tondella ML, Dezoysa A, Efstratiou A, Sing A, Zasada A, Bernard K, Guiso N, Badell E, Rosso ML, Baldwin A, Dowson C (2010) Multilocus sequence typing identifies evidence for recombination and two distinct lineages of *Corynebacterium diphtheriae*. J Clin Microbiol 48(11):4177–4185

Bonmarin I, Guiso N, Le Flèche-Matéos A, Patey O, Patrick AD, Levy-Bruhl D (2009) Diphtheria: a zoonotic disease in France? Vaccine 27(31):4196–4200

Bostock AD, Gilbert FR, Lewis D, Smith DC (1984) *Corynebacterium ulcerans* infection associated with untreated milk. J Infect 9(3):286–288

Butterworth A, Abbott JD, Simmons LE, Ironside AG, Williams RF, Mandal BK, Brennard J, Mann NM, Simon S (1974) Diphtheria in the Manchester area, 1967–1971. Lancet 304(7896):1558–1561

Carter HS, Wilson W (1949) Note on the recognition of toxigenic strains of *C. diphtheriae in vitro*. Glasgow Med J 30(1):43–48

Cassiday PK, Pawloski LC, Tiwari T, Sanden GN, Wilkins PP (2008) Analysis of toxigenic *Corynebacterium ulcerans* strains revealing potential for false-negative real-time PCR results. J Clin Microbiol 46(1):331–333

Cetinkaya B, Karahan M, Atil E, Kalin R, De Baere T, Vaneechoutte M (2002) Identification of *Corynebacterium pseudotuberculosis* isolates from sheep and goats by PCR. Vet Microbiol 88(1):75–83

Chang DN, Laughren GS, Chalvardjian NE (1978) Three variants of *Corynebacterium diphtheriae* subsp. *mitis* (Belfanti) isolated from a throat specimen. J Clin Microbiol 8(6):767–768

Colman G, Weaver E, Efstratiou A (1992) Screening tests for pathogenic corynebacteria. J Clin Pathol 45(1):46–48

Coyle MB, Minshew BH, Bland JA, Hsu PC (1979) Erythromycin and clindamycin resistance in *Corynebacterium diphtheriae* from skin lesions. Antimicrob Agents Chemother 16(4):525–527

Coyle MB, Nowowiejski DJ, Russell JQ, Groman NB (1993) Laboratory review of reference strains of *Corynebacterium diphtheriae* indicates mistyped *intermedius* strains. J Clin Microbiol 31(11):3060–3062

De Zoysa AS, Efstratiou A (1999) PCR typing of *Corynebacterium diphtheriae* by random amplification of polymorphic DNA. J Med Microbiol 48(4):335–340
De Zoysa AS, Efstratiou A (2000) Use of amplified fragment length polymorphisms for typing *Corynebacterium diphtheriae*. J Clin Microbiol 38(10):3843–3845
De Zoysa AS, Efstratiou A, George RC, Jahkola M, Vuopio-Varkila J, Deshevoi S, Tseneva G, Rikushin Y (1995) Molecular epidemiology of *Corynebacterium diphtheriae* from northwestern Russia and surrounding countries studied by using ribotyping and pulsed-field gel electrophoresis. J Clin Microbiol 33(5):1080–1083
De Zoysa A, Hawkey PM, Engler K, George R, Mann G, Reilly W, Taylor D, Efstratiou A (2005) Characterization of toxigenic *Corynebacterium ulcerans* strains isolated from humans and domestic cats in the United Kingdom. J Clin Microbiol 43(9):4377–4381
De Zoysa A, Hawkey P, Charlett A, Efstratiou A (2008) Comparison of four molecular typing methods for characterization of *Corynebacterium diphtheriae* and determination of transcontinental spread of *C. diphtheriae* based on BstEII rRNA gene profiles. J Clin Microbiol 46(11):3626–3635
Dias AA, Santos LS, Sabbadini PS, Santos CS, Silva FC Jr, Napoleão F, Nagao PE, Villas-Bôas MH, Hirata R Jr, Guaraldi AL (2011) *Corynebacterium ulcerans* diphtheria: an emerging zoonosis in Brazil and worldwide. Rev Saude Publica 45(6):1176–1191
Dixon JMS (1984) Diphtheria in North America. J Hyg 93(3):419–432
Dorella FA, Pacheco LG, Oliveira SC, Miyoshi A, Azevedo V (2006) *Corynebacterium pseudotuberculosis*: microbiology, biochemical properties, pathogenesis and molecular studies of virulence. Vet Res 37(2):201–218
Efstratiou A, George RC (1996) Microbiology and epidemiology of diphtheria. Rev Med Microbiol 7(1):31–42
Efstratiou A, George RC (1999) Laboratory guidelines for the diagnosis of infections caused by *Corynebacterium diphtheriae* and *C. ulcerans*. Commun Dis Public Health 2(4):250–257
Efstratiou A, Maple PAC (1994) Manual for the Laboratory Diagnosis of Diphtheria. The Expanded Programme on Immunisation in the European Region of WHO. ICP/EPI 038(C)
Efstratiou A, Engler KH, Dawes CS, Sesardic D (1998) Comparison of phenotypic and genotypic methods for detection of diphtheria toxin among isolates of pathogenic corynebacteria. J Clin Microbiol 36(11):3173–3177
Elek SD (1948) The recognition of toxigenic bacterial strains *in vitro*. Br Med J 1(4549):493–496
Elek SD (1949) The plate virulence test for diphtheria. J Clin Pathol 2(4):250–258
Engler KH, Efstratiou A (2000) Rapid enzyme immunoassay for determination of toxigenicity among clinical isolates of corynebacteria. J Clin Microbiol 38(4):1385–1389
Engler KH, Glushkevich T, Mazurova IK, George RC, Efstratiou A (1997) A modified Elek test for detection of toxigenic corynebacteria in the diagnostic laboratory. J Clin Microbiol 35(2):495–498
Engler KH, Kozlov RS, Copping SJ, Efstratiou A (2001a) International external quality assessment scheme for the laboratory diagnosis of diphtheria. J Med Microbiol 50(11):1006–1012
Engler KH, Warner M, George RC (2001b) *In vitro* activity of ketolides HMR 3004 and HMR 3647 and seven other antimicrobial agents against *Corynebacterium diphtheriae*. J Antimicrob Chemother 47(1):27–31
Engler KH, Efstratiou A, Norn D, Kozlov RS, Selga I, Glushkevich TG, Tam M, Melnikov VG, Mazurova IK, Kim VE, Tseneva GY, Titov LP, George RC (2002) Immuno-chromatographic strip test for rapid detection of diphtheria toxin: description and multicenter evaluation in areas of low and high prevalence of diphtheria. J Clin Microbiol 40(1):80–83
European Centre for Disease Prevention and Control (2011) External quality assurance scheme for diphtheria diagnostics 2010. As part of the European Diphtheria Surveillance Network (EDSN). ECDC; 2011, Stockholm
Fraser DT (1931) The technique of a method for quantitative determination of diphtheria antitoxin by skin test in rabbits. Trans R Soc Can (Sec V) 25 (2):175–181
Fraser DT (1937) The intracutaneous virulence test for *Corynebacterium diphtheriae*. Am J Public Health 27(3):121–124

Fraser DT, Weld CB (1926) The intracutaneous virulence test for *Corynebacterium diphtheriae*. Trans R Soc Can (Sec V) 20:343

Freeman VJ (1950) Influence of type and concentration of antitoxin on the *in vitro* toxigenicity test for *C. diphtheria*. Public Health Rep 65(28):875–882

Freney J, Duperron MT, Courtier C, Hansen W, Allard F, Boeufgras JM, Monget D, Fleurette J (1991) Evaluation of API Coryne in comparison with conventional methods for identifying coryneform bacteria. J Clin Microbiol 29(1):38–41

Frobisher M Jr, King EO, Parsons EI (1951) A test *in vitro* for virulence of *Corynebacterium diphtheriae*. Am J Clin Pathol 21(3):282–285

Funke G, Bernard KA (2007) Coryneform gram-positive rods. In: Murray PR, Baron EJ, Jorgensen JH, Landry ML (eds). Manual of Clinical Microbiology. American Society of Microbiology Press, 9th ed. Washington, DC, pp 485–514

Funke G, von Graevenitz A, Clarridge JE 3rd, Bernard KA (1997a) Clinical microbiology of coryneform bacteria. Clin Microbiol Rev 10(1):125–159

Funke G, Renaud FN, Freney J, Riegel P (1997b) Multicenter evaluation of the updated and extended API (RAPID) Coryne database 2.0. J Clin Microbiol 35(12):3122–3126

Funke G, Peters K, Aravena-Roman M (1998) Evaluation of the RapID CB Plus System for identification of coryneform bacteria and *Listeria* spp. J Clin Microbiol 36(9):2439–2442

Gavin SE, Leonard RB, Briselden AM, Coyle MB (1992) Evaluation of the rapid CORYNE identification system for *Corynebacterium* species and other coryneforms. J Clin Microbiol 30(7):1692–1695

Grimont PA, Grimont F, Efstratiou A, De Zoysa A, Mazurova I, Ruckly C, Lejay-Collin M, Martin-Delautre S, Regnault B (2004) International nomenclature for *Corynebacterium diphtheriae* ribotypes. Res Microbiol 155(3):162–166

Groman N, Schiller J, Russell J (1984) *Corynebacterium ulcerans* and *Corynebacterium pseudotuberculosis* responses to DNA probes derived from corynephage beta and *Corynebacterium diphtheriae*. Infect Immun 45(2):511–517

Hall AJ, Cassiday PK, Bernard KA, Bolt F, Steigerwalt AG, Bixler D, Pawloski LC, Whitney AM, Iwaki M, Baldwin A, Dowson CG, Komiya T, Takahashi M, Hinrikson HP, Tondella ML (2010) Novel *Corynebacterium diphtheriae* in domestic cats. Emerg Infect Dis 16(4):688–691

Hallas G, Harrison TG, Samuel D, Colman G (1990) Detection of diphtheria toxin in culture supernates of *Corynebacterium diphtheriae* and *C. ulcerans* by immunoassay with monoclonal antibody. J Med Microbiol 32(4):247–253

Hatanaka A, Tsunoda A, Okamoto M, Ooe K, Nakamura A, Miyakoshi M, Komiya T, Takahashi M (2003) *Corynebacterium ulcerans* diphtheria in Japan. Emerg Infect Dis 9(6):752–753

Hauser D, Popoff MR, Kiredjian M, Boquet P, Bimet F (1993) Polymerase chain reaction assay for diagnosis of potentially toxinogenic *Corynebacterium diphtheriae* strains: correlation with ADP-ribosylation activity assay. J Clin Microbiol 31(10):2720–2723

Health Protection Agency (2008) National Standard Method: Identification of *Corynebacterium* species, BSOP ID2i3. www.evaluations-standards.org.uk

Hoy CS, Sesardic D (1994) *In vitro* assays for detection of diphtheria toxin. Toxicol In Vitro 8(4):693–695

Hudspeth MK, Hunt Gerardo S, Citron DM, Goldstein EJ (1998) Evaluation of the RapID CB Plus system for identification of *Corynebacterium* species and other gram-positive rods. J Clin Microbiol 36(2):543–547

Jalgaonkar SV, Saoji AM (1993) Coagglutination for rapid testing of toxin producing *Corynebacterium diphtheriae*. Indian J Med Res 97:35–36

Janda MW (1999) The corynebacteria revisited: new species, identification kits, and antimicrobial susceptibility testing. Clinical Microbiological Newsletter 21(22):175–182

Jellard CH (1971) Comparison of Hoyle's medium and Billings' modification of Tinsdale's medium for the bacteriological diagnosis of diphtheria. J Med Microbiol 4(3):366–369

Jellard CH (1980) Toxigenic and non-toxigenic *Corynebacterium diphtheriae*. Lancet 315(8169):650

Katsukawa C, Kawahara R, Inoue K, Ishii A, Yamagishi H, Kida K, Nishino S, Nagahama S, Komiya T, Iwaki M, Takahashi M (2009) Toxigenic *Corynebacterium ulcerans* isolated from the domestic dog for the first time in Japan. Jpn J Infect Dis 62(2):171–172

Katsukawa C, Komiya T, Yamagishi H, Ishii A, Nishino S, Nagahama S, Iwaki M, Yamamoto A, Takahashi M (2012) Prevalence of *Corynebacterium ulcerans* in dogs in Osaka, Japan. J Med Microbiol 61:266–273

Khamis A, Raoult D, La Scola B (2004) *rpoB* gene sequencing for identification of *Corynebacterium* species. J Clin Microbiol 42(9):3925–3931

Khamis A, Raoult D, La Scola B (2005) Comparison between *rpoB* and 16S rRNA gene sequencing for molecular identification of 168 clinical isolates of *Corynebacterium*. J Clin Microbiol 43(4):1934–1936

Kneen R, Pham NG, Solomon T, Tran TM, Nguyen TT, Tran BL, Wain J, Day NP, Tran TH, Parry CM, White NJ (1998) Penicillin vs. erythromycin in the treatment of diphtheria. Clin Infect Dis 27(4):845–450

Kobaidze K, Popovic T, Nakao H, Quick L (2000) Direct polymerase chain reaction for detection of toxigenic *Corynebacterium diphtheriae* strains from the Republic of Georgia after prolonged storage. J Infect Dis 181(Suppl 1):S152–S155

Komiya T, Seto Y, De Zoysa A, Iwaki M, Hatanaka A, Tsunoda A, Arakawa Y, Kozaki S, Takahashi M (2010) Two Japanese *Corynebacterium ulcerans* isolates from the same hospital: ribotype, toxigenicity and serum antitoxin titre. J Med Microbiol 59(12):1497–1504

Konrad R, Berger A, Huber H, Boschert V, Hörmansdorfer S, Busch U, Hogardt M, Schubert S, Sing A (2010) Matrix-assisted laser desorption/ionisation time-of-flight (MALDI-TOF) mass spectrometry as a tool for rapid diagnosis of potentially toxigenic *Corynebacterium* species in the laboratory management of diphtheria-associated bacteria. Euro Surveill 15(43):pii=19699

Krech T (1994) Epidemiological typing of *Corynebacterium diphtheriae*. Med Microbiol Lett 3(1):1–8

Krech T, Wittelsbürger C (1987) Immunologische Methoden zum Nachweis von Diphtherietoxin (Passive Haemagglutination und ELISA zum Toxinnachweis aus Kulturen und im Serum). Zentralbl Bakteriol Mikrobiol Hyg A 265(1–2):124–135

Laird W, Groman N (1973) Rapid, direct tissue culture test for toxigenicity of *Corynebacterium diphtheriae*. Appl Microbiol 25(2):709–712

Lennox ES, Kaplan AS (1957) Action of diphtheria toxin on cells cultivated *in vitro*. Proc Soc Exp Biol Med 95(4):700–702

Lindenmann K, von Graevenitz A, Funke G (1995) Evaluation of the Biolog system for the identification of asporogenous, aerobic gram-positive rods. Med Microbiol Lett 4(6):287–296

Lucchini GM, Gruner E, Altwegg M (1992) Rapid detection of diphtheria toxin by the polymerase chain reaction. Med Microbiol Lett 1:276–283

Martínez-Martínez L, Ortega MC, Suárez AI (1995) Comparison of E-test with broth microdilution and disk diffusion for susceptibility testing of coryneform bacteria. J Clin Microbiol 33(5):1318–1321

Mattos-Guaraldi AL, Sampaio JL, Santos CS, Pimenta FP, Pereira GA, Pacheco LG, Miyoshi A, Azevedo V, Moreira LO, Gutierrez FL, Costa JL, Costa-Filho R, Damasco PV, Camello TC, Hirata Jr R (2008) First detection of *Corynebacterium ulcerans* producing a diphtheria-like toxin in a case of human with pulmonary infection in the Rio de Janeiro metropolitan area Brazil. Mem Inst Oswaldo Cruz 103(4):396–400

Maximescu P, Fîciu S (1980) Slide agglutination assay in the *C. diphtheriae* identification. Arch Roum Pathol Exp Microbiol 39(3):205–212

Maximescu P, Oprişan A, Pop A, Potorac E (1974) Further studies on *Corynebacterium* species capable of producing diphtheria toxin (*C. diphtheriae, C. ulcerans, C. ovis*). J Gen Microbiol 82(1):49–56

McNamara PJ, Cuevas WA, Songer JG (1995) Toxic phospholipases D of *Corynebacterium pseudotuberculosis, C. ulcerans* and *Arcanobacterium haemolyticum*: cloning and sequence homology. Gene 156(1):113–118

Meers PD (1979) A case of classical diphtheria, and other infections due to *Corynebacterium ulcerans*. J Infect 1(1):139–142

Mikhailovich VM, Melnikov VG, Mazurova IK, Wachsmuth IK, Wenger JD, Wharton M, Nakao H, Popovic T (1995) Application of PCR for detection of toxigenic *Corynebacterium diphtheriae* strains isolated during the Russian diphtheria epidemic, 1990 through 1994. J Clin Microbiol 33(11):3061–3063

Mina NV, Burdz T, Wiebe D, Rai JS, Rahim T, Shing F, Hoang L, Bernard K (2011) Canada's first case of a multidrug-resistant *Corynebacterium diphtheriae* strain, isolated from a skin abscess. J Clin Microbiol 49(11):4003–4005

Mokrousov I (2009) *Corynebacterium diphtheriae*: genome diversity, population structure and genotyping perspectives. Infect Genet Evol 9(1):1–15

Mokrousov I, Narvskaya O, Limeschenko E, Vyazovaya A (2005) Efficient discrimination within a *Corynebacterium diphtheriae* epidemic clonal group by a novel macroarray-based method. J Clin Microbiol 43(4):1662–1668

Mothershed EA, Cassiday PK, Pierson K, Mayer LW, Popovic T (2002) Development of a real-time fluorescence PCR assay for rapid detection of the diphtheria toxin gene. J Clin Microbiol 40(12):4713–4719

Mouton RP (1960) Isolation of non-toxigenic diphtheria bacteria and phages during a local outbreak of diphtheria. Antonie Van Leeuwenhoek 26(1):297–304

Murphy JR, Bacha P, Teng M (1978) Determination of *Corynebacterium diphtheriae* toxigenicity by a colorimetric tissue culture assay. J Clin Microbiol 7(1):91–96

Nakao H, Popovic T (1997) Development of a direct PCR assay for detection of the diphtheria toxin gene. J Clin Microbiol 35(7):1651–1655

Nakao H, Popovic T (1998) Use of random amplified polymorphic DNA for rapid molecular subtyping of *Corynebacterium diphtheriae*. Diagn Microbiol Infect Dis 30(3):167–172

Neal SE, Efstratiou A (2009) International external quality assurance for laboratory diagnosis of diphtheria. J Clin Microbiol 47(12):4037–4042

Nielsen PB, Koch C, Friis H, Heron I, Prag J, Schmidt J (1987) Double-antibody sandwich enzyme-linked immunosorbent assay for rapid detection of toxin-producing *Corynebacterium diphtheriae*. J Clin Microbiol 25(7):1280–1284

Noble WC, Dixon JMS (1990) *Corynebacterium* and other coryneform bacteria. In: Parker MT, Duerden BI (eds) Topley Wilson's principles of bacteriology, virology and immunity, vol. 2. Edward Arnold, London, pp 103–118

Ouchterlony O (1948) In vitro method for testing the toxin-producing capacity of diphtheria bacteria. Acta Pathol Microbiol Scand 25(1–2):186–191

Pacheco LG, Pena RR, Castro TL, Dorella FA, Bahia RC, Carminati R, Frota MN, Oliveira SC, Meyer R, Alves FS, Miyoshi A, Azevedo V (2007) Multiplex PCR assay for identification of *Corynebacterium pseudotuberculosis* from pure cultures and for rapid detection of this pathogen in clinical samples. J Med Microbiol 56(4):480–486

Pallen MJ (1991) Rapid screening for toxigenic *Corynebacterium diphtheriae* by the polymerase chain reaction. J Clin Pathol 44(12):1025–1026

Pallen MJ, Hay AJ, Puckey LH, Efstratiou A (1994) Polymerase chain reaction for screening clinical isolates of corynebacteria for the production of diphtheria toxin. J Clin Pathol 47(4):353–356

Pascual C, Lawson PA, Farrow JA, Gimenez MN, Collins MD (1995) Phylogenetic analysis of the genus *Corynebacterium* based on 16S rRNA gene sequences. Int J Syst Bacteriol 45(4):724–728

Pereira GA, Pimenta FP, Santos FR, Damasco PV, Hirata R Jr, Mattos-Guaraldi AL (2008) Antimicrobial resistance among Brazilian *Corynebacterium diphtheriae* strains. Mem Inst Oswaldo Cruz 103(5):507–510

Petrie GF, Steabben D (1943) Specific identification of the chief pathogenic clostridia of gas gangrene. Brit Med J 1(4290):377–379

Pietrzak J, Muehlestein S, Gasser M (1990) Sandwich-dot immunobinding assay (Sandwich-DIA), a new immunological method for the detection of diphtheria toxin. Zentralbl Bakteriol 274(1):61–69

Pike RM (1976) Laboratory-associated infections: summary and analysis of 3921 cases. Health Lab Sci 13(2):105–114

Pike RM, Sulkin SE, Schulze ML (1965) Continuing importance of laboratory-acquired infections. Am J Public Health Nations Health 55(2):190–199

Pimenta FP, Matias GA, Pereira GA, Camello TC, Alves GB, Rosa AC, Hirata R Jr, Mattos-Guaraldi AL (2008a) A PCR for *dtxR* gene: application to diagnosis of non-toxigenic and toxigenic *Corynebacterium diphtheriae*. Mol Cell Probes 22(3):189–192

Pimenta FP, Souza MC, Pereira GA, Hirata R Jr, Camello TC, Mattos-Guaraldi AL (2008b) DNase test as a novel approach for the routine screening of *Corynebacterium diphtheriae*. Lett Appl Microbiol 46(3):307–311

Popovic T, Kombarova SY, Reeves MW, Nakao H, Mazurova IK, Wharton M, Wachsmuth IK, Wenger JD (1996) Molecular epidemiology of diphtheria in Russia, 1985–1994. J Infect Dis 174(5):1064–1072

Reinhardt DJ, Lee A, Popovic T (1998) Antitoxin-in-membrane and antitoxin-in-well assays for detection of toxigenic *Corynebacterium diphtheriae*. J Clin Microbiol 36(1):207–210

Rennie RP, Brosnikoff C, Turnbull L, Reller LB, Mirrett S, Janda W, Ristow K, Krilcich A (2008) Mulitcenter evaluation of the Vitek 2 anaerobe and *Corynebacterium* identification card. J Clin Microbiol 46(8):2646–2651

Riegel P, Ruimy R, de Briel D, Prévost G, Jehl F, Christen R, Monteil H (1995) Taxonomy of *Corynebacterium diphtheriae* and related taxa, with recognition of *Corynebacterium ulcerans* sp. nov. nom. rev. FEMS Microbiol Lett 126(3):271–276

Ruimy R, Riegel P, Boiron P, Monteil H, Christen R (1995) Phylogeny of the genus *Corynebacterium* deduced from analyses of small-subunit ribosomal DNA sequences. Int J Syst Bacteriol 45(4):740–746

Schubert JH, Bickham ST, Wiggins GL (1968) Tissue culture method for toxigenicity testing of *Corynebacterium diphtheriae*. Appl Microbiol 16(11):1748–1752

Schuhegger R, Kugler R, Sing A (2008a) Classical diphtheria caused by *Corynebacterium ulcerans* in Germany: amino acid sequence differences between diphtheria toxins from *Corynebacterium diphtheriae* and *C. ulcerans*. Clin Infect Dis 47(2):288

Schuhegger R, Lindermayer M, Kugler R, Heesemann J, Busch U, Sing A (2008b) Detection of toxigenic *Corynebacterium diphtheriae* and *Corynebacterium ulcerans* strains by a novel real-time PCR. J Clin Microbiol 46(8):2822–2823

Schuhegger R, Schoerner C, Dlugaiczyk J, Lichtenfeld I, Trouillier A, Zeller-Peronnet V, Busch U, Berger A, Kugler R, Hörmansdorfer S, Sing A (2009) Pigs as source for toxigenic *Corynebacterium ulcerans*. Emerg Infect Dis 15(8):1314–1315

Segal E, Eylan E, Imre Z (1973) Cystine-serum-tellurite: a differential medium for *Corynebacterium diphtheriae*. Med Microbiol Immun 158(3):165–169

Seto Y, Komiya T, Iwaki M, Kohda T, Mukamoto M, Takahashi M, Kozaki S (2008) Properities of corynephage attachment site and molecular epidemiology of *Corynebacterium ulcerans* isolated from humans and animals in Japan. Jpn J Infect Dis 61(2):116–122

Simmons LE, Abbott JD, Macaulay ME, Jones AE, Ironside AG, Mandal BK, Stanbridge TN, Maximescu P (1980) Diphtheria carriers in Manchester: simultaneous infection with toxigenic and non-toxigenic *mitis* strains. Lancet 315(8163):304–305

Sing A, Hogardt M, Bierschenk S, Heesemann J (2003) Detection of differences in the nucleotide and amino acid sequences of diphtheria toxin from *Corynebacterium diphtheriae* and *Corynebacterium ulcerans* causing extrapharyngeal infections. J Clin Microbiol 41(10):4848–4851

Sing A, Bierschenk S, Heesemann J (2005) Classical diphtheria caused by *Corynebacterium ulcerans* in Germany: amino acid sequence differences between diphtheria toxins from *Corynebacterium diphtheriae* and *C. ulcerans*. Clin Infect Dis 40(2):325–326

Sing A, Berger A, Schneider-Brachert W, Holzmann T, Reischl U (2011) Rapid detection and molecular differentiation of toxigenic *Corynebacterium diphtheriae* and *Corynebacterium ulcerans* strains by LightCycler PCR. J Clin Microbiol 49(7):2485–2489

Soto A, Zapardiel J, Soriano F (1994) Evaluation of API Coryne system for identifying coryneform bacteria. J Clin Pathol 47(8):756–975

Strauss N, Hendee ED (1959) The effect of diphtheria toxin on the metabolism of HeLa cells. J Exp Med 109(2):145–163

Sulakvelidze A, Kekelidze M, Gomelauri T, Deng Y, Khetsuriani N, Kobaidze K, De Zoysa A, Efstratiou A, Morris JG Jr, Imnadze P (1999) Diphtheria in the Republic of Georgia: use of molecular typing techniques for characterization of *Corynebacterium diphtheriae* strains. J Clin Microbiol 37(10):3265–3270

Synder JW (2004) *Corynebacterium diphtheriae* cultures. In: Isenberg HD (ed), Clinical Microbiology Procedures Handbook, 2nd ed. ASM Press, Washington, DC, pp 3.11.7.1–3.11.7.9.

Tang YW, von Graevenitz A, Waddington MG, Hopkins MK, Smith DH, Li H, Kolbert CP, Montgomery SO, Persing DH (2000) Identification of coryneform bacterial isolates by ribosomal DNA sequence analysis. J Clin Microbiol 38(4):1676–1678

Taylor I, Tomlinson AJH, Davies JR (1962) Diphtheria control in the 1960's. J R Soc Promot Health 82(3):158–164

Thilo W, Kiehl W, Geiss HK (1997) A case report of laboratory-acquired diphtheria. Euro Surveill 2(8):67–68

Thompson NL, Ellner PD (1978) Rapid determination of *Corynebacterium diphtheriae* toxigenicity by counterimmunoelectrophoresis. J Clin Microbiol 7(5):493–494

Thompson JS, Gates-Davis DR, Yong DC (1983) Rapid microbiochemical identification of *Corynebacterium diphtheriae* and other medically important corynebacteria. J Clin Microbiol 18(4):926–929

Tinsdale GFW (1947) A new medium for the isolation and identification of *Corynebacterium diphtheriae* based on the production of hydrogen sulphide. J Pathol Bacteriol 59(3):461–466

Tiwari TS, Golaz A, Yu DT, Ehresmann KR, Jones TF, Hill HE, Cassiday PK, Pawloski LC, Moran JS, Popovic T, Wharton M (2008) Investigations of 2 cases of diphtheria-like illness due to toxigenic *Corynebacterium ulcerans*. Clin Infect Dis 46(3):395–401

Toma C, Sisavath L, Iwanaga M (1997) Reversed passive latex agglutination assay for detection of toxigenic *Corynebacterium diphtheriae*. J Clin Microbiol 35(12):3147–3149

Vaneechoutte M, Riegel P, de Briel D, Monteil H, Verschraegen G, De Rouck A, Claeys G (1995) Evaluation of the applicability of amplified rDNA-restriction analysis (ARDRA) to identification of species of the genus *Corynebacterium*. Res Microbiol 146(8):633–641

von Graevenitz A, Funke G (1996) An identification scheme for rapidly and aerobically growing gram-positive rods. Zentralblatt Bakteriol Parasitenkd Infekt Hyg 284(2–3):246–254

von Graevenitz A, Pünter-Streit V, Riegel P, Funke G (1998) Coryneform bacteria in throat cultures of healthy individuals. J Clin Microbiol 36(7):2087–2088

von Hunolstein C, Scopetti F, Efstratiou A, Engler K (2002) Penicillin tolerance amongst nontoxigenic Corynebacterium diphtheriae isolated from cases of pharyngitis. J Antimicrob Chemother 50 (1):125–128

Wagner KS, Stickings P, White JM, Neal S, Crowcroft NS, Sesardic D, Efstratiou A (2009) A review of the international issues surrounding the availability and demand for diphtheria antitoxin for therapeutic use. Vaccine 28(1):14–20

Wagner KS, White JM, Crowcroft NS, De Martin S, Mann G, Efstratiou A (2010) Diphtheria in the United Kingdom, 1986–2008: the increasing role of *Corynebacterium ulcerans*. Epidemiol Infect 138(11):1519–1530

Weiss K, Laverdière M, Rivest R (1996) Comparison of antimicrobial susceptibilities of *Corynebacterium* species by broth microdilution and disk diffusion methods. Antimicrob Agents Chemother 40(4):930–933

Wellinghausen N, Sing A, Kern WV, Perner S, Marre R, Rentschler J (2002) A fatal case of necrotizing sinusitis due to toxigenic *Corynebacterium ulcerans*. Int J Med Microbiol 292(1):59–63

Wirsing von Koenig CH, Krech T, Finger H, Bergmann M (1988) Use of fosfomycin disks for isolation of diphtheroids. Eur J Clin Microbiol Infect Dis 7(2):190–193

Young PS, Smith DD (1976) The detection of toxin production by *Corynebacterium diphtheriae* in diagnostic laboratories. Pathology 8(2):101–104

Zapardiel J, Nieto E, Gegúndez MI, Gadea I, Soriano F (1994) Problems in minimum inhibitory concentration determinations in coryneform organisms. Comparison of an agar dilution and the Etest. Diagn Microbiol Infect Dis 19(3):171–173

Chapter 10
Diphtheria Surveillance

Karen S. Wagner, Katherina Zakikhany, Joanne M. White, Gayatri Amirthalingam, Natasha S. Crowcroft and Androulla Efstratiou

Abstract Diphtheria is uncommon in developed countries but when cases do arise they are often severe with high mortality. This disease has demonstrated its potential to re-emerge to epidemic proportions in areas where it was previously thought to be under control. Ongoing monitoring and surveillance is therefore essential and in general follows the principles of surveillance utilised for most other vaccine preventable infections, but with some specific adaptations relevant to a disease that is close to elimination. Surveillance across countries and regions is complicated by a number of factors including the use of different case definitions and the variation in laboratory policy and expertise. International networks have been valuable in improving knowledge and skills in this area.

A. Efstratiou (✉)
Public Health England, Microbiology Services Division: Colindale,
61 Colindale Avenue, NW9 5EQ, London, UK
e-mail: Androulla.Efstratiou@phe.gov.uk

K. S. Wagner · J. M. White · G. Amirthalingam
Centre for Infectious Disease Surveillance and Control, Public Health England,
61 Colindale Avenue, NW9 5EQ London, UK
e-mail: karen.wagner@phe.gov.uk

K. Zakikhany
Swedish Institute for Communicable Disease Control, Nobels vag 18, Solna, Sweden
e-mail: k.zakikhany@gmail.com

The European Programme for Public Health Microbiology Training (EUPHEM),
European Centre for Disease Prevention and Control (ECDC), Stockholm, Sweden

J. M. White
e-mail: joanne.white@phe.gov.uk

G. Amirthalingam
e-mail: gayatri.amirthalingham@phe.gov.uk

N. S. Crowcroft
Infectious Diseases, Public Health Ontario, 480 University avenue,
Suite 300. Toronto, Ontario, Canada
e-mail: natasha.crowcroft@pahpp.ca

A. Burkovski (ed.), *Corynebacterium diphtheriae and Related Toxigenic Species*,
DOI 10.1007/978-94-007-7624-1_10,

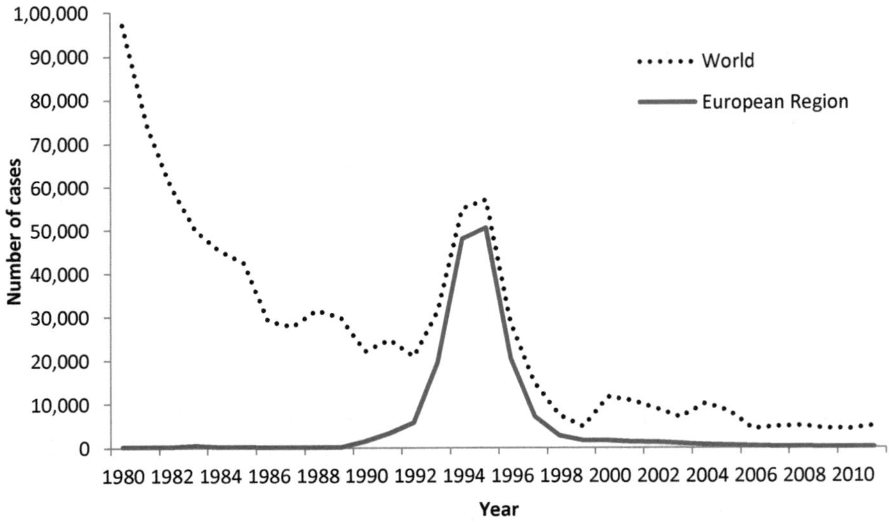

Fig. 10.1 Diphtheria cases reported to the World Health Organization (Data downloaded 24/07/2012), 1980–2011. (Source: WHO 2012)

Keywords Case definitions · Diphtheria · Epidemiology · Monitoring · Surveillance

10.1 Introduction

In the pre-vaccine era, diphtheria was a common cause of morbidity and a leading cause of childhood death. This led to the development of diphtheria antitoxin for treatment, the provision of which was associated with heroic stories such as that of Balto the dog (one of the sleigh dogs who carried antitoxin to an isolated community in Nome, Alaska during an outbreak in 1925) whose statue stands in Central Park, New York. By the middle of the twentieth century, comprehensive routine childhood vaccination schemes were implemented across Europe and North America and marked the beginning of a global decline in incidence (Fitzgerald et al. 1932; Galazka et al. 1995; Vitek and Wharton 1998). Furthermore, in 1974 the Expanded Programme on Immunization (EPI) was established to introduce and improve immunisation coverage in developing countries (Keja et al. 1988). By the beginning of the 1980s in Europe, cases of clinical diphtheria were extremely uncommon and many countries were progressing towards elimination (Galazka and Robertson 2004; Vitek and Wharton 1998). However, by the end of this decade a resurgence of epidemic diphtheria occurred in the Russian Federation, affecting all newly independent states of the former Soviet Union (Vitek and Wharton 1998). Between 1990 and 1998, more than 157,000 cases and 5,000 deaths were reported from the region, accounting for more than 80 % of diphtheria cases reported worldwide (see Fig. 10.1) (Dittmann et al. 2000; Markina et al. 2000; Vitek and Wharton

1998). One of the striking features of this large-scale diphtheria outbreak was the high proportion of adult cases, reported in all affected countries (Dittmann et al. 2000; Galazka 2000; Galazka and Robertson 1995).

Many factors were considered to have contributed to the resurgence, including increased adult susceptibility due to waning immunity, low vaccination coverage among children in the 1980s and early 1990s, disruptions to health service infrastructure and large-scale population movements as a consequence of the break up of the former Soviet Union, as well as the introduction of toxigenic strains (Dittmann et al. 2000). Although now largely under control this epidemic highlighted the continuing danger posed by diphtheria and the need for vigilance and rapid control procedures even in areas where the number of reported cases is usually low.

Although diphtheria is now uncommon in Europe, it is present in every WHO Region (WHO 2012), and there are new and ongoing epidemics (for example in Africa and South East Asia (FluTrackers 2012; WHO 2012), notably India, Indonesia and the Sudan (WHO 2012)) being reported regularly. The risk of importation of *Corynebacterium diphtheriae* from endemic countries, particularly those beyond the European Region, remains. In addition, indigenous infections caused by toxigenic *Corynebacterium ulcerans* (associated with consumption of raw dairy products (Barrett 1986; Bostock et al. 1984), contact with cattle (Hart 1984), and increasingly domestic pets (De Zoysa et al. 2005; Hogg et al. 2009; Lartigue et al. 2005)) further emphasise the need to maintain high levels of vaccination coverage. Only a few countries regularly report toxigenic isolates of *C. ulcerans* (Tiwari et al. 2008; Wagner et al. 2012), however, it is likely that this organism is also circulating in other countries but remains undetected due to a lack of robust surveillance and/or laboratory diagnostics.

10.2 Surveillance

10.2.1 Definition of Surveillance

The purpose of surveillance is to provide sufficient information to enable countries to take timely public health action in the prevention and control of diphtheria. The International Health Regulations 2005 define surveillance as 'the systematic ongoing collection, collation and analysis of data for public health purposes and the timely dissemination of public health information for assessment and public health response as necessary' (WHO 2008).

Surveillance practices vary according to the systems, health structures and resources in place within different countries. In addition the level of surveillance needed will depend to a certain extent on the incidence of diphtheria; where the disease is common diagnosis based on symptoms for a proportion of cases may be reliable, but as a country approaches the elimination phase the need for laboratory confirmation and detailed follow-up of all clinically suspected cases becomes essential. In addition to disease surveillance, other important types of surveillance include vaccine coverage and assessment of population immunity, as for all vaccination programmes.

Objectives of Diphtheria Surveillance The World Health Organization (WHO) outlines the rationale for the surveillance of diphtheria as follows (WHO 2003): 'Diphtheria is a widespread severe infectious disease that has the potential for causing epidemics. Surveillance data can be used to monitor levels of coverage and disease as a measure of the impact of control programmes. Recent epidemics have highlighted the need for adequate surveillance and epidemic preparedness.'

The objectives of a diphtheria surveillance system are:

a. To estimate the burden of disease in different populations

In the first instance, a surveillance system should enable timely recognition and response to increases in cases (or individual cases as is most relevant for diphtheria in non-endemic countries) and clusters (clusters may not be detected at local level if they span geographical boundaries) through both early detection and communication of information through appropriate channels such that management and control procedures can be rapidly implemented.

A picture of the overall disease burden in a country will generally be based on data from several sources. Incidence rates for different age groups, geographical areas and time periods can indicate particular risk groups and periods. However, the accuracy of the estimated burden of disease will depend on the sensitivity and specificity of the surveillance systems in place.

b. To assess the public health threat posed by the disease

This is dependent on:

- transmissibility of the organism
- opportunities for spread:
  - living conditions e.g. overcrowding
  - population movements nationally and internationally
  - availability of containment facilities
  - infrastructure, external threats e.g. conflict, breakdown of immunisation programmes
- susceptibility of the population:
  - degree of natural immunity from infection
  - current and historical vaccine coverage
  - vaccine schedule, type of vaccine
  - existence of risk groups e.g. injecting drug users, homeless, alcoholics, immunocompromised individuals, those in contact with animals
- diagnosis and treatment options:
  - awareness of health staff to recognise the disease and enable early diagnosis and treatment
  - laboratory diagnostic facilities
  - health service factors e.g. access to antitoxin/antibiotics/vaccine

c. To identify risk factors for carriage, transmission and complications

Analysis of data collected through enhanced surveillance can give indications of risk factors and provide a starting point for further exploratory studies. For example, recent individual case based data has been used to build up an evidence base for understanding risk factors relating to *C. ulcerans* and its association with domestic animals (for example: Berger et al. 2011; Hatanaka et al. 2003; Hogg et al. 2009; Schuhegger et al. 2009). Where case numbers are low, pooling datasets across several countries is of value. However, this requires standardisation of case definitions and surveillance systems across countries.

d. To monitor the effectiveness of control programmes and inform policy
Reviewing vaccination history on all cases through enhanced surveillance is a valuable means of monitoring the effectiveness of the vaccine. Seroprevalence studies can also be used to assess whether or not the vaccination programme is sufficiently protective, as well as monitoring the degree of waning immunity and the duration of protection (Edmunds et al. 2000; Di Giovine et al. 2012). A high case fatality ratio may indicate problems with late diagnosis, case management, or the availability or quality of antibiotics and/or antitoxin. In addition, a process of review and debrief of incident teams after a case or cluster has occurred enables 'lessons learned' to be identified and can inform guidance.

e. To monitor phenotypic and genotypic changes in the causative organism
The application of molecular epidemiological tools is essential for monitoring the spread of epidemic clones and to allow for distinction between epidemic, endemic and imported cases. This also has major implications for timely and adequate preventative measures. It is facilitated by close partnerships between public health microbiology and epidemiology. Although ribotyping is still used as the 'gold standard' for molecular epidemiological studies, novel typing methods such as MLST (multi locus sequence typing) are being explored with promising results. In addition, standardised protocols have led to more rapid and accurate detection of these 'clones' globally along with the establishment of an online international database for automatic recognition of genotypes. The use of state of the art genome sequencing has also been undertaken and has provided invaluable information on novel targets for rapid typing and information on the pathogenicity of the organisms. Databases for ribotyping and MLST are available within the public domain. The current molecular typing ((http://www.dipnet.org/ribo.public.php, http://pubmlst.org/cdiphtheriae/)) database for diphtheria, which was built using ribotype data from a diverse collection of strains, was imported from Taxotron® to Bionumerics. This has allowed access to laboratories worldwide for *C. diphtheriae* and *C. ulcerans* (via the WHO Collaborating Centre) pattern analysis and has significantly contributed towards the molecular epidemiology of diseases caused by these organisms (De Zoysa et al. 2008; De Zoysa et al. 2005). An EU consensus towards typing at different levels is being developed according to country laboratory capabilities.

f. To disseminate public health information appropriately
Alongside the collection and analysis of data from the sources described above, surveillance includes the need for timely dissemination of information so that front line staff can update their knowledge with respect to the latest findings. Updat-

Table 10.1 Comparison of two case-definitions currently in use within Europe

	1994 WHO case definition for diphtheria (Begg 1994)	2012 EU case definition for diphtheria (European Commission 2012)
Suspected/ possible case	Laryngitis or nasopharyngitis or tonsillitis plus pseudomembrane	Any person meeting the clinical criteria for classical respiratory diphtheria *Classical respiratory diphtheria* An upper respiratory tract illness with laryngitis or nasopharyngitis or tonsillitis AND an adherent membrane/pseudomembrane
Probable case	Suspected case plus one of the following: Recent (<2 weeks) contact with a confirmed case Diphtheria epidemic currently in the area Stridor Swelling/edema of neck Submucosal or skin petechial hemorrhages Toxic circulatory collapse Acute renal insufficiency Myocarditis and/or motor paralysis 1-6 weeks after onset Death	Any person meeting the clinical criteria for diphtheria (classic respiratory diphtheria, mild respiratory diphtheria, cutaneous diphtheria, diphtheria of other sites) with an epidemiological link to a human confirmed case or with an epidemiological link to animal to human transmission Clinical criteria *Classical respiratory diphtheria—see above* *Mild respiratory diphtheria* An upper respiratory tract illness with laryngitis or nasopharyngitis or tonsillitis WITHOUT an adherent membrane/pseudomembrane *Cutaneous diphtheria* Skin lesion *Diphtheria of other sites* Lesion of conjunctiva or mucous membranes Epidemiological criteria At least one of the following epidemiological links Human to human transmission Animal to human transmission

Table 10.1 (continued)

	1994 WHO case definition for diphtheria (Begg 1994)	2012 EU case definition for diphtheria (European Commission 2012)
Confirmed case	Probable case plus isolation of a toxigenic strain of *C. diphtheriae* from a typical site (nose, throat, skin ulcer, wound, conjunctiva, ear, vagina) or fourfold or greater rise in serum antitoxin, but only if both serum samples were obtained before the administration of diphtheria toxoid or antitoxin Confirmed cases should be classified as indigenous or imported (infection acquired abroad) Note: demonstration of toxin production is recommended but not required in typical cases. Microscopic examination of a direct smear of a clinical specimen is not sufficiently accurate to substitute as a culture	Any person meeting the laboratory criteria AND at least one of the clinical forms Laboratory criteria Isolation of toxin-producing *C. diphtheriae, C. ulcerans*, or *C. pseudotuberculosis* from a clinical specimen

ing national and international guidance for clinicians and laboratories allows new recommendations to be clearly communicated. In addition, since cases of diphtheria are uncommon in most European countries, publishing individual case reports in widely read journals can be a valuable means of enabling others to learn from recent experience. The development of surveillance networks which include both epidemiologists and microbiologists allows more direct sharing of information and expertise in a timely manner (see Sect. 10.6).

10.3 Case Definitions

Currently two different case definitions are applied in public health settings across the WHO European region: the 2012 EU case definition, which considers disease caused by *C. diphtheriae, C. ulcerans* and *C. pseudotuberculosis*, and the 1994 WHO case definition, which only considers classical respiratory diphtheria cases caused by *C. diphtheriae* ('epidemic diphtheria'). Both definitions are detailed in Table 10.1 and differences discussed below. In addition, it is standard practice that case definitions may be modified during specific incidents or outbreaks.

The definitions differ with respect to both the clinical presentations and organisms included.

Clinical Presentation Both definitions include classic respiratory diphtheria. The EU case definition also includes alternative presentations such as cutaneous disease and milder respiratory symptoms. The detection of mild cases and unusual presentations can serve as an indicator of a sensitive surveillance system. Loss of laboratory expertise increases the risk of missing mild, subclinical or unusual manifestations. Whilst epidemic diphtheria is associated with respiratory disease, cutaneous infections can be a source of respiratory infection (Koopman and Campbell 1975) so early identification of such presentations can be valuable in assisting with the prevention and control of diphtheria. Information concerning asymptomatic carriers of toxigenic strains, although not included in the case definitions, may be recorded as part of the follow-up of contacts of a case.

Organism Both definitions include toxigenic *C. diphtheriae* infections, the cause of epidemic diphtheria. In addition, the EU definition includes toxigenic *C. ulcerans* infections which can have the same clinical presentations. The reservoirs and transmission routes of *C. ulcerans* are still not fully understood and since only a few cases are reported each year within Europe it is useful to be able to consolidate surveillance data from several countries to try to improve understanding. It has also been suggested that the capability to detect *C. ulcerans* isolates could be interpreted as an indicator for a well functioning surveillance system.

In addition the EU definition includes toxigenic *C. pseudotuberculosis*, a rare infection in humans, typically associated with contact with sheep or goats (Dorella et al. 2006).

Neither of the definitions includes non-toxigenic organisms, which do not require any public health action (antibiotic treatment only). A subset of non-toxigenic *C. diphtheriae* strains are known to carry the *tox* gene (non-toxigenic toxin-bearing [NTTB] or toxin-carrying [NTTC] strains) (Groman 1984; INTAS 2004). During and after the diphtheria epidemic in the former Soviet Union, the circulation of NTTBs was observed (Melnikov et al. 2000), and two NTTBs were detected in Lithuania in a recent screening study (Wagner et al. 2011), but in general the prevalence of NTTB strains is ill-defined. NTTB strains are often isolated alongside other organisms (Berger et al. 2012; Lowe et al. 2011; Reacher et al. 2000). Although the potential of phage conversion to transform NTTBs into toxigenic strains is considered low and rarely observed *in vivo*, circulation of NTTBs may act as a repository for *tox* gene sequences and therefore pose a risk for disease. However, the public health relevance of these strains is yet to be examined and further studies are needed in order to assess the true burden of NTTB strains in Europe and globally.

Sensitivity and Specificity In general, the broader the definition is, the more sensitive the surveillance system will be. However, countries must balance the need for surveillance with financial constraints, and for a rare disease such as diphtheria, it may be sensible to consider broadening the definition once elimination is approaching. With appropriate laboratory expertise the specificity of the surveillance system for diphtheria in a country where the disease is uncommon should be high as all cases would be laboratory confirmed. In countries where diphtheria is more widespread cases may be reported on the basis of symptoms and an epidemiological link only, which is less specific, though this can be reliable when symptoms are typical and the patient has had close contact with another case.

10.4 Sources of Surveillance Data

10.4.1 Case-based Surveillance Data

Early Reporting of Probable Individual Cases to Public Health Authorities on the Basis of Symptoms The frequency of reporting to central level will depend on the incidence of infection (see sect. 10.5). Identification of a probable case of diphtheria necessitates immediate reporting to national level in most countries. It may be sensible to modify the criteria for notifying cases to national level based on the prevalence of disease in a particular country to ensure that the notification system is appropriate and able to give a realistic picture of the current situation. In countries with a high incidence of diphtheria, reports relying on symptoms in the first instance are likely genuine cases, but in countries where diphtheria is rare, cases can typically be notified on the basis of symptoms and later discovered to be due to other cause(s) hence looking at such notifications alone may be misleading.

Laboratory Confirmed Cases Screening policies, laboratory expertise, and also the availability of laboratory reagents differ between countries and will influence the number of cases detected, as well as the ability to characterise strains.

Screening Policies Diagnostics relating to diphtheria are complex and require specialist expertise and media because *C. diphtheriae* and *C.ulcerans* are easily obscured by the normal throat flora. Laboratory guidelines and flow-diagrams have been published to support microbiological identification (Efstratiou et al. 2000; Efstratiou et al. 1998; Efstratiou and George 1999). There are a number of different policy options:

- No screening (where resources and/or expertise are lacking; in many countries there is complacency and screening is not considered to be cost effective).
- Screening only specimens with a specific clinician request and/or risk factor such as a history of travel to an endemic region (this is the practice in many non-endemic countries).
- Sentinel screening of all throat swabs by particular laboratories (undertaken for example in Denmark, Ireland, UK) (Wagner et al. 2011).
- Routine screening of all throat swabs (this practice is no longer undertaken by any countries within Europe).

Laboratory Diagnostics A recent external quality assurance evaluation (EQA) assessing microbiological procedures for diphtheria across the WHO European Region revealed that less than 20% (6/34) of participating international centres were fully capable of diagnosing the specimen correctly (Neal and Efstratiou 2009). This indicated the significant challenges that need to be overcome in terms of developing and maintaining laboratory expertise. Although originally described in the 1940s, the Elek test (Elek 1949; Engler et al. 1997), conventional (24h-48h incubation time) or modified (16h-24h incubation time), is still the gold standard method for the detection of toxigenic strains as it detects expression of the active toxin. In times of austerity and financial constraints, many laboratories do not maintain the laborious laboratory infrastructure or cannot afford to stock the specialised media and diphtheria antitoxin (needed for the Elek test) (Neal and Efstratiou 2009) and therefore, use PCR alone for the detection of the *tox* gene. However, PCR cannot distinguish between toxigenic and non-toxigenic toxin gene-bearing strains (NTTB) and should therefore only be used in combination with the Elek test (Efstratiou et al. 2000; Efstratiou and George 1999).

Strain Characterisation In addition to identifying the organism, specialised reference laboratories also offer procedures for further characterisation; these typing methods allow the identification of clonal groups and (epidemiologically) closely related strains and can provide information on the geographic origin of the strain.

A. Biotyping (C. diphtheriae only) The first stage in characterisation to species level (to distinguish between biovars *gravis, mitis*, *intermedius* and *belfanti*), undertaken by most clinical microbiology laboratories.

B. Ribotyping The currently widely recognised gold standard method for typing of *Corynebacterium spp*. During the diphtheria epidemic in the former Soviet Union in the 1990s ribotyping enabled the differentiation between endemic and epidemic strains (Damian et al. 2002; De Zoysa et al. 1995; Kolodkina et al. 2006; Popovic et al. 1996; Skogen et al. 2002; von Hunolstein et al. 2003). Nowadays ribotyping is used during outbreak investigations for *C. diphtheriae*, and for *C. ulcerans* to identify possible sources of infection and to investigate suspected transmission from domestic animals to humans (Bonmarin et al. 2009; De Zoysa et al. 2005; Lartigue et al. 2005). However, ribotyping is a subjective, band-matching based system using continuous values for classification; it is therefore prone to generate ambiguous data, which, together with a required rigid standardisation procedure, negatively affects reproducibility and portability of the method and no evolutionary information can be collected.

C. MLST (multi locus sequence typing) MLST has been established as a promising successor to ribotyping (Bolt et al. 2010); overcoming problems with ambiguity encountered with ribotyping by indexing nucleotide variations within core metabolic (*housekeeping*) genes. Selected genes are directly sequenced and sequential differences within each gene (allele) provide an allelic profile. The allelic profiles and sequence types are unambiguous, meaning strains can readily be compared between laboratories and numerical values are easily stored in databases, thereby, providing reproducible and portable data appropriate for the epidemiological and evolutionary investigation of diphtheria (Bolt et al. 2010; Maiden et al. 1998). However, the discriminatory power of MLST, and also other sequencing based methods (e.g. VNTR [variable tandem repeat analysis]) which are currently under investigation, is slightly lower compared to ribotyping and not directly comparable; therefore, more systematic studies are required to evaluate and assess the method for the purposes needed.

Enhanced Surveillance This is typically initiated in response to an epidemiological or laboratory report and should involve collaboration with epidemiology and laboratory colleagues to ascertain detailed information regarding at a minimum:

- the patient
  - date of birth, sex, geographical area of residence
- clinical information
  - onset of symptoms
  - description of symptoms
  - duration of illness
  - outcome
- laboratory confirmation
  - organism *(C. diphtheriae, C. ulcerans, C. pseudotuberculosis)*
  - toxigenicity test result
- vaccination history (dates and doses received)
- travel history (for cases of *C. diphtheriae*)
- animal contact (for cases of *C. ulcerans, C. pseudotuberculosis)*

- case management
  - antibiotics (date, name, dose)
  - antitoxin (date, dose)
  - diphtheria vaccine (dates and doses received)
- management of contacts

Hospitalisations Where hospitalisations for diphtheria or symptoms associated with diphtheria recorded in hospital databases may act as a further source of surveillance data for severe cases. However, ideally these cases would have already been reported to relevant authorities via the channels above.

Death Registrations Where diphtheria is recorded as a cause of death provide a further source of data which can be useful in ensuring severe case reports have not been missed. International Classification of Diseases (ICD) coding allows for both the main cause and any underlying causes of death to be recorded, such as a complication from diphtheria in childhood contributing to the death of an adult; therefore full information on any death certification mentioning diphtheria should be sought.

Medical Literature In some instances case reports may be detected in the medical literature that have not been reported through the standard national channels (Wagner et al. 2010).

10.4.2 Population-Level Surveillance Data

Seroprevalence Studies Measurement of population immunity by age group can demonstrate the effectiveness of national immunisation programmes. They can also highlight population groups susceptible to infection, and indicate the need to adapt a country's vaccination schedule. To achieve elimination of diphtheria, a minimum immunity rate of 90 % in children and 75 % in adults is recommended (Begg 1994).

Age-Specific Vaccine Coverage at National and Sub-National Level Reduced vaccination coverage may indicate the need for heightened surveillance in particular age groups or geographical areas, and/or new public health campaigns. The targets proposed by the WHO expert group in 1992 include achieving 95 % coverage of both the primary immunisation series (DTP3) by 2 years of age, and a booster dose in school age children in every district (Begg 1994).

Screening Studies Occasional studies may be of value to detect the presence of toxigenic and non-toxigenic *C. diphtheriae* and/or *C. ulcerans* in throat swabs and can provide reassurance that cases are not being missed. Currently, very few cases of diphtheria are reported within the EU however, results of a recent pan-European screening study revealed that toxigenic organisms are still circulating in Latvia and Lithuania. At least one of the toxigenic organisms circulating in

Lithuania would have not been detected in the absence of this screening study (Wagner et al. 2012).

10.5 Frequency of Reporting

WHO recommended standards for surveillance of diphtheria are as follows (WHO 2003):

- Routine monthly reporting of aggregated data on probable or confirmed cases is recommended from the peripheral level to the intermediate and central levels.
- Designated reporting sites at all levels should report at a specified frequency (e.g. weekly or monthly) even if there are zero cases (often referred to as "zero reporting").
- All outbreaks should be investigated immediately and case-based data should be collected.
- In countries achieving low incidence (usually where coverage is >85-90%[1]), immediate reporting of case-based data of probable or confirmed cases is recommended from the peripheral level to the intermediate and central levels.

10.6 The Need for a Cross-Country Approach to Surveillance: European Surveillance Networks

While national approaches within a country might be advisable for high incidence and high prevalence diseases, a combined European approach to diphtheria surveillance, into which national approaches can feed, allows skills and resources to be shared and increases the ability of a country and its neighbours to detect and respond to epidemics that are small, widespread, and/or at an early stage. With the increase in international travel and migration and the steady increase of the number of countries in the European Union, the benefits of a European-wide surveillance system become even more relevant.

At the peak of the diphtheria epidemic in the former Soviet Union in 1993, the European Laboratory Working Group on Diphtheria (ELWGD) was established as an initiative of the WHO Regional Office for Europe in response to the urgent need to develop laboratory techniques for diphtheria diagnosis and analysis (Efstratiou and Roure 2000). In 2001, and as a response to the epidemic, this approach was taken one step further and the European Commission (EC) (Directorate-General for Health and Consumers (DG SANCO)) funded a feasibility study which led to the

[1] Population coverage of approximately 85% is required for elimination, based on a basic reproductive number (R_0; the average number of secondary cases produced by one primary case in a wholly susceptible population) of 6–7 (Fine 1993).

Diphtheria Surveillance Network (DIPNET). This included both the epidemiological and microbiological aspects of diphtheria and other infections caused by potentially toxigenic corynebacteria. The network was officially recognised in 2006 as an EC Dedicated Surveillance Network, bringing together 25 EU partner countries and 21 collaborating countries from the WHO European Region (Neal and Efstratiou 2007) with the objectives to:

- Harmonise and enhance surveillance of *C. diphtheriae* and *C. ulcerans* within the WHO European Region.
- Determine the disease prevalence and characteristics of toxigenic and non-toxigenic *C. diphtheriae* and *C. ulcerans* in a variety of populations with emphasis upon higher risk countries.
- Expand the DIPNET external quality assurance schemes for laboratory diagnosis to include epidemiological typing and serological immunity (Neal and Efstratiou 2009).
- Develop novel tools for integrated molecular epidemiological characterisation so as to gain a clearer understanding of the spread of epidemic clones throughout the WHO European Region (Efstratiou et al. 2009).
- Undertake serological immunity studies within 'high risk countries' and assessment of serological methodologies across all EU Member States (Di Giovine et al. 2010).

The research findings of DIPNET, which included highlighting a lack of laboratory expertise across Europe (Neal and Efstratiou 2009), identification of toxigenic *C. diphtheriae* during a screening study in a country that had not reported diphtheria in the previous five years (Wagner et al. 2011), and the analysis of aggregated data across member countries (Wagner et al. 2012) emphasised the benefit of a coordinated approach, as well as further training and studies to assist with monitoring progress across Europe.

From February 2010 the responsibility for the activities of DIPNET were transferred to the European Centre for Disease Control (ECDC), based in Stockholm, Sweden, in the form of the European Diphtheria Surveillance Network (EDSN) (ECDC 2012). The activities of the EDSN are aimed at integrating the epidemiology and laboratory surveillance of diseases caused by *C. diphtheriae* and *C. ulcerans*.

At the global level, WHO undertakes surveillance in each of its regional units, and together with the United Nations Children's Fund (UNICEF) collates global data at the national level on diphtheria incidence and vaccine coverage (WHO 2012).

10.7 Future Challenges

Remarkable advances have been made since the epidemic of the 1990s with respect to reducing diphtheria case numbers, case management and laboratory diagnostics. However, considerable challenges remain in terms of the surveillance of this disease.

It is vital that cases can be identified and treated in a timely manner. Accurate microbiological and epidemiological surveillance is therefore essential. This entails ensuring that clinicians are aware of the various clinical presentations of diphtheria as well as risk factors for infection including those specific to *C. ulcerans*, and that microbiologists have sufficient skills and resources for microbiological diagnosis. Maintaining this expertise in the face of low prevalence of disease is one of the key challenges for diphtheria surveillance.

Furthermore, as the table included earlier in this chapter demonstrates, consistent case definitions are not currently used across different countries and world regions. This presents challenges when analysing data and limits the ability to pool data across countries, reducing the opportunities to understand risk factors, for example, for *C. ulcerans*.

A further challenge lies in accurately monitoring vaccine coverage in all age groups and maintaining consistently high vaccination coverage across Europe. Owing to the rarity of this disease, the fear of diphtheria and the consequent demand for vaccination is lessened. Furthermore anti-immunisation sentiment may actively discourage vaccination in some countries. Clear public health messages and strong efforts towards achieving the minimum 95 % coverage recommended by the WHO are essential.

Maintaining good inter-country communication, particularly during times of austerity and financial constraints is an additional challenge to overcome as part of a concerted cross-country commitment to achieving and maintaining diphtheria elimination.

Acknowledgements We gratefully acknowledge the helpful comments from Dr Shona Neal on the final draft.

References

Barrett NJ (1986) Communicable disease associated with milk and dairy products in England and Wales: 1983–1984. J Infect 12(3):265–272

Begg NT (1994) Diphtheria: manual for the management and control of diphtheria in the European Region. World Health Organization, Copenhagen

Berger A, Huber I, Merbecks S-S, Ehrhard I, Konrad R, Hörmansdorfer S, Hogardt M, Sing A (2011) Toxigenic *Corynebacterium ulcerans* in woman and cat. Emerg Infect Dis 17(9):1767–1769

Berger A, Lensing C, Konrad R, Huber I, Sing A (2012) Sexually transmitted diphtheria. Sex Transm Infect, published online May 23, 2012 doi: 10.1136/sextrans-2011-050418

Bisgard KM, Rhodes P, Hardy IR, Litkina IL, Filatov NN, Monisov AA, Wharton M (2000) Diphtheria toxoid vaccine effectiveness: A case-control study in Russia. J Infect Dis 181(Suppl 1):S184–S187

Bolt F, Cassiday P, Tondella ML, De Zoysa A, Efstratiou A, Sing A, Zasada A, Bernard K, Guiso N, Badell E, Rosso ML, Baldwin A, Dowson C (2010) Multilocus sequence typing identifies evidence for recombination and two distinct lineages of *Corynebacterium diphtheriae*. J Clin Microbiol 48(11):4177–4185

Bonmarin I, Guiso N, Fleche-Mateos A, Patey O, Patrick AD, Levy-Bruhl D (2009) Diphtheria: a zoonotic disease in France? Vaccine 27(31):4196–4200

Bostock AD, Gilbert FR, Lewis D, Smith DC (1984) *Corynebacterium ulcerans* infection associated with untreated milk. J Infect 9(3):286–288

Damian M, Grimont F, Narvskaya O, Straut M, Surdeanu M, Cojocaru R, Mokrousov I, Diaconescu A, Andronescu C, Melnic A, Mutoi L, Grimont PA (2002) Study of *Corynebacterium diphtheriae* strains isolated in Romania, northwestern Russia and the Republic of Moldova. Res Microbiol 153(2):99–106

De Zoysa A, Efstratiou A, George RC, Jahkola M, Vuopio-Varkila J, Deshevoi S, Tseneva G, Rikushin Y (1995) Molecular epidemiology of *Corynebacterium diphtheriae* from northwestern Russia and surrounding countries studied by using ribotyping and pulsed-field gel electrophoresis. J Clin Microbiol 33(5):1080–1083

De Zoysa A, Hawkey P, Charlett A, Efstratiou A (2008) Comparison of four molecular typing methods for characterization of *Corynebacterium diphtheriae* and determination of transcontinental spread of *C. diphtheriae* based on BstEII rRNA gene profiles. J Clin Microbiol 46(11):3626–3635

De Zoysa A, Hawkey PM, Engler K, George R, Mann G, Reilly W, Taylor D, Efstratiou A (2005) Characterization of toxigenic *Corynebacterium ulcerans* strains isolated from humans and domestic cats in the United Kingdom. J Clin Microbiol 43(9):4377–4381

Di Giovine P, Kafatos G, Nardone A, Andrews N, Olander RM, Alfarone G, Broughton K, Cohen D, Kriz B, Mikova I, O'Flanagan D, Schneider F, Selga I, Valinsky L, Velicko I, Karacs I, Pebody R, von Hunolstein C (2012) Comparative seroepidemiology of diphtheria in six European countries and Israel. Epidemiol Infect 141(1):132–142 http://dx.doi.org/10.1017/S0950268812000210

Di Giovine P, Pinto A, Olander RM, Sesardic D, Stickings P, Berbers G, Neal S, Efstratiou A, Paberza R, Dauksiene S, Bujko M, Detcheva A, Jõks U, Levent B, von Hunolstein C (2010) External quality assessment for the determination of diphtheria antitoxin in human serum. Clin Vaccine Immunol 17(8):1282–1290

Dittmann S, Wharton M, Vitek C, Ciotti M, Galazka A, Guichard S, Hardy I, Kartoglu U, Koyama S, Kreysler J, Martin B, Mercer D, Ronne T, Roure C, Steinglass R, Strebel P, Sutter R, Trostle M (2000) Successful control of epidemic diphtheria in the states of the Former Union of Soviet Socialist Republics: lessons learned. J Infect Dis 181(Suppl 1):S10–S22

Dorella FA, Pacheco LG, Oliveira SC, Miyoshi A, Azevedo V (2006) *Corynebacterium pseudotuberculosis*: microbiology, biochemical properties, pathogenesis and molecular studies of virulence. Vet Res 37(2):201–218

Edmunds WJ, Pebody RG, Aggerback H, Baron S, Berbers G, Conyn-van Spaendonck MAE, Hallander HO, Olander R, Maple PAC, De Melker HE, Olin P, Fievret-Groyne F, Rota C, Salmaso S, Tischer A, von-Hunolstein C, Miller E, on behalf of the ESEN Project (2000) The seroepidemiology of diphtheria in Western Europe. Epidemiol Infect 125(1):113–125

Efstratiou A, Engler KH, De Zoysa A (1998) Diagnosis and epidemiology of diphtheria. Methods Mol Med 15:191–212

Efstratiou A, Engler KH, Mazurova IK, Glushkevich T, Vuopio-Varkila J, Popovic T (2000) Current approaches to the laboratory diagnosis of diphtheria. J Infect Dis 181(Suppl 1):S138–S145

Efstratiou A, George RC (1999) Laboratory guidelines for the diagnosis of infections caused by *Corynebacterium diphtheriae* and *C. ulcerans*. Commun Dis Public Health 2(4):250–257

Efstratiou A, Neal SE, and on behalf of all DIPNET members (2009) Diphtheria Surveillance Network: DIPNET; European Commission-DGSANCO. Final Report. Agreement No: 2005210DIPNET

Efstratiou A, Roure C (2000) The European Laboratory Working Group on Diphtheria: A global microbiologic network. J Infect Dis 181(Suppl 1):S146–S151

Elek D (1949) The plate virulence test for diphtheria. J Clin Pathol 2(4):250–258

Engler KH, Glushkevich T, Mazurova IK, George RC, Efstratiou A (1997) A modified Elek test for detection of toxigenic corynebacteria in the diagnostic laboratory. J Clin Microbiol 35(2):495–498

European Centre for Disease Prevention and Control (ECDC) (2012) European Diphtheria Surveillance Network (EDSN). http://www.ecdc.europa.eu/en/activities/surveillance/EDSN/Pages/index.aspx. Accessed 1 Oct 2012

European Commission (2012) Diphtheria. In: Commission implementing decision of 8 August 2012 amending Decision 2002/253/EC laying down case definitions for reporting communicable diseases to the Community network under Decision No 2119/98/EC of the European Parliament and of the Council (notified under document C(2012) 5538). Official Journal of the European Union (L series)

Fine PE (1993) Herd immunity: history, theory, practice. Epidemiol Rev 15(2):265–302

Fitzgerald JG, Defries RD, Fraser DT, Moloney PJ, McKinnon NE (1932) Experiences with diphtheria toxoid in Canada. Am J Public Health Nations Health 22(1):25–28

FluTrackers (2012) http://www.flutrackers.com. Accessed 15 Oct 2012

Galazka A (2000) The changing epidemiology of diphtheria in the vaccine era. J Infect Dis 181(Suppl 1): S2–S9

Galazka A, Robertson SE (2004) Diphtheria. In: Murray CJL, Lopez AD, Mathers CD (eds) Global Epidemiology of Infectious Diseases. World Health organisation (WHO), Geneva, pp 55–74

Galazka AM, Robertson SE (1995) Diphtheria: changing patterns in the developing world and the industrialized world. Eur J Epidemiol 11(1):107–117

Galazka AM, Robertson SE, Oblapenko GP (1995) Resurgence of diphtheria. Eur J Epidemiol 11(1):95–105

Groman NB (1984) Conversion by corynephages and its role in the natural history of diphtheria. J Hyg (Lond) 93(3):405–417

Hart RJ (1984) *Corynebacterium ulcerans* in humans and cattle in North Devon. J Hyg (Lond) 92(2):161–164

Hatanaka A, Tsunoda A, Okamoto M, Ooe K, Nakamura A, Miyakoshi M, Komiya T, Takahashi M (2003) *Corynebacterium ulcerans* diphtheria in Japan. Emerg Infect Dis 9(6):752–753

Hogg RA, Wessels J, Hart J, Efstratiou A, De Zoysa A, Mann G, Allen T, Pritchard GC (2009) Possible zoonotic transmission of toxigenic *Corynebacterium ulcerans* from companion animals in a human case of fatal diphtheria. Vet Rec 165(23):691–692

INTAS (2004) Study of the role of non-toxigenic *Corynebacterium diphtheriae* in the epidemiology of diphtheria (2002–2004). INTAS reference number 01-2289. Final report

Keja K, Chan C, Hayden G, Henderson RH (1988) Expanded programme on immunization. World Health Stat Q 41(2):59–63

Kolodkina V, Titov L, Sharapa T, Grimont F, Grimont PA, Efstratiou A (2006) Molecular epidemiology of *C. diphtheriae* strains during different phases of the diphtheria epidemic in Belarus. BMC Infect Dis 15(6):129

Koopman JS, Campbell J (1975) The role of cutaneous diphtheria infections in a diphtheria epidemic. J Infect Dis 131(3):239–244

Lartigue MF, Monnet X, Le Fleche A, Grimont PA, Benet JJ, Durrbach A, Fabre M, Nordmann P (2005) *Corynebacterium ulcerans* in an immunocompromised patient with diphtheria and her dog. J Clin Microbiol 43(2):999–1001

Lowe CF, Bernard KA, Romney MG (2011) Cutaneous diphtheria in the urban poor population of Vancouver, British Columbia, Canada: a 10-year review. J Clin Microbiol 49(7):2664–2666

Maiden MC, Bygraves JA, Feil E, Morelli G, Russell JE, Urwin R, Zhang Q, Zhou J, Zurth K, Caugant DA, Feavers IM, Achtman M, Spratt BG (1998) Multilocus sequence typing: a portable approach to the identification of clones within populations of pathogenic microorganisms. Proc Natl Acad Sci USA 95(6):3140–3145

Markina SS, Maksimova NM, Vitek CR, Bogatyreva EY, Monisov AA (2000) Diphtheria in the Russian Federation in the 1990s. J Infect Dis 181(Suppl 1):S27–S34

Melnikov VG, De Zoysa A, Mazurova I, Kombarova SI, Borisova OI, Engler K, Efstratiou A (2000) Molecular screening for the identification of non-toxigenic tox bearing strains of Corynebacterium diphtheriae from the Russian Federation. Abstract in: Proceedings of the Sixth International Meeting of the European Laboratory Working Group on Diphtheria, Brussels, Belgium, June 2000, p 60

Neal SE, Efstratiou A (2009) International external quality assurance for laboratory diagnosis of diphtheria. J Clin Microbiol 47(12):4037–4042

Neal S, Efstratiou A (2007) DIPNET—establishment of a dedicated surveillance network for diphtheria in Europe. Euro Surveill 12(12)

Popovic T, Kombarova SY, Reeves MW, Nakao H, Mazurova IK, Wharton M, Wachsmuth IK, Wenger JD (1996) Molecular epidemiology of diphtheria in Russia, 1985–1994. J Infect Dis 174(5):1064–1072

Reacher M, Ramsay M, White J, De Zoysa A, Efstratiou A, Mann G, Mackay A, George RC (2000) Nontoxigenic *Corynebacterium diphtheriae*: an emerging pathogen in England and Wales? Emerg Infect Dis 6(6):640–645

Schuhegger R, Schoerner C, Dlugaiczyk J, Lichtenfeld I, Trouillier A, Zeller-Peronnet V, Busch U, Berger A, Kugler R, Hörmansdorfer S, Sing A (2009) Pigs as a source for toxigenic *Corynebacterium ulcerans*. Emerg Infect Dis 15(8):1314–1315

Skogen V, Cherkasova VV, Maksimova N, Marston CK, Sjursen H, Reeves MW, Olsvik O, Popovic T (2002) Molecular characterization of *Corynebacterium diphtheriae* isolates, Russia, 1957-1987. Emerg Infect Dis 8(5):516–518

Tiwari TS, Golaz A, Yu DT, Ehresmann KR, Jones TF, Hill HE, Cassiday PK, Pawloski LC, Moran JS, Popovic T, Wharton M (2008) Investigations of 2 cases of diphtheria-like illness due to toxigenic *Corynebacterium ulcerans*. Clin Infect Dis 46(3):395–401

Vitek CR, Wharton M (1998) Diphtheria in the former Soviet Union: reemergence of a pandemic disease. Emerg Infect Dis 4(4):539–550

von Hunolstein C, Alfarone G, Scopetti F, Pataracchia M, La Valle R, Franchi F, Pacciani L, Manera A, Giammanco A, Farinelli S, Engler K, De Zoysa A, Efstratiou A (2003) Molecular epidemiology and characteristics of *Corynebacterium diphtheriae* and *Corynebacterium ulcerans* strains isolated in Italy during the 1990s. J Med Microbiol 52(Pt 2):181–188

Wagner KS, White JM, Crowcroft NS, De Martin S, Mann G, Efstratiou A (2010) Diphtheria in the United Kingdom, 1986-2008: the increasing role of *Corynebacterium ulcerans*. Epidemiol Infect 138(11):1519–1530

Wagner KS, White JM, Lucenko I, Mercer D, Crowcroft NS, Neal S, Efstratiou A (2012) Diphtheria in the postepidemic period, Europe, 2000-2009. Emerg Infect Dis 18(2):217–225

Wagner KS, White JM, Neal S, Crowcroft NS, Kuprevičiene N, Paberza R, Lucenko I, Jõks U, Akbas E, Alexandrou-Athanassoulis H, Detcheva A, Vuopio J, Hunolstein C von, Murphy PG, Andrews N, Efstratiou A (2011) Screening for *Corynebacterium diphtheriae* and *Corynebacterium ulcerans* in patients with upper respiratory tract infections 2007–2008: a multicentre European study. Clin Microbiol Infect 17(4):519–525

World Health Organization (WHO) (2003) WHO-recommended standards for surveillance of selected vaccine-preventable diseases. World Health Organization, Geneva

World Health Organization (WHO) (2008) International Health Regulations (2005), 2nd edn. World Health Organization, Geneva

World Health Organization (WHO) (2012) Immunization surveillance, assessment and monitoring: Data, statistics and graphics. http://www.who.int/immunization_monitoring/data/en/. Accessed 8 Oct 2012

Chapter 11
History of Diphtheria Vaccine Development

Enrico Malito and Rino Rappuoli

Abstract Diphtheria was a major cause of death among children until the beginning of the nineteenth century. The introduction in the 1920s of formaldehyde-detoxified diphtheria toxin for mass immunization, led to the control of diphtheria that is considered to be the first conquest of vaccination. Pioneering scientific discoveries were the isolation of *Corynebacterium diphtheriae* and of the secreted toxin responsible for disease, diphtheria toxin. The simultaneous progress on bacteriology and immunology brought to the development of detoxified toxins as safe and efficacious subunit vaccines that are still used in current vaccines. In this chapter we will review the main scientific breakthroughs that led to the development of the first vaccine against diphtheria and to the control of this disease.

Keywords Diphtheria · Immunization · Toxin · Toxoid · Vaccine

11.1 Introduction

The scientific progress on diphtheria and the ensuing development of the anti-diphtheria vaccine are milestones in the annals of medicine and immunology (English 1985). The control of diphtheria, once one of the most feared childhood diseases, is considered the first conquest of immunization (Pappenheimer 1984). The causative agent of diphtheria, *Corynebacterium diphtheriae*, was one of the first bacterial pathogens to be isolated and grown in pure culture, and the toxin responsible for disease, diphtheria toxin (DT), was among the first toxins discovered (Collier 2001).

After the introduction of passive immunization at the beginning of the twentieth century using toxin-neutralizing antibodies isolated from animals previously injected with sub-lethal doses of DT, the vaccine against diphtheria was developed in the 1920s using formaldehyde-detoxified DT, which was named toxoid. The

R. Rappuoli (✉) · E. Malito
Novartis Vaccines and Diagnostics, Siena, Italy
e-mail: rino.rappuoli@novartis.com

A. Burkovski (ed.), *Corynebacterium diphtheriae and Related Toxigenic Species*,
DOI 10.1007/978-94-007-7624-1_11,

widespread use of the toxoid vaccine for active immunization that began in the 1930s in the United States, and later in Europe, led to the significant reduction in the number of cases of diphtheria and to the almost complete elimination of the disease from industrialized countries. In the 1980s, diphtheria was considered as the only example of bacterial disease almost completely eliminated as a direct result of active immunization (Pappenheimer 1984), and the consensus was that the disease could not re-emerge in the same way as before the advent of immunization (Christenson 1986; Kleinman 1992). Remarkably, the same toxoid vaccine against diphtheria that was discovered in the 1920s is still in use today.

Although toxoid immunization has been very successful, rare outbreaks of diphtheria still occur today in poorly immunized groups (Murphy 1996; Eskola et al. 1998), and diphtheria remains a threat in developing countries (Mattos-Guaraldi et al. 2003). At the beginning of the 1990s, diphtheria reemerged in the Russian Federation and in the Newly Independent States (NIS) (Galazka 2000; Mattos-Guaraldi et al. 2003). The epidemic, which originated because of the low vaccination coverage in many areas as a result of the dissolution of the Soviet Union and lack of adequate public health measures, affected mainly adolescents and adults (Centers for Disease Control and Prevention (CDC) 1995). The shift in age distribution away from childhood presented a new paradigm for understanding the epidemiology of diphtheria in the vaccine era (Galazka 2000). This recent epidemic confirms the role and importance of vaccination coverage among both children and adults to reduce diphtheria (Bisgard et al. 2000; Hadfield et al. 2000; Bergamini et al. 2000).

11.2 Epidemiology of Diphtheria

Epidemics of diphtheria or of its typical symptoms (sore throat, membrane production, and death by suffocation) were already described during Egyptian and Greek times (Vitek 2006). The first reports of epidemics of what at the time was called "throat distemper" are from the sixteenth century (English 1985; Caulfield 1939). Several epidemics of diphtheria were reported in 1700s in England, and throughout the eighteenth and nineteenth century. By the late nineteenth century, diphtheria became endemic in many countries and caused extremely high morbidity and mortality among children, both in Europe and in North America (Vitek 2006). In the 1880s, mortality rates during outbreaks of diphtheria reached 50 % in certain areas of Europe and the US (World Health Organization 2006). After the introduction of antitoxin treatment at the end of the eighteenth century, the incidence of diphtheria in Europe during the First World War decreased to below 15 %. Active immunization with diphtheria toxoid that begun between 1930s and 1940s led to a further decrease of reported cases in most Western countries, with only 40 cases reported in the 1980s in the US (Eskola et al. 1998). After a major epidemic occurred during the Second World War in Europe, with about a million cases reported, the first major epidemic of the vaccination era is from the 1990s in Russia and in the Newly Independent States (NIS) of the former Soviet Union (Eskola et al. 1998). The incidence of this last epidemic ranged from 64–82 % in people older than 15, and by 1999 more than 157,000 cases and 5,000 deaths were reported with almost half being adults between 40 and 49 years old (Mattos-Guaraldi et al. 2003). Also, rare outbreaks in countries where the disease is endemic, which include Africa, India, Bangladesh, Vietnam, the tropics and certain areas of South America, are still reported (Mattos-Guaraldi et al. 2003; Galazka and Robertson 1995; Galazka 2000). This indicates that diphtheria is still a dangerous disease, and that it will be important to maintain high vaccination coverage in the population.

11.3 The Discovery of Diphtheria and of Diphtheria Toxin

The first of a series of breakthroughs in the understanding of diphtheria dates back to 1821, during an epidemic in France, when Pierre Bretonneau was the first to define diphtheria based on the clinical observations of the membranes it formed in the throat (Nezelof 2002). In 1883, Klebs observed under the microscope cocci and bacilli in sections of diphtheritic membranes, and in 1884 Loeffler isolated the diphtheria bacillus (Loeffler 1884). This became known as the Klebs-Loeffler bacillus, and was only later named *Corynebacterium diphtheriae*. Loeffler also isolated avirulent (nontoxigenic) *C. diphtheriae* from healthy individuals, and noticed that it was identical to the virulent (toxigenic) bacterium of strains isolated from patients with diphtheria. This discovery was crucial in establishing the carrier state as important in the maintenance and spread of the disease. It is now well known that avirulent strains of *C. diphtheriae* may be converted to the virulent phenotype following infection and lysogenization by corynebacteriophages that carry the structural gene for DT, *tox* (Murphy 1996).

The observation by Loeffler that the bacillus could only be isolated from the membranes in the throat and did not spread to other tissues opened the question of how the bacteria could cause damage throughout the body. Loeffler hypothesized that the bacteria were able to release a soluble poison, or toxin, capable of inducing damage away from the respiratory tract. This was demonstrated in 1888, by the revolutionary work of Roux and Yersin, which grew in the laboratory bacilli isolated from the throat of children with diphtheria (Roux and Yersin 1888). By filtering the liquid broth where the bacteria grew, Roux and Yersin isolated the toxin responsible for diphtheria, called diphtheria toxin (DT), and later injected these bacteria-free filtrates into healthy guinea pigs. The guinea pigs soon became sick with symptoms resembling those caused by diphtheria in man. Importantly, Roux suggested that to combat diphtheria it was necessary to target specifically the newly-discovered toxin.

11.4 Diphtheria Antitoxin

The cornerstone of a therapy for diphtheria emerged in the 1890s, when Behring and Kitasato discovered the "antitoxic" effect of serum from animals challenged with either diphtheria or tetanus bacilli (Grundbacher 1992; Linton 2005). In a revolutionary publication that is considered the first major therapeutic breakthrough of the bacteriological era, Behring and Kitasato demonstrated that sub-lethal doses of either diphtheria or tetanus toxins induced mammals to form substances in the blood that would neutralize the activity of the injected toxin (Behring 1890). In addition, Behring and Kitasato showed that the antitoxic capacity found in the blood of rabbits immunized against tetanus could be transferred to other animals. The substances responsible for protection (antitoxins) were shown to be protective antibodies. This research represents the basis of serum therapy and passive immunization against both diphtheria and tetanus, and led to the discovery of humoral immunity.

Previously, Koch observed how blood from animals surviving diphtheria protected susceptible animals from infection, and that serum only was able to provide protection (English 1985; Lévy 1975). Behring injected sub-lethal doses of diphtheria bacilli previously treated with iodine trichloride in guinea pigs, rabbits, sheep, goats and horses. The elucidation of the specific substance of the blood serum that was responsible for immunity, thus the recognition of the existence of antibodies, was described in a second seminal paper that was published 1 week after the previous one written with Kitasato, and signed by Behring only (Linton 2005). Here, Behring focused exclusively on diphtheria, and reported five methods to immunize guinea pigs, but recognized that these methods could not be used in humans. In discussing his theory that a special property of the immune animals' serum was responsible for the immunizing power, rather than some sort of "habituation", Behring also hypothesized the therapeutic potential of antitoxin. These studies earned Behring the first Nobel Prize for medicine in 1901 (Grabenstein 2010), "for his work on serum therapy, especially its application against diphtheria, by which he has opened a new road in the domain of medical science and thereby placed in the hands of the physician a victorious weapon against illness and deaths" (Behring 1901).

Reports of the first serum therapy with antitoxin prepared in sheep and horses, and conducted on children, appeared in 1891–1892. At the University of Leipzig (Germany), children under the care of Heubner were treated with antitoxin; while in 1893 Roux and Martin conducted a clinical trial in two hospitals in Paris (Grabenstein 2010). Also in 1892, several enterprises started the commercial production of antitoxin, and in 1894 Roux showed that horses were the most efficient at producing antitoxin (Dolman 1973). Both Heubner's and Roux's clinical trials showed great reduction of mortality rates, and the final proof of the efficacy of the diphtheria antitoxin therapy came in 1898 from the work of the Danish physician Fibiger, who performed "the best-known nineteenth century example of a controlled therapeutic experiment", or in other words the first random clinical trials in medical history (Gachelin 2007; Chalmers 2001). Fibiger's trial is a milestone in the history of clinical trials, the first one to introduce random allocation, and the first one to be planned and reported with high accuracy even by modern standards (Hróbjartsson et al. 1998).

Diphtheria antitoxin quickly became international news, and in 1895 became widely available. Hermann Biggs, chief bacteriologist of the New York City Health Department, was visiting the European laboratories working on diphtheria at the time of Roux's results on serum therapy, and positively impressed he cabled to

William Park in New York: "Diphtheria antitoxin is a success; begin to produce it" (de Kruif 1926). The diphtheria antitoxin became available in the United States between 1894 and 1895, and its use for immunization was strongly supported by public campaigns sponsored by newspapers. The New York Herald organized a public fund-raising drive to support the production of antitoxin. The results of immunizing with antitoxin were striking. The number of diphtheria deaths reported in 1894 in New York City was 2,870, and by 1900 it was down to 1,400. In addition, this number declined steadily in the following decades (Hammonds 1999).

To solve problems of potencies with antitoxin preparations, in 1894 Williams and Park isolated a strain with high toxigenicity that was named Park-Williams No. 8 (PW#8) (Lampidis and Barksdale 1971; Pappenheimer 1984). This strain is still in use today to make DT for the preparation of toxoid. At the same time, Ehrlich developed a technique to standardize the potency of diphtheria antitoxin, and this became the first international standard reference preparation (Ehrlich 1897). Ehrlich's studies on toxin-antitoxin represent the foundation of modern immunology, explaining interactions of antigens and antibodies by the receptor theory, and earned him the Nobel Prize for Medicine in 1908 (Grabenstein 2010). Ehrlich wrote in his Nobel prize lecture: "My quantitative investigations of this process [of production of specific antitoxin in the animal body] have shown that the toxins, especially the solutions of diphtherial toxin, will—either spontaneously if left standing for some time, or through the action of thermal influences or certain chemicals (iodine)—change in such a way that they are more or less deprived of their toxicity but retain their ability to produce antibodies" (Ehrlich 1908).

A method to diagnose immunity to diphtheria was introduced in 1913 by Schick (Schick 1913), and consisted of intradermic injections of a dilution of diphtheria. This was a simple means of testing for antitoxic immunity, as susceptible individuals would develop erythema following the injection.

11.5 Toxin-Antitoxin Immunization

The main impediment to passive immunizations by use of antitoxin was the short-term immunity that generally lasted only for few months. For this reason, antitoxin was mainly used during epidemics, and to treat acute infections, but could not be used as a prophylactic measure. Mixtures of toxin-antitoxin were found to be more efficient at producing immunity in animals by Babes in 1895 (Park 1922). Previously, Behring and Wernicke, in 1892, had studied the combination of antitoxin and bacilli for immunizing animals. In 1895, Wernicke observed immunity in the offspring of guinea pigs previously immunized with a mixture of toxin-antitoxin (Smith 1907). In 1896, Park immunized horses that were used for the production of antitoxin with mixtures of toxin-antitoxin (Park 1922). Also, in 1909, Smith reported the use of mixtures of DT and antitoxin for producing active immunity (Smith 1909). In 1913, Behring injected for the first time toxin-antitoxin in humans, and published his results in 1914 discussing also why passive immunization with antitoxin alone never succeeded in raising long-term protection. Furthermore, in this publication Behring described his clinical studies on the new toxin-antitoxin (TA) vaccine (Linton 2005; Park et al. 1914). The first toxin-antitoxin immunizations were organized in New York, where Park and Zingher injected toxin-antitoxin mixtures in non-immune children as identified by the use of the Schick test (Park 1931; Baker 2000). Park played an influential role in promoting toxin–antitoxin in New York, and by 1933, 1 million children were immunized.

Clearly, immunization with toxin-antitoxin mixtures was at the time the most promising therapy for diphtheria. As Glenny wrote in 1921: "The Schick reaction and diphtheria prophylaxis by active immunisation with toxin-antitoxin mixtures together constitute the greatest advance in our knowledge of the problems associated with diphtheria since the introduction of antitoxin. It is probable that when public opinion is sufficiently familiarized with these two methods and their significance, they will come into universal use and there will follow the same reduction in the occurrence of diphtheria that has been brought about by Jennerian vaccination for smallpox" (Glenny and Brien 1921). However, the risk of using toxin-antitoxin mixtures in humans was too high, with the main problem being the difficulty to carefully control the ratio between toxin and antitoxin. In addition, because very small doses of toxin had to be used several injections were required per subject.

11.6 Toxoid Vaccine and Active Immunization

To obviate the intrinsic risks of using crude toxin for injections, researchers were studying possible nontoxic forms of DT that could be used for safe and effective mass vaccination. In 1922, Park wrote: "It was early realized that if we ever attempted actively to immunize children, the advantage would be great if we could replace pure toxin with an altered toxin which would be equally effective and without danger" (Park 1922). Ehrlich's work on the titration of anti-diphtheric sera from the 1890s introduced the concept of toxoid, defined as a nontoxic product of the toxin but still capable of interaction with the antitoxin. Fraenkel used toxin after heat treatment at 60 °C in 1890 (Pope 1963), while Behring tested iodine trichloride to inactivate DT (Linton 2005). Between 1898 and 1904, Salkowski and Loewenstein studied the effect of formaldehyde on soluble toxins from both tetanus and diphtheria using a concentration of formaldehyde of 0.15–0.2 % (FitzGerald 1927; Loewenstein 1909). At the same time, Smith immunized guinea pigs with toxoid demonstrating a durable immunity, and suggested a potential role for toxoids in prophylaxis (Smith 1907). In 1921, Glenny and Südmersen reported on the immunogenicity of DT rich in toxoid prepared by previous treatment with formaldehyde (Glenny and Südmersen 2009), and in 1923 Glenny and Hopkins published their results on modifying toxin with 0.1 % formaldehyde and exposing it for 4 weeks at 37 °C (FitzGerald 1927; Glenny and Hopkins 1923). In both cases, although the toxoid did not kill the animals, it was not completely nontoxic, as the guinea pigs developed from small to large swelling at the site of injection depending on the dose of toxoid injected. As such, the toxoid prepared by Glenny was useful only if used in combination with antitoxin. As Glenny and Hopkins wrote: "It is hoped by further reduction in toxicity and by concentration to produce an immunising agent of far greater efficiency than any yet employed for protection against diphtheria." (Glenny and Hopkins 1923).

The breakthrough discovery that allowed the preparation of a completely nontoxic diphtheria toxoid is due to Gaston Ramon, who in 1922 introduced the flocculation method for measuring the dosage of toxin and antitoxin (Ramon 1922; Ramon 1923a; Ebisawa 1996). This technique allowed Ramon to measure the antigenic and immunogenic activity of toxins by observing their flocculation properties once mixed, and it eliminated the use of animals as a means of titrating DT and antitoxin (Bayne-Jones 1924). Also, the formation of a flocculent precipitate allowed Ramon to measure the activities of toxin modified by formaldehyde and treated by heat. In 1924, Ramon treated DT with a 1:2,000 dilution of formaldehyde (which at the time was used as antiseptic), and the toxoid so generated had the same antigenic behavior as the original toxin (Ebisawa 1987). By increasing the amount of formaldehyde to 3–4 ml per 1 L of toxin, and by heating the preparation at 40–42 °C, Ramon was the first to produce a completely nontoxic substance (Relyveld 1996; Ebisawa 1996). This formaldehyde-treated toxin was immunogenic in guinea pigs and horses (Ramon 1923b), and Ramon attempted to demonstrate the safety of the toxoid by injecting himself with 0.25 ml (Ramon 1924; Ebisawa 1987). The toxoid prepared by Ramon was named anatoxin by Roux, and this could now be used for safe active immunization in humans. Vaccinations using toxoid in Schick-positive children began in 1924, and the efficacy of this prophylaxis was later unambiguously confirmed (Bloomberg and Fleming 1927; Keller and Harris 1934; Ebisawa 1987; Kjeldsen et al. 1985), and recommended immunization procedures established (Emerson et al. 1935).

Arguably, the discovery of anatoxin by Ramon represents the beginning of the modern era of toxoid vaccines (Pizza et al. 2003). As all vaccines in use at the time were made of living virus with attenuated virulence originally introduced by Jenner and Pasteur, toxoid vaccines were the first example of what were to be called "chemical vaccines", or what we call today "subunit vaccines." Therefore, Ramon's discovery of diphtheria toxoid was truly revolutionary in the field of vaccinology.

After Ramon's discovery of its detoxifying properties, formaldehyde has been used to inactivate several toxins, whole bacteria cells, and viruses (Rappuoli 1994). Although it is well established that formaldehyde treatment has a great effect on the toxicity, antigenicity, and immunogenicity of DT, the exact location and mechanism of detoxification are not fully understood. In all likelihood, formaldehyde induces the formation of intramolecular cross-links in several regions of DT, with those located in the NAD^+-binding cavity and/or the receptor-binding site being the most disruptive for toxicity (Metz et al. 2005). In addition to its inactivation properties, formaldehyde acts also to stabilize antigens, which is useful for their long-term storage (Rappuoli 1994).

11.7 Preparation of Diphtheria Vaccine

Today, diphtheria vaccine (diphtheria toxoid) is manufactured in a large number of countries. The original procedure to produce antigenic toxoids as first reported by Ramon in 1923 is still used with only small refinements (Plotkin 2011). The hypertoxinogenic strain of *C. diphtheriae* PW8 is grown in fermentors for 36–48 h, using a medium containing enzymatic digests of beef (Linggood 1941). When the concentration of toxin in the supernatant reaches 150–250 Lf/mL (the flocculation units, Lf, being defined as the amount of toxin that flocculates 1 unit of an international reference antitoxin) bacteria are removed by centrifugation and 0.75% of formaldehyde is added to the supernatant. The complete detoxification of DT is obtained by storing the supernatant for up to 6 weeks at 37 °C (Linggood et al. 1963; Rappuoli 1990). Once fully detoxified, the toxoid is concentrated by ultrafiltration and purified by a combination of salt fractionation and chromatographic methods, and it is then tested for potency (Keller 2011), toxicity, sterility, and reversibility (toxoid potency is measured in international units, IU, as determined by measuring the amount of neutralizing antitoxin in previously immunized guinea-pigs). For vaccine preparation the toxoids are absorbed onto aluminum hydroxide, aluminum or calcium phosphate.

11.8 Adjuvants

Another crucial development in vaccinology due to Ramon is the use of adjuvants as enhancer of the immune response (Prigge 1955; Ebisawa 1987). While studying the immunization of animals, Ramon observed that horses that were developing abscesses at the site of injection produced antitoxin of higher titers than horses that did not develop abscesses (Relyveld 1996). Ramon subsequently tested several organic and inorganic substances as adjuvants, and observed that tapioca was enhancing antibody production (Ramon 1925a; Ramon 1925b).

Already Roux and Yersin, in 1889, observed that filtrates of diphtheriae cultures formed antigen-containing precipitates that were insoluble (Prigge 1955). In 1926, Glenny used aluminum potassium, known as alum, to precipitate diphtheria toxoid and to delay absorption from the site of injection, observing: "toxin or toxoid may be precipitated by the addition of varying quantities of potassium alum. An emulsion of such a precipitate has high antigenic properties" (Glenny et al. 1926). In summary, the toxoid adjuvanted by precipitation with aluminum potassium persisted in the tissue and elicited a durable protection. In 1930, Glenny also showed that both diphtheria and tetanus toxoids absorbed on alum had better immunity (Glenny 1930). Today, aluminum salts are the most widely used vaccine adjuvants in human vaccines (Clements and Griffiths 2002; Lindblad 2004).

11.9 Diphtheria and Tetanus as Precursors of Combination Vaccines

At the same time of the discovery of diphtheria anatoxin, the tetanus anatoxin was also discovered (Smith 1969). Ramon and Zoeller investigated the possibility of performing simultaneous vaccination against diphtheria and tetanus using the respective anatoxins (Ramon and Zoeller 1926). This combination of anatoxins resulted in no negative interference and in an enhanced antibody response; the era of "combined vaccination" was thus born. Combinations of antigens from different pathogens are used in many current vaccines, and their benefits have been documented (Halsey 2001). Since the 1940s, children in industrialized countries are routinely immunized with a combined diphtheria, tetanus, and pertussis (DTP) vaccine, which became among the most widely used vaccines and has been part of the WHO Expanded Programme on Immunization (EPI) since 1974. Diphtheria toxoid vaccine is also available in combination with tetanus toxoid only, in two formulations with higher or lower diphtheria antigen for immunizing children or adolescents and adults, respectively. Since the 1990s, the combination vaccine DTaP (diphtheria, tetanus, and acellular-pertussis) replaced the DTwP (diphtheria, tetanus, and whole-cell pertussis) for childhood immunization. The average duration of protection from the DTP vaccine is about 10 years, and protective immunity may be boosted naturally, through exposure to circulating strains of toxigenic *C. diphtheriae*, or by use of booster doses beyond infancy and early school age (World Health Organization 2006). The WHO recommendation for diphtheria immunization is to apply an effective primary immunization in infancy and to maintain immunity throughout life (World Health Organization 2009).

11.10 Toxoids as Carriers for Conjugate Vaccines

Since its discovery, DT quickly became one of the most extensively studied bacterial toxins, and it occupied a central role in the field of toxicology in the last century (Collier 2001). At the time of its first use in mixtures with antitoxin (1914), and later as formaldehyde-treated toxoid (1924), the chemical nature of DT was unknown. Pivotal contributions to the understanding of the nature and fundamental properties of DT came from the work of Pappenheimer Jr. (Keith 2011; Collier 1996; Collier 2001; Pappenheimer 1993; Pappenheimer and Gill 1973; Ratts and Murphy 2005), who in 1937 was able to purify DT in crystalline form (Pappenheimer 1937). This was the first bacterial toxin to be purified. Later, Pappenheimer and Collier showed that DT inhibited protein synthesis in HeLa cells and in HeLa cell extracts (Collier and Pappenheimer 1964), and Collier showed that DT was able to inactivate elongation factor 2 (EF-2) in the presence of NAD (Collier 1967). In 1968, Pappenheimer and Gill showed that the inactivation of EF-2 was the result of ADP-ribosylation catalyzed by DT (Gill et al. 1969).

In 1972, again in the laboratory of Pappenheimer, Uchida used nitroguanidine to mutagenize the phage containing the gene for DT, isolating a number of phages that encoded mutated and nontoxic forms of DT (Uchida et al. 1973). These mutant proteins were named CRMs (cross-reacting materials), and they played a crucial role for understanding the mechanism of action of DT. Despite their potential for vaccine development, being naturally nontoxic, CRMs did not replace the classical diphtheria toxoid because of higher susceptibility to proteases, lower immunogenicity and low production yields with respect to DT (Pizza et al. 2003). However, one of the CRMs, CRM197, which is an enzymatically inactive and nontoxic form containing a single amino acid substitution from Glycine to Glutamate at position 52 (Giannini et al. 1984), has been extensively used as carrier for conjugate vaccines against encapsulated bacteria (Bröker et al. 2011; Ada and Isaacs 2003; Rappuoli 1983). Vaccines containing CRM197 as a carrier and currently used to immunize hundreds of millions of children annually include Menveo® against serogroups A-C-W135-Y of *Neisseria meningitidis*, Menjugate® and Meningitec® (against serotype C of *N. meningitidis*), Vaxem-Hib® and HibTITER® (against *Haemophilus influenzae* type B, Hib) and the multivalent pneumococcal conjugate Prevnar™ (Shinefield 2010; Bröker et al. 2011; Costantino et al. 2011). In addition to its role as a carrier, it was also shown that CRM197 found in conjugate vaccines can boost the anti-diphtheria vaccine (Bröker et al. 2011; Podda et al. 1991).

In 1992, the crystal structure of DT was solved revealing the composition of the toxin in three discrete domains (Choe et al. 1992), and very recently the crystal structure of CRM197 has also been solved, providing the structural and molecular bases for the lack of toxicity of this mutant (Malito et al. 2012).

11.11 Conclusions

The current diphtheria toxoid vaccine, which is almost exclusively formulated in combination with tetanus and pertussis antigens, is among the oldest, safest, and most affordable vaccine available (World Health Organization 2006). Only slight refinements to the original toxoid formulations of Ramon to obtain higher purity and better immunogenicity by using adjuvants have been implemented for the production of the current diphtheria vaccines (Rappuoli 1994; World Health Organization 2009). Advantages of this refined diphtheria vaccine are the absence of anaphylactic and severe reactions, as well as its relatively simple composition and production, which is a very well established and safe process performed in many countries (World Health Organization 2006). Further developments of the DTP vaccine, have seen the inclusion of additional antigens in order to comply with immunization schedules and reduce the number of injections. In addition to the DTP components, new combination vaccines include antigens from inactivated poliomyelitis (IPV), *H. influenzae* type B (Hib), and/or hepatitis B virus (HBV), for tetra-, penta-, or hexa-valent products (World Health Organization 2009; Kitchin 2011; Borrow et al. 2011).

Although the control of diphtheria between the 1930s and 1980s can also be partially ascribed to factors such as new public health policies and progress in medicine practices, advances in bacteriology and immunology between the end of the nineteenth and the beginning of the twentieth century unquestionably provided the crucial knowledge to develop a safe and efficient vaccine. However, as the epidemics of the 1990s in the Russian Federation and in the Newly Independent States (NIS) show, diphtheria can still be considered a dangerous disease. Measures to ensure continued control of diphtheria through immunization and high vaccination coverage among children, adolescents, and adults are therefore strongly advisable.

Acknowledgements We thank Francesco Berti, Paolo Costantino, and Matthew Bottomley for useful discussions, and Catherine Mallia for editorial assistance.

References

Ada G, Isaacs D (2003) Carbohydrate–protein conjugate vaccines. Clin Microbiol Infec 9(2):79–85

Baker JP (2000) Immunization and the American way: 4 childhood vaccines. Am J Public Health 90(2):199–207

Bayne-Jones S (1924) The titration of diphtheria toxin and antitoxin by Ramon's flocculation method. J Immunol 9(6):481–504

Behring E (1890) Untersuchungen uber das zustandekommen der diphtherie-immunitat bei theiren. Deut Med Wochenschr 16:1145–1148

Behring E (1901) The Nobel prize in physiology or medicine 1901. Nobelprize.org. Nobel Media AB 2013. http://www.nobelprize.org/nobel_prizes/medicine/laureates/1901/

Bergamini M, Fabrizi P, Pagani S, Grilli A, Severini R, Contini C (2000) Evidence of increased carriage of Corynebacterium spp. in healthy individuals with low antibody titres against diphtheria toxoid. Epidemiol Infect 125(1):105–112

Bisgard KMK, Rhodes PP, Hardy IRI, Litkina ILI, Filatov NNN, Monisov AAA, Wharton MM (2000) Diphtheria toxoid vaccine effectiveness: a case-control study in Russia. J Infect Dis 181(Suppl 1):S184–S187

Bloomberg MW, Fleming AG (1927) Diphtheria immunization with diphtheria toxoid (anatoxin-ramon). Can Med Assoc J 17(7):801–803

Borrow R, Dagan R, Zepp F, Hallander H, Poolman J (2011) Glycoconjugate vaccines and immune interactions, and implications for vaccination schedules. Expert Rev Vaccines 10(11):1621–1631

Bröker M, Costantino P, DeTora L, McIntosh ED, Rappuoli R (2011) Biochemical and biological characteristics of cross-reacting material 197 (CRM197), a non-toxic mutant of diphtheria toxin: use as a conjugation protein in vaccines and other potential clinical applications. Biologicals 39(4):195–204

Caulfield E (1939) The "Throat Distemper" of 1735–1740: part II. Yale J Biol Med 11(4):277–335

Centers for Disease Control and Prevention (CDC) (1995) Diphtheria epidemic–New independent states of the former Soviet Union, 1990–1994. MMWR. Morb Mortal Wkly Rep 44(10):177–181

Chalmers I (2001) Comparing like with like: some historical milestones in the evolution of methods to create unbiased comparison groups in therapeutic experiments. Int J Epidemiol 30(5):1156–1164

Choe S, Bennett MJ, Fujii G, Curmi PM, Kantardjieff KA, Collier RJ, Eisenberg D (1992) The crystal structure of diphtheria toxin. Nature 357(6375):216–222

Christenson B (1986) Is diphtheria coming back? Ann Clin Res 18(2):69–70

Clements C, Griffiths E (2002) The global impact of vaccines containing aluminium adjuvants. Vaccine 20:S24–S33

Collier RJ (1967) Effect of diphtheria toxin on protein synthesis: inactivation of one of the transfer factors. J Mol Biol 25(1):83–98

Collier RJ (1996) A tribute to Pap. In: Frandsen PL, Alouf JE, Falmagne P et al (ed) Bacterial Protein Toxins. Gustav Fischer Verlag-Stuttgart-Jena, New York, pp 1–4

Collier RJ (2001) Understanding the mode of action of diphtheria toxin: a perspective on progress during the 20th century. Toxicon 39(11):1793–1803

Collier RJ, Pappenheimer AM (1964) Studies on the mode of action of diphtheria toxin. II. Effect of toxin on amino acid incorporation in cell-free systems. J Exp Med 120:1019–1039

Costantino P, Rappuoli R, Berti F (2011) The design of semi-synthetic and synthetic glycoconjugate vaccines. Expert Opin Drug Discov 6(10):1045–1066

de Kruif P (1926) Microbe hunters, 1996 Ed. Harcourt, Inc

Dolman C (1973) The Donald T. Fraser Memorial Lecture, 1973. Landmarks and pioneers in the control of diphtheria. Can J Public Health 64(4):317–336

Ebisawa I (1987) The encounter of Gaston Ramon (1886–1963) with formalin: a biographical study of a great scientist. Kitasato Arch Exp Med 60(3):55–70

Ebisawa I (1996) Three to four instead of one millilitre of formalin. Vaccine 14(3):247

Ehrlich P (1897) Die wertbemessung des diphtherieheilserums und deren theoretische grundlagen. In, vol 6. Klinisches Jahrbuch, pp 299–326

Ehrlich P (1908) The Nobel prize in physiology or medicine 1908. Nobelprize.org. Nobel Media AB 2013. http://www.nobelprize.org/nobel_prizes/medicine/laureates/1908/

Emerson H, Freeman AW, Chapin CV, Dublin LI, Rice JL, Vaughan HF, Mountin JW, Walker WF (1935) Recommended procedures for diphtheria immunization: sub-committee on evaluation of administrative practices of the committee on administrative practice. Am J Publ Health Nation's Health 25(6):712–714

English PC (1985) Diphtheria and theories of infectious disease: centennial appreciation of the critical role of diphtheria in the history of medicine. Pediatrics 76(1):1–9

Eskola J, Lumio J, Vuopio-Varkila J (1998) Resurgent diphtheria—are we safe? Br Med Bull 54(3):635–645

FitzGerald JG (1927) Diphtheria toxoid as an immunizing agent. Can Med Assoc J 17(5):524–529

Gachelin G (2007) The designing of anti-diphtheria serotherapy at the Institut Pasteur (1888–1900): the role of a supranational network of microbiologists. Dynamis (Granada, Spain) 27:45–62

Galazka AM (2000) The changing epidemiology of diphtheria in the vaccine era. J Infect Dis 181(Suppl 1):S2–S9

Galazka AM, Robertson SE (1995) Diphtheria: changing patterns in the developing world and the industrialized world. Eur J Epidemiol 11(1):107–117

Giannini G, Rappuoli R, Ratti G (1984) The amino-acid sequence of two non-toxic mutants of diphtheria toxin: CRM45 and CRM197. Nucleic Acids Res 12(10):4063–4069

Gill DM, Pappenheimer AM, Brown R, Kurnick JT (1969) Studies on the mode of action of diphtheria toxin. VII. Toxin-stimulated hydrolysis of nicotinamide adenine dinucleotide in mammalian cell extracts. J Exp Med 129(1):1–21

Glenny AT (1930) Insoluble precipitates in diphtheria and tetanus immunization. Bri Med J 2(3632):244

Glenny AT, Brien R (1921) The Schick reaction and diphtheria prophylactic immunization with toxin-antitoxin mixture. Lancet 197(5102):1236–1237

Glenny AT, Hopkins BE (1923) Diphtheria toxoid as an immunizing agent. Br J Exp Pathol 4:283–288

Glenny AT, Pope CG, Waddington H, Wallace U (1926) Immunological notes. XVII–XXIV. J Pathol Bacteriol 29(1):31–40

Glenny AT, Südmersen HJ (2009) Notes on the production of immunity to diphtheria toxin. J Hygiene 20(02):176–220

Grabenstein JD (2010) Toxoid vaccines. In: Artenstein A (ed) Vaccines: a biography. Springer, pp 105–124

Grundbacher F (1992) Behring's discovery of diphtheria and tetanus antitoxins. Immunol Today 13(5):188–190

Hadfield TL, McEvoy P, Polotsky Y, Tzinserling VA, Yakovlev AA (2000) The pathology of diphtheria. J Infect Dis 181(Suppl 1):S116

Halsey NA (2001) Combination vaccines: defining and addressing current safety concerns. Rev Infect Dis 33(Suppl 4):S312–S318
Hammonds EM (1999) Childhood's deadly scourge: the campaign to control diphtheria in New York City, 1880–1930. The Johns Hopkins University Press
Hróbjartsson A, Gøtzsche PC, Gluud C (1998) The controlled clinical trial turns 100 years: fibiger's trial of serum treatment of diphtheria. BMJ 317(7167):1243–1245
Keith JM (2011) Bacterial protein toxins used in vaccines. In: Rappuoli R, Bagnoli F (eds) Vaccine design: innovative approaches and novel strategies. Caister Academic Press, Norfolk, UK,pp 109–137
Keller AE, Harris S (1934) The use of diphtheria toxoid in immunization of medical students and nurses. J Am Med Assoc 102(26):2163
Keller JE (2011) Overview of currently approved serological methods with a focus on diphtheria and tetanus toxoid potency testing. Procedia in Vaccinology 5:192–199
Kitchin NRE (2011) Review of diphtheria, tetanus and pertussis vaccines in clinical development. Expert Rev Vaccines 10(5):605–615
Kjeldsen K, Simonsen O, Heron I (1985) Immunity against diphtheria 25–30 years after primary vaccination in childhood. Lancet 325(8434):900–902
Kleinman LC (1992) To end an epidemic. Lessons from the history of diphtheria. New Engl J Med 326(11):773–777
Lampidis T, Barksdale L (1971) Park-Williams number 8 strain of Corynebacterium diphtheriae. J Bacteriol 105(1):77–85
Lévy F (1975) The fiftieth anniversary of diphtheria and tetanus immunization. Prev Med 4:226–237
Lindblad EB (2004) Aluminium compounds for use in vaccines. Immunol Cell Biol 82(5):497–505
Linggood FV (1941) Purification of diphtheria toxin and toxoid made from tryptic digest broth. Bri J Exp Path 22:255–261
Linggood FV, Stevens MF, Fulthorpe A, Woiwod A, Pope CG (1963) The toxoiding of purified diphtheria toxin. Bri J Exp Path 44(2):177
Linton DS (2005) Emil von Behring: infectious disease, immunology, serum therapy. Amer Philosophical Society
Loeffler F (1884) Untersuchungen über die bedeutung der mikroorganismen für die entstehung der diphtherie beim menschen, bei der taube und beim kalbe. 2:421–499
Loewenstein E (1909) Über aktive schutzimpfung bei tetanus durch toxoide. Med Microbiol Immunol 62(1):491–508
Malito E, Bursulaya B, Chen C, Surdo PL, Picchianti M, Balducci E, Biancucci M, Brock A, Berti F, Bottomley MJ, Nissum M, Costantino P, Rappuoli R, Spraggon G (2012) Structural basis for lack of toxicity of the diphtheria toxin mutant CRM197. Proc Natl Acad Sci U S A 109(14):5229–5234
Mattos-Guaraldi AL, Moreira LO, Damasco PV, Hirata Júnior R (2003) Diphtheria remains a threat to health in the developing world: an overview. Memórias do Instituto Oswaldo Cruz 98(8):987–993
Metz B, Kersten GFA, de Jong A, Meiring H, ten Hove J, Hennink WE, Crommelin DJA, Jiskoot W (2005) Identification of formaldehyde-induced modifications in proteins: reactions with diphtheria toxin. In: Structural characterisation of diphtheria toxoid. vol 7. Proefschrift Universiteit Utrecht, pp 139–154
Murphy JR (1996) Corynebacterium diphtheriae. In: Baron S (ed) Medical Microbiology, 4th ed. vol 32. NCBI Bookshelf, Galveston (TX)
Nezelof C (2002) Pierre Fidèle Bretonneau 1778–1862. A pioneer in understanding infectious diseases. Ann Diagn Pathol 6(1):74–82
Pappenheimer AM (1937) Diphtheria toxin. I. Isolation and characterization of a toxic protein from Corynebacterium diphtheriae filtrates. J Biol Chem 120(2):543–553
Pappenheimer AM (1984) Diphtheria. In: Germanier R (ed) Bacterial vaccines. vol 1. Academic Press, Inc, pp 1–35
Pappenheimer AM (1993) The story of a toxic protein, 1888–1992. Protein science 2(2):292–298
Pappenheimer AM, Gill DM (1973) Diphtheria. Science (New York, NY) 182(110):353–358
Park WH (1922) Toxin-antitoxin immunization against diphtheria. J Am Med Assoc 79(19): 1584–1591

Park WH (1931) The history of diphtheria in New York City. Arch Pediatr Adolesc Med 42(6):1439–1445
Pizza M, Masignani V, Rappuoli R (2003) Toxins as vaccines and adjuvants. In: Burns D, Barbieri J, Iglewski B, Rappuoli R (eds). Bacterial protein toxins vol 22. ASM Press, Washington, D. C., pp 311–325
Plotkin S (2011) History of vaccine development. Springer, New York, NY
Podda A, Vescia N, Donati D, Marsili I, Volpini G, Nencioni L, Mastroeni I, Rappuoli R, Fara GM (1991) A phase-I clinical trial of a new antitetanus/antidiphtheria vaccine for adults. Ann Ig 3(2):79–84
Pope CG (1963) Development of knowledge of antitoxins. Bri Med Bull 19:230–234
Prigge R (1955) The development of diphtheria vaccines. Bull World Health Organ 13(3):473–478
Ramon G (1922) Floculation dans un melange neutre de toxine-antitoxine diphteriques. R Soc Biol (Paris) 86:661–663.
Ramon G (1923a) La floculations dans les melanges de toxine et de serum antidiphtherique. Ann Inst Pasteur 37:1001–1011
Ramon G (1923b) Sur le pouvoir floculant et sur les propriétés immunisantes d'une toxine diphtérique rendue anatoxique (anatoxine). CR Acad Sci 177:1338–1340
Ramon G (1924) Sur la toxine et sur l'anatoxine diphtériques. Pouvoir floculant et propriétés immunisantes. Ann Inst Pasteur 48:1–10
Ramon G (1925a) Sur la production de l'antitoxine diphtérique. CR Soc Biol 93:506–507
Ramon G (1925b) Sur la production des antitoxines. CR Soc Biol 181:157–159
Ramon G, Zoeller C (1926) Les "vaccins associés" par union d'une anatoxine et d'un vaccin microbien (T.A.B.) ou par mélange d'anatoxines. CR Soc Biol 13:155–165
Rappuoli R (1983) Isolation and characterization of Corynebacterium diphtheriae nontandem double lysogens hyperproducing CRM197. Appl Environ Microbiol 46(3):560–564
Rappuoli R (1990) New and improved vaccines against diphtheria and tetanus. In: Woodrow G, Levine M (eds) New generation vaccines. vol 17. Marcel Dekker Inc., New York, pp 417–435
Rappuoli R (1994) Toxin inactivation and antigen stabilization: two different uses of formaldehyde. Vaccine 12(7):579–581
Ratts R, Murphy JR (2005) Diphtheria toxin, diphtheria-related fusion protein toxins, and the molecular mechanism of their action against eukaryotic cells. In: Schmitt M, Schaffrath R (eds) Microbial protein toxins. Topics in current genetics, vol 11. Springer Berlin Heidelberg, Heidelberg, pp 1–20
Relyveld EH (1996) A history of toxoids. In: Plotkin S, Fantini B (eds) Vaccinia, vaccination and vaccinology: Jenner, Pasteur and their successors. Springer, Vaccinology Elsevier Paris, pp 95–105
Roux E, Yersin A (1888) Contribution a l'etude de la diphtherie. Annales de l'institut pasteur 2(12):629–661
Schick B (1913) Die diphtherietoxin—Hautreaktion des menschen als vorprobe der prophylaktischen diphtherieheilseruminjektion. Münch Med Woch 60(47):2608–2610
Shinefield HR (2010) Overview of the development and current use of CRM(197) conjugate vaccines for pediatric use. Vaccine 28(27):4335–4339
Smith JW (1969) Diphtheria and tetanus toxoids. Bri Med Bull 25(2):177–182
Smith T (1907) The degree and duration of passive immunity to diphtheria toxin transmitted by immunized female guinea-pigs to their immediate offspring. J Med Res 16(2):359–379
Smith T (1909) Active immunity produced by so called balanced or neutral mixtures of diphtheria toxin and antitoxin. J Exp Med 11(2):241–256
Uchida T, Pappenheimer AM, Greany R (1973) Diphtheria toxin and related proteins. I. Isolation and properties of mutant proteins serologically related to diphtheria toxin. J Biol Chem 248(11):3838–3844
Vitek CR (2006) Diphtheria. Curr Top Microbiol Immunol 304:71–94
World Health Organization (2006) Diphtheria vaccine. Wkly Epidemiol Rec 81(3):24–32
World Health Organization (2009) The Immunological Basis for Immunization Series. Module 2: Diphtheria Update 2009. http://apps.who.int/iris/handle/10665/44094?mode=full&submit_simple=Show+full+item+record

Chapter 12
Antimicrobial Susceptibility and Treatment

Aleksandra Anna Zasada

Abstract Antibiotic treatment is essential in all—toxigenic and nontoxigenic—*Corynebacterium diphtheriae* infections as well as in close contacts of diphtheria patients. The chapter presents methods for antimicrobial susceptibility testing of *C. diphtheriae* and recommendations with the focus on discrepancies concerning testing conditions and interpretive breakpoints. Resistance of *C. diphtheriae* isolated in different regions of the world has been described including multidrug-resistant strains.

Keywords Antibiotic resistance · Antimicrobial susceptibility testing · Diphtheria antitoxin · Multidrug-resistant

12.1 Treatment of *C. diphtheriae* Infections

In diphtheria cases, the most effective treatment is early administration of diphtheria antitoxin (DAT), along with appropriate antimicrobial therapy to eliminate the corynebacteria from the site of infection, stop exotoxin production, and reduce communicability (Wagner et al. 2009; Tiwari 2012). DAT is a preparation of immunoglobulins or immunoglobulin $F(ab)_2$ fractions produces from immunization of horses. Antitoxin will only neutralize circulating toxin that is not yet bound to tissue, it is therefore critical that DAT is administrated as soon as presumptive diagnosis has been made without waiting for bacteriological confirmation. The dose of DAT to be administrated depends upon the site and extent of the diphtheritic membrane, the degree of toxicity, and the duration of illness. Recommendations are described in the WHO manual for the management and control—document ICP/EPI 038 (B) (Begg 1994; Wagner et al. 2009).

A. A. Zasada (✉)
National Institute of Public Health—National Institute of Hygiene,
Department of Bacteriology, Chocimska 24, 00-791 Warsaw, Poland
e-mail: azasada@pzh.gov.pl

A. Burkovski (ed.), *Corynebacterium diphtheriae and Related Toxigenic Species*,
DOI 10.1007/978-94-007-7624-1_12,

Antimicrobial prophylaxis should also be applied for close contacts of diphtheria patients (Tiwari 2012). The recommended drugs for the antimicrobial treatment and control of diphtheria are penicillin and erythromycin (Tiwari 2012; Begg 1994). Penicillin also represents the first antibiotic used to treat systemic *C. diphtheriae* infections, usually in combination with an aminoglycoside (Pereira et al. 2008; Tiley et al. 1993).

12.2 *C. diphtheriae* Susceptibility to Antibiotics

12.2.1 Penicillin and Erythromycin

Penicillin is a drug of choice for diphtheria patients and carriers. Strains resistant to penicillin have been isolated from invasive infections (Farfour et al. 2012) and from diphtheria cases (Gladin et al. 1999; Pereira et al. 2008,). However, resistance of *C. diphtheriae* to this antimicrobial agent have been notified very rare. Nevertheless, failures to eliminate *C. diphtheriae* from cases and carriers treated with penicillin have been reported for many years (eg. Zalma et al. 1970; McLaughlin et al. 1971; Wilson 1995; von Hunolstein et al. 2002). von Hunolstein et al. (2002) analyzed MIC and MBC (Minimal Bactericidal Concentration) of penicillin for non-toxigenic *C. diphtheriae* and found that for 71 % of the strains MBC/MIC ratio was ≥ 32, which has been suggested as indicating tolerance. The results were compared with therapy outcomes and revealed that a high percentage of non-toxigenic *C. diphtheriae* isolates are tolerant to penicillin and that there are no differences in the MIC (minimum inhibitory concentration) between susceptible and tolerant strains. Therefore, the authors recommended treatment of non-toxigenic *C. diphtheriae* infections with erythromycin. On the other hand, Kneen et al. (1998) showed that median time to fever clearance in diphtheria cases was significantly shorter for penicillin-treated patients than for the erythromycin-treated patients and that penicillin was better tolerated. The authors recommended penicillin as the drug of choice for treatment patients with diphtheria and suggested that erythromycin should be reserved for treatment of penicillin-allergic patients.

Resistance of *C. diphtheriae* to erythromycin is recorded more often than to penicillin. Erythromycin-resistant strains were identified in Vietnam (Kneen et al. 1998), France (Patey et al. 1995), USA (Coyle et al. 1979), Russia (Gladin et al. 1999). Resistance to erythromycin could be induced by subinhibitory levels of the antibiotic. Interestingly, subinhibitory concentrations of erythromycin induced also resistance to the chemically unrelated antibiotic clindamycin (Coyle et al. 1979; Kneen et al. 1998).

The erythromycin resistance of *C. diphtheriae* was found to be plasmid-meditated. Schiller et al. (1983) and Serwold-Davis and Groman (1986) showed that the plasmid and plasmid elements that correlate with erythromycin resistance in *C. diphtheriae* are closely related to plasmid sequences in skin coryneforms. Thus it is supposed that the normal skin flora could serve as a reservoir of resistance plasmids for the pathogen. However, despite the high frequency of plasmid carriage by skin coryneforms, plasmid carriage by *C. diphtheriae* is uncommon (Schiller et al. 1983; Pereira et al. 2008; Zasada et al. 2010).

12.2.2 Other Antibiotics

Resistance of *C. diphtheriae* to antimicrobial drugs is different in different countries. Variations in biotype and toxigenicity of the isolates have no effect on the susceptibility of the strains (Engler et al. 2001; Pereira et al. 2008). It is worth to mention a high level of resistance among *C. diphtheriae* to trimethoprim-sulphamethoxazole in India achieving over 83 % (Sharma et al. 2007). A high level (86 %) of tetracycline resistance of *C. diphtheriae* was reported in Indonesia (Rockhill et al. 1982), but also in Western Europe (Funke et al. 1999) and Brasil (Pereira et al. 2008), despite the fact that the resistance was regarded as typical for the isolates from Swiss injecting drug users (Gruner et al. 1992). Reduced susceptibility to cefotaxime was recorded in Poland (Zasada et al. 2010) and Italy (von Hunolstein et al. 2003). Generally, *C. diphtheriae* isolates described in international literature could be regarded as susceptible to most of antibiotics, with some exception. Nonetheless, in must be kept in mind that multidrug-resistant strains have also been identified.

12.2.3 Multidrug-Resistant **C. diphtheriae**

Multidrug-resistant *C. diphtheriae* strains have been recognized only very rarely in recent global literature. The most multiple resistance phenotypes were observed among *C. diphtheriae* isolates in Brazil where 95.7 % of strains were resistant to between four and seven drugs tested. 8.5 % of isolates had multiple resistance to seven drugs: mupirocin, penicillin and/or ampicillin, oxacillin, ceftazidime, aztreonam, tetracycline, and/or lincomycin, clindamycin and erythromycin (Pereira et al. 2008). Studies conducted on 130 *C. diphtheriae* strains (toxigenic and non-toxigenic) isolated in St. Petersburg (Russia) revealed 2.3 % of isolates resistant to eight drugs: benzylpenicillin, ampicillin, oxacillin, chloramphenicol, erythromycin, lincomycin, trimethoprim, and nitroxolin (Gladin et al. 1999). The first case of multidrug-resistant toxigenic *C. diphtheriae* strain in Canada was isolated from skin abscess in 2011. The strain was resistant to clindamycin, erythromycin, tetracycline and trimethoprim-sulfamethoxazole and intermediate to ceftriaxone and cefotaxime (Mina et al. 2011). Multidrug-resistant *C. diphtheriae* were also identified in Vietnam (Kneen et al. 1998) and France (Patey et al. 1995).

12.3 Antimicrobial Susceptibility Testing—Methods and Recommendations

The performance of antimicrobial susceptibility testing of bacterial isolates is an important task of the clinical microbiology laboratory. Traditionally, three methods for antimicrobial susceptibility testing have been used: a disk diffusion method, a broth dilution or microdilution method and an E-test method called also a gradient method.

Disk diffusion is one of the oldest approaches to antimicrobial susceptibility testing and remains one of the most widely used antimicrobial susceptibility testing methods in routine clinical laboratories. It is a qualitative method and categorizes a bacterial isolate as sensitive, intermediate or resistant to a particular antibiotic. The test is performed by applying a standardized bacterial inoculum to the surface of an appropriate agar growth medium. Then, paper disks containing fixed concentration of antibiotics are placed on the inoculated agar surface. After incubation, the zones of growth inhibition around each of the antibiotic disk are measured to the nearest millimeter. The diameter of the zone is related to the susceptibility of the isolate and to the diffusion rate of the drug through the agar medium.

The broth dilution or microdilution method is a quantitative method. It enables to determine the minimum amount of antibiotic that inhibits the visible growth of an isolate or minimum inhibitory concentration (MIC). This procedure involves preparing two-fold dilutions of antibiotics in a liquid growth medium dispensed in test tubes or small reagent wells of plastic trays. Then, the tubes or wells are inoculated with a standardized bacterial suspension. Following incubation the tubes or wells are examined for visible bacterial growth as evidenced by turbidity.

The E-test could serve as an alternative to microdilution. In this method the standardized bacterial inoculum is applied to the surface of an appropriate agar growth medium and plastic strips impregnated with a dried antibiotic concentration gradient are placed on the agar surface. The test plates are evaluated after appropriate incubation time given in applying recommendations—for details see Table 12.1 (Jorgensen and Ferraro 2009).

Comparison of these three methods for susceptibility testing of *Corynebacterium* species revealed a good overall agreement between them (Weiss et al. 1996; Martinez-Martinez et al. 1995; von Hunolstein et al. 2002) and all the three methods have been used for susceptibility testing for *C. diphtheriae.* However, there is no unified recommendation for antimicrobial susceptibility testing of *Corynebacterium* species, including *C. diphtheriae.* Recommendations apply in different countries differ in testing conditions, like medium, inoculum and incubation, as well as in interpretive breakpoints (Table 12.1). Whereas density of the inoculum seems not to influence results of the test, different results could be obtained when different media are used (McLaughlin et al. 1971; Zasada et al. 2010).

The Clinical and Laboratory Standard Institute (CLSI), which recommendations are broadly used in many countries, for the first time published the recommendations for antimicrobial susceptibility testing of *Corynebacterium* ssp. (in-

Table 12.1 Comparison of selected recommendations for antimicrobial susceptibility testing of *Corynebacterium* spp. applied in some countries

Source of recommendations	Method	Medium	Inoculum	Incubation
Clinical and Laboratory Standard Institute (CLSI). Methods for antimicrobial dilution and disk susceptibility testing of infrequently isolated or fastidious bacteria; approved guideline—second edition. CLSI document M45-A2. Clinical and Laboratory Standard Institute, Wayne, PA USA, 2010	MIC (broth microdilution)	Cation-adjusted Mueller-Hinton broth with 2.5–5 % lysed horse blood	0.5 McFarland	35 °C, ambient air, 24–48 h
British Society for Antimicrobial Chemotherapy (BSAC). BSAC Methods for Antimicrobial Susceptibility Testing. Version 10.2 May 2011	Disk diffusion and MIC	Iso-Sensitest agar (ISA) with 5 % horse blood and 20 mg/L NAD	0.5 McFarland diluted 1/10	35–37 °C, 4–6 % CO_2 in air, 18–20 h
Comite de l'antibiogramme de la Societe Francaise de Microbiologie (CA-SFM). Recommandations 2011 [in French][a]	Disk diffusion and MIC	Mueller-Hinton agar with 5 % sheep blood	0.5 McFarland diluted 1/10	35–37 °C, 18–24 h
Krajowy Ośrodek Referencyjny ds. Lekowrażliwości Drobnoustrojów (KORDL). Rekomendacje doboru testów do oznaczania wrażliwości bakterii na antybiotyki i chemioterapeutyki 2010. Oznaczanie wrażliwości pałeczek Gram-dodatnich z rodzaju *Corynebacterium* spp. Krajowy Ośrodek Referencyjny ds. Lekowrażliwości Drobnoustrojów, Narodowy Instytut Leków Centralny Ośrodek Badan Jakości w Diagnostyce Mikrobiologicznej, 2010 [in Polish]	MIC	Mueller-Hinton agar with 5 % sheep blood	1.0 McFarland	35 °C, 5 % CO_2 in air or ambient air, 24–48 h or longer if slow grower
BioMerieux, Etest Application Guide—16273A-2010/11	MIC	Mueller-Hinton agar with 5 % blood	1.0 McFarland	35 °C, 5 % CO_2 in air, 20–24 h (48 h if required)

[a] CA-SFM suggests use of recommendations for *Streptococcus* spp

cluding *C. diphtheriae*) in May 2006 in the document M45-A "*Methods for the Antimicrobial Dilution and Disk Susceptibility Testing of Infrequently Isolated or Fastidious Bacteria*". On the other hand, the European Committee on Antimicrobial Susceptibility Testing (EUCAST), which has been harmonized most antimicrobial MIC breakpoints in Europe, is still working on the recommendations for *Corynebacterium* ssp.

Before the recommendations of CLSI for *Corynebacterium* spp. were published the criteria for *Streptococcus* ssp. or less often for *Listeria* sp. were used for *C. diphtheriae* antimicrobial susceptibility testing and they are still used in some countries. Moreover, some researchers applied breakpoints for *Staphylococcus aureus* to penicillin, ampicillin and oxacillin (Funke et al. 1996; Gomes et al. 2009). Weiss et al. (1996) conducted analysis concerning comparison of usefulness of the criteria for the determination of the susceptibility of streptococci and *L. monocytogenes* to penicillin and revealed that in order to avoid errors, the *Streptococcus* instead of the *Listeria* interpretative criteria should be used when assessing the activity of penicillin against *Corynebacterium* spp.

It is also worth to mention automated systems as well as genetic methods for antimicrobial susceptibility testing. Currently, several automated instrument systems for antimicrobial susceptibility testing are available on the market, for example Vitek and Vitek 2 Systems (BioMerieux), BD Phoenix Automated Microbiology System (Becton Dickinson Diagnostics), MicroScan WalkAway (Siemens Healthcare Diagnostics). Automated systems often produce susceptibility test results in a shorter period than manual readings because sensitive optical detection systems allow detection of subtle changes in bacterial growth. However, there is no published information concerning the usefulness of the systems for *C. diphtheriae* so far.

As resistance to antibiotics is very often related to presence of resistance genes in the bacterial genome, PCR and hybridization analysis are common methods used to detect antibiotic resistance genes in bacteria, currently. In recent years, microarray technology has been of growing interest to microbiologists. DNA microarrays offer an alternative method for screening for the presence of a wide diversity of genes. In this format, probes specific to each gene are deposited onto a solid substrate (usually glass) in a lattice pattern. DNA is then labeled and hybridized to the array, and specific target-probe duplexes are detected with a reporter molecule (Call et al. 2003). Parreten et al. (2005) proposed the use of microarray technology for routine microbial investigations that allows rapid and efficient screening of gram-positive bacteria for the presence of up to 90 of the most prevalent and transferable antibiotic resistance genes. Unfortunately, the method has not been tested for *Corynebacterium* spp.

12.4 Summary

Penicillin and erythromycin are the drugs of choice in treatment of *C. diphtheriae* infections. But emerging resistant and multidrug-resistant isolates forces testing of antimicrobial susceptibility of strains isolated form clinical samples. For these reasons it seems to be necessary to establish consistent international recommendations for antimicrobial susceptibility testing of *C. diphtheriae*. Usefulness of automated systems for antimicrobial susceptibility testing of *C. diphtheriae* also should be evaluated as many clinical laboratories are using such systems routinely.

References

Begg N (1994) Manual for the management and control of diphtheria in the European region. WHO Europe, ICP/EPI 038 (B) Copenhagen

Call DR, Bakko MK, Krug MJ, Roberts MC (2003) Identifying antimicrobial resistance genes with DNA microarrays. Antimicrob Agents Chemother 47(10):3290–3295

Coyle MB, Minshew BH, Bland JA, Hsu PC (1979) Erythormycin and clindamycin resistance in *Corynebacterium diphtheriae* from skin lesions. Antimicrob Agents Chemother 16(4):525–527

Engler KH, Warner M, George RC (2001) In vitro activity of ketolides HMR 3004 and HMR 3647 and seven other antimicrobial agents against *Corynebacterium diphtheriae*. J Antimicrob Chemother 47(1):27–31

Farfour E, Badell E, Zasada A, Hotzel H, Tomaso H, Guillot S, Guiso N (2012) Characterization and comparison of invasive *Corynebacterium diphtheriae* isolates from France and Poland. J Clin Microbiol 50(1):173–175

Funke G, Altwegg M, Frommelt L, von Graevenitz A (1999) Emergence of related nontoxigenic *Corynebacterium diphtheriae* biotype *mitis* strains in Western Europe. Emerg Infect Dis 5(3):477–480

Funke G, Punter V, von Graevenitz A (1996) Antimicrobial susceptibility patterns of some recently established coryneform bacteria. Antimicrob Agents Chemother 40(12):2874–2878

Gladin DP, Kozlova NS, Zaitseva TK, Cherednichenko AS, Khval SA (1999) Sensitivity of *Corynebacterium diphtheriae* isolated in Saint-Petersburg to antibacterial drugs. Antibiot Khimioter 44(5):17–21

Gomes DLR, Martins CAS, Faria LMD, Sanots LS, Santos CS, Sabbadini PS, Souza MC, Alves GB, Rosa ACP, Nagao PE, Pereira GA, Hirata Jr R, Mattos-Guaraldi AL (2009) *Corynebacterium diphtheriae* as an emerging pathogen in nephrosomy catheter-related infection: evaluation of traits associated with bacterial virulence. J Med Microbiol 58(11):1419–1427

Gruner E, Zuber PLF, Martinelli-Lucchini G, von Graevenitz A, Altwegg M (1992) A cluster of non-toxigenic *Corynebacterium diphtheriae* infections among Swiss intravenous drug abusers. Med Microbiol Lett 1:160–167

Jorgensen JH, Ferraro MJ (2009) Antimicrobial susceptibility testing: a review of general principles and contemporary practices. Clin Infect Dis 49(11):1749–1755

Kneen R, Pham NG, Solomon T, Tran TM, Nguyen TT, Tran BL, Wain J, Day NP, Tran TH, Parry CM, White NJ (1998) Penicillin vs. erythromycin in the treatment of diphtheria. Clin Infect Dis 27(4):845–850

Martinez-Martinez L, Ortega MC, Suarez AI (1995) Comparison of E-test with broth microdilution and disk diffusion for susceptibility testing of coryneform bacteria. J Clin Microbiol 33(5):1318–1321

McLaughlin JV, Bickham ST, Wiggins GL, Larsen SA, Balows A, Jones WL (1971) Antibiotic susceptibility patterns of recent isolates of *Corynebacterium diphtheriae*. Appl Microbiol 21(5):844–851

Mina NV, Burdz T, Wiebe D, Rai JS, Rahim T, Shing F, Hoang L, Bernard K (2011) Canada's first case of a multidrug-resistant *Corynebacterium diphtheriae* strain, isolated from a skin abscess. J Clin Microbiol 49(11):4003–4005

Parreten V, Vorlet-Fawer L, Slickers P, Ehricht R, Kuhnert P, Frey J (2005) Microarray-based detection of 90 antibiotic resistance genes of gram-positive bacteria. J Clin Microbiol 43(5):2291–2302

Patey O, Bimet F, Emond JP, Estrangin E, Riegel Ph, Halioua B, Dellion S, Kiredjian M (1995) Antibiotic susceptibilities of 38 non-toxigenic strains of *Corynebacterium diphtheriae*. J Antimicrob Chemother 36(6):1108–1110

Pereira GA, Pimeta FP, Santos FR, Damasco PV, Hirata R Jr, Mattos-Guaraldi AL (2008) Antimicrobial resistance among Brazilian *Corynebacterium diphtheriae* strains. Mem Inst Oswaldo Cruz 103(5):507–510

Rockhill RC, Hadiputranto SH, Siregar SP, Muslihun B (1982) Tetracycline resistance of *Corynebacterium diphtheriae* isolated from diphtheria patients in Jakarta, Indonesia. Antimicrob Agents Chemother 21(5):842–843

Serwold-Davis TM, Groman NB (1986) Mapping and cloning of *Corynebacterium diphtheriae* plasmid pNG2 and characterization of its relatedness to plasmids from skin coryneforms. Antimicrob Agents Chemother 30(1):69–72

Sharma NC, Banavaliker JN, Ranjan R, Kumar R (2007) Bacteriological and epidemiological characteristics of diphtheria cases in and around Delhi—a retrospective study. Indian J Med Res 126(6):545–552

Schiller J, Storm M, Groman N, Coyle M (1983) Relationship between pNG2, an Em^r plasmid in *Corynebacterium diphtheriae*, and plasmids in aerobic skin coryneforms. Antimicrob Agents Chemother 24(6):892–901

Tiley SM, Kociuba KR, Heron LG, Munro R (1993) Infective endocarditis due to non-toxigenic *Corynebacterium diphtheriae*: report of seven cases and review. Clin Infect Dis 16(2):435–439

Tiwari TSP (2012) Diphtheria. In: Brunette GW (ed) CDC health information for international travel 2012. The yellow book. Oxford Iniversity Press, Inc. New York, pp 162–165.

von Hunolstein C, Scopetti F, Efstratiou A, Engler K (2002) Penicillin tolerance amongst non-toxigenic *Corynebacterium diphtheriae* isolated from cases of pharyngitis. J Antimicrob Chemother 50(1):125–128

von Hunolstein C, Alfarone G, Scopetti F, Pataracchia M, La Valle R, Franchi F, Pacciani L, Manera A, Giammanco A, Farinelli S, Engler K, De Zoysa A, Efstratiou A (2003) Molecular epidemiology and characteristics of *Corynebacterium diphtheriae* and *Corynebacterium ulcerans* strains isolated in Italy during the 1990s. J Med Microbiol 52(2):181–188

Wagner KS, Stickings P, White JM, Neal S, Crowcroft NS, Sesardic D, Efstratiou A (2009) A review of the international issues surrounding the availability and demand for diphtheria antitoxin for therapeutic use. Vaccine 28(1):14–20

Weiss K, Laverdiere M, Rivest R (1996) Comparison of antimicrobial susceptibilities of *Corynebacterium* species by broth microdilution and disk diffusion methods. Antimicrob Agents Chemother 40(4):930–933

Wilson APR (1995) Treatment of infection caused by toxigenic and non-toxigenic strains of *Corynebacterium diphtheriae*. J Antimicrob Chemother 35(6):717–720

Zalma VM, Older JJ, Brooks GF (1970) The Austin, Texas, diphtheria outbreak. J Amer Med Assoc 211(13):2125–2129

Zasada AA, Baczewska-Rej M, Wardak S (2010) An increase in non-toxigenic *Corynebacterium diphtheriae* infections in Poland—molecular epidemiology and antimicrobioal susceptibility of strains isolated from past outbreaks and those currently circulation in Poland. Int J Infect Dis 14(10):e907–e912

Chapter 13
Sialidases of Corynebacteria and their Biotechnological Applications

Seonghun Kim, Doo-Byoung Oh and Ohsuk Kwon

Abstract Sialidase (EC 3.2.1.18) is a glycosyl hydrolase which hydrolyzes terminal sialic acid residues from the glycans of glycoproteins, glycolipids, and polysaccharides. *Corynebacterium diphtheriae* harbors an extracellular exo-α-sialidase, NanH, which can cleave terminal sialic acids α(2,3)- or α(2,6)-linked to glycoconjugates. These catalytic activities of the sialidase can be used potentially for the enzymatical production of sialylated complex glycans using regioselective hydrolysis reactions. They can also be used for sialylation via transglycosylation. This chapter focuses on the biochemical properties and the structural features of *C. diphtheriae* NanH sialidase and its homologous proteins to synthesize sialyloligosaccharides through chemoenzymatic approaches. In addition, the chapter describes potential applications of NanH, including a putative vaccine candidate as a virulence factor and an exoglycosidase for analyses of the glycan structure.

Keywords *Corynebacterium diphtheriae* · *In vitro* trans-sialylation · Sialic acid · Sialidase · Sialoglycoconjugate

Abbreviations

CMP	cytidine monophosphate
Gal	galactose
Glc	glucose

O. Kwon (✉) · D.-B. Oh
Biochemicals and Synthetic Biology Research Center,
Korea Research Institute of Bioscience and Biotechnology (KRIBB),
125 Gwahak-ro, Yuseong-gu, Daejeon 305–806, Korea
e-mail: oskwon@kribb.re.kr

S. Kim
Jeonbuk Branch Institute, Korea Research Institute of Bioscience
and Biotechnology (KRIBB), 181 Ipsin-gil, Jeongeup 580–185, Korea
e-mail: seonghun@kribb.re.kr

D.-B. Oh
e-mail: dboh@kribb.re.kr

A. Burkovski (ed.), *Corynebacterium diphtheriae and Related Toxigenic Species*,
DOI 10.1007/978-94-007-7624-1_13, © Springer Science+Business Media Dordrecht
(outside the USA) 2014

MU-Gal	4-methylumbelliferyl-α-D-galactopyranoside
Neu5Ac	neuraminic acid (sialic acid)
*p*NP-	para-nitrophenyl-

13.1 Introduction

Sialic acids are a family of nine carbon α-keto aldonic acids, which are often occupied at the non-reducing end of oligosaccharide chains on glycoproteins, glycolipids, and polysaccharides. Sialic acid-containing structures naturally appear in diverse forms with different sialic acid linkages along with several functional group modifications. Four sialic acids, *N*-acetylneuraminic acid, *N*-glycolylneuraminic acid, deaminated neuraminic acid, and neuraminic acid, are major monosaccharide sugars. These major sialic acids are diversified by modification with additional substitutions at the hydroxyl group with *O*-acetyl, *O*-phosphate, *O*-sulfate, and *O*-methyl groups (Varki 1999; Chen and Varki 2010). More than 50 sialic acid structure derivatives have been detected and found to be widely distributed in various organisms, from bacteria to animals. The structural diversity of sialic acids reflects their involvement in the mediation and/or modulation of many biological processes, including intercellular interaction, cellular trafficking, intracellular adhesion, cell development, and microbial attachment (Varki 2007).

For microbial infections, the terminal sialic acids displayed on the surface of vertebrate cells, including erythrocytes and other blood cells and serum glycoproteins, mediate the adhesion of bacteria to host tissues as an initial and essential step in the infection process (Lehmann et al. 2006; Varki 2007). Several pathogenic bacteria have been shown to produce specific surface-adhesion proteins which harbor high affinity toward the sialic acids of oligosaccharide exposed on the host cell surface (Vimr et al. 2004; Lehmann et al. 2006). Some pathogenic bacteria, fungi, and protozoa can utilize the sialic acids of hosts as one type of carbon source for cell growth. In other cases, a form of pathogenic microorganism utilizes the sialic acids of host cell for the decoration of their cell surface to escape the host's immune system (Vimr et al. 2004). In the pathogenic machinery used to mediate specific interactions with sialic-acid-containing glycoconjugates on host cells, sialidases contribute to the recognition of sialic acids exposed on host cell surfaces.

Sialidase, or neuraminidase (E.C. 3.2.1.18), is an exo-α-glycosidase which hydrolyzes terminal sialic acids from a variety of sialoglycoconjugates. Sialidases are widely distributed in diverse organisms, including viruses, bacteria, fungi, protozoa, and vertebrate animals, but not plants. At least, 70 different microorganisms capable of sialidase activities have been reported (Kim et al. 2011). Several gram-positive and gram-negative bacteria-producing sialidases are commonly in close contact with mammals, acting as commensals or pathogens. Sialidases in several pathogenic microorganisms are considered to be a potential virulence factor.

Corynebacterium diphtheriae, a gram-positive pathogenic bacterium, causes diphtheria. The bacterium colonizes the mucosal surface of the respiratory tract in humans during the early stage of infection and then secretes the diphtheria toxin, inducing necrosis and injury to epithelial cells. Sialidase activity in *C. diphtheriae* has been identified in a toxin preparation (Blumberg and Warren 1961), and exo-sialidase induced in an iron-enriched culture of the bacterium was later characterized (Warren and Spearing 1963). In addition, sialidase production and the cell surface-glycan contents of *C. diphtheriae* were found to be affected by the iron ion concentration in the culture (Mattos-Guaraldi et al. 1999; Moreira et al. 2003). However, the cellular regulatory mechanisms with three different effectors, the production of extracellular sialidase, the cell surface-glycan content and the iron ion concentration, in *C. diphtheriae* have yet to be clarified. Although no detailed studies of these cellular mechanisms in the bacterium have been reported, *C. diphtheriae* sialidase is a useful enzyme to apply to the chemoenzymatic synthesis of glycoconjugates. Recently, we identified an extracellular sialidase of *C. diphtheriae*, NanH, characterized its biochemical properties, and investigated its potential catalytic activity for sialidase-mediate transglycosylation reactions (Kim et al. 2010a; Kim et al. 2010b).

This chapter will focus on *C. diphtheriae* sialidase and its homologous proteins, with special emphasis placed on the biochemical features and structures of sialidases that are potentially useful for the chemoenzymatic synthesis of sialoglycoconjugates and/or other biotechnological applications.

13.2 *Corynebacterium diphtheriae* Sialidase and its Homologous Proteins

C. diphtheriae secrets an extracellular sialidase that can hydrolyze the sialic acid at the terminal position of glycans on glycoproteins and oligosaccharides (Warren and Spearing 1963; Kim et al. 2010a). Several microorganisms, including *Arthrobacter nicotianae*, *Arthrobacter ureafaciens*, *Bacteroides fragilis*, *Clostridium perfringens*, *Pasteurella multocida*, and *Streptococcus pneumoniae*, harbor more than one sialidase with different catalytic activities (Kim et al. 2011). Although the cellular functions of these isoenzymes remain unidentified, they vary in their hydrolysis activity towards various linkages of sialic acids as well as in their expression patterns (Corfield et al. 1983; Tanaka et al. 1994; Iwamori et al. 1997; Iwamori et al. 2005). These isoenzymes may play important roles in the interaction with other organisms or in the infection of a specific tissue by the reorganization of different sialic acid-linkages (King et al. 2006; Manco et al. 2006; Uchiyama et al. 2009).

C. diphtheriae NCTC13129 genome data also shows that this bacterium possesses two putative sialidases, NanH (DIP0543, protein accession no. NP_938919) and NanI (DIP0330 protein accession no. NP_938718) (Cerdeño-Tárraga et al. 2003). We partially purified NanH (protein accession no. ACS34893), a secreted protein of *C. diphtheriae* KCTC3075, and cloned the corresponding gene *nanH* together with the putative *nanI* gene encoding another sialidase (protein accession no. ACS34894) from its genomic DNA (Kim et al. 2010a). The amino acid sequence of *C. diphtheriae* KCTC3075 NanH and NanI demonstrates 75% and 100% identity levels with the DIP0543 and DIP0330 proteins of *C. diphtheriae* NCTC13129, respectively.

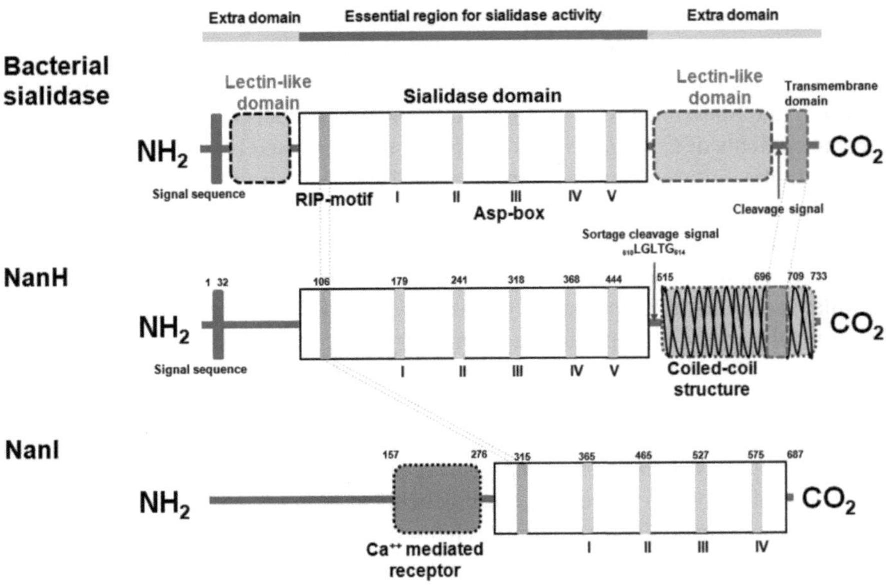

Fig. 13.1 Schematic representation and comparison of *C. diphtheriae* NanH and NanI sialidases with a typical bacterial sialidase structure. The location of the signal sequence, the RIP (Arg-Ile/Leu-Pro) motif, the Asp-boxes (I–V) lectin-like domain, and the transmembrane domain are indicated. The numbers indicate the position of each domain on the amino acid sequences of the *C. diphtheriae* NanH and NanI proteins. (Reprinted with minor adaptation from Kim et al. 2011. With permission)

Bioinformatics analysis reveals that the *C. diphtheriae* KCTC3075 NanH protein (733 amino acids) contains a putative signal sequence of 32 amino acids (Met$_1$-Ala$_{32}$) at its N-terminus and a hydrophobic transmembrane domain of 13 amino acids (Gly$_{696}$-Phe$_{709}$) at its C-terminus (Fig. 13.1). These predictions indicate that NanH is a membrane protein belonging to a typical type-Ia transmembrane protein and that its mature form would have the N$_{out}$-C$_{in}$ orientation. On the other hand, *C. diphtheriae* KCTC3075 NanI does not have any signal sequence or membrane-anchored domain. It appears to be an intracellular protein in the cytoplasm. The amino acid sequences of both NanH and NanI of *C. diphtheriae* KCTC3075 possess the conserved motifs found in the bacterial sialidase family of a nonviral origin, i.e., four or five copies of an aspartate (Asp)-box (Ser/Thr-x-Asp-x-Gly-x-Thr-Trp/Phe; where x represents any amino acid) and the RIP (Arg-Ile/Leu-Phe)-motif observed upstream of the first Asp-box in the sialidase catalytic domain (Fig. 13.1). However, the NanH and NanI proteins do not contain any of the lectin-like domains occasionally observed at the N-terminal or C-terminal region of other bacterial sialidases as extra domains. Interestingly, the C-terminus of NanH contains a unique alanine-rich domain (Asp$_{515}$-Gln$_{733}$) homologous to a putative adhesion protein of *Haemophilus somnus* 129PT. This domain is predicted to be composed of α-helical coiled-coil structures, which may serve as a secreted virulence factor or as an adhesion protein of pathogenic bacteria (Jedrzejas 2001; Delahay and Frankel 2002). In addition, the C-terminal region of NanH includes a putative sortase cleavage site, $L_{510}GLTG_{514}$, in front of a potential coiled-coil structure. These sequence analysis results indicate that the NanH protein can be localized on the cell surface or released extracellularly via sortase cleavage, playing a putative role as a virulence factor, like the sialidases of *S. pneumonia* and *Propionibacterium acnes* (Tai 2006; Nakatsuji et al. 2008).

On the other hand, although several features and functions of the *C. diphtheriae* KCTC3075 NanH protein are predictable, those of the NanI remain unclear. Although NanI includes sialidase motifs as well as potential active site residues, the recombinant protein did not show any activity toward sialoglycoconjugates when tested (Kim et al. 2010a). It may be one of the products encoding non-expressed pseudo-genes present in the cytoplasm. However, an analysis of its amino acid sequence showed that NanI may also include a putative calcium-ion-mediated receptor domain (a laminin-binding domain) containing a potential binding site for adhesion to laminin and the cell surface. This suggests that NanI is involved in adhesion, migration and differentiation through interaction with cell adhesion molecules, although further studies are required to understand its physiological roles (Magdesian et al. 2001; Tonelli et al. 2010).

13.3 Structural Properties of *C. diphtheriae* NanH Sialidase for a Catalytic Activity

Generally, sialidases are categorized according to their origins and according to the similarities of their amino acid sequences in glycoside hydrolase families in the *C*arbohydrate-*A*ctive Enzymes database (CAZy): GH33 (bacterial and eukaryotic enzymes), GH34 (influenza virus derivative enzymes), and GH83 (other virus originated enzymes) for exo-α-sialidase and GH58 (bacteriophage endosialidase) for endosialidase families are grouped (http://www.cazy.org/Glycoside-Hydrolases.html). Although *C. diphtheriae* NanH and NanI sialidases, both belonging to the GH33 family, share less than 30 % homology in their overall amino sequences, the topology of their sialidase catalytic domains described above are well conserved and share the same motifs and residues in their structures.

In a protein structure model predicted by PHYRE (Bennett-Lovsey et al. 2008) based on the homologous 3-D structure of bacterial sialidases, *C. diphtheriae* NanH displays a six-bladed β-propeller fold (Fig. 13.2a). The Asp-boxes of five copies are observed at topologically identified positions in the β-sheet folding. Although the precise NanH protein structure is still not available to elucidate the sialidase catalytic mechanism, we can predict the catalytic mechanism of *C. diphtheriae* NanH based on the structure models of other bacterial sialidases, as the motifs and special residues observed in the catalytic center are highly conserved in the bacterial enzymes (Luo et al. 1999; Newstead et al. 2008; Xu et al. 2008).

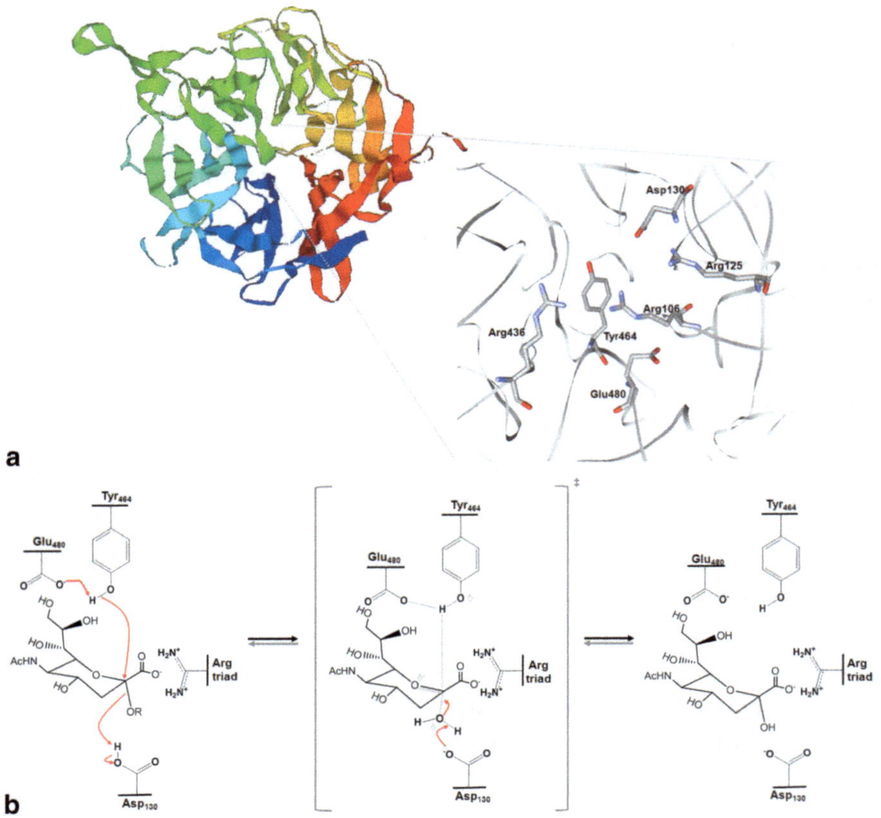

Fig.13.2 Structural model of *C. diphtheriae* NanH sialidase. **a** Overall structure of the NanH sialidase and zoomed-in view of its putative active site. The protein structure was generated by the PHYRE Protein fold recognition server (http://www.sbg.bio.ic.ac.uk/~phyre/) based on the homologous sialidase structures in the PDB database and via a PSI-blast search. **b** Proposed catalytic mechanism of the *C. diphtheriae* NanH sialidase based on bacterial sialidase structure models. (Reprinted with minor adaptation from Kim et al. 2011. With permission)

When comparing the active sites of the bacterial sialidases, several common features of *C. diphtheriae* NanH sialidase could be proposed. The highly conserved catalytic center of *C. diphtheriae* NanH consists of a tyrosine (Tyr_{464})/ glutamate (Glu_{480}) residue as a potential nucleophilic pair, an aspartic acid residue (Asp_{130}) as the acid/base catalyst, and an arginine triad (Arg_{106}-Arg_{125}-Arg_{436}) clustered with the Arg residue at the 106 position of the $R_{106}I_{107}P_{108}$ motif with two other Arg residues for stabilization through interaction with the carboxylate group of sialic acid. One of the conserved arginine residues in the active site can be stabilized by the glutamate residue (Glu_{480}). The hydroxyl group of a tyrosine residue (Tyr_{464}) is close to the C1 and C2 carbons of sialic acid (Fig. 13.2a). The catalytic mechanism for the hydrolysis reaction is initiated by the glutamate (Glu_{480}) residue, which facilitates the nucleophilic attack of the tyrosine (Tyr_{464}) residue. Once destabilized, the bond-breaking energy is offset by the formation of a delocalized π orbital between the positively charged C2 and the electron-rich O6, and 2-carboxylate can attack the proton of the carboxyl group on the aspartate (Asp_{130}) residue in an electrophilic addition reaction (Fig. 13.2b). In the transient state of the enzyme-substrate-complex, the complex sialic acid (Neu5Ac) with NanH would result in an oxocarbenium ion intermediate, with the positive charge delocalized between the anomeric carbon and the endocyclic oxygen. The structure of the covalent complex with NanH shows that the intermediate can be stabilized through a covalent bond with the nucleophilic Tyr_{464}. Asp_{130} can then activate an incoming water molecule, a nucleophile, to attack the positive charge in the anomeric carbon, creating a protonated alcoholic intermediate. The loss of H^+ from this protonated alcohol back to Asp_{130} generates the hydrolyzed Neu5Ac.

Despite the fact that the catalytic residues involved in the enzyme activity are commonly shared with others bacterial sialidases, the preference for a specific linkage, the catalytic efficiency and the enzyme kinetics of *C. diphtheriae* NanH toward sialic acid-linked substrates represent different features. This implies that other amino acid residues around the active site or the substrate binding pocket in the protein influence these properties of *C. diphtheriae* NanH rather than any conserved residues.

13.4 Application of *C. diphtheriae* NanH Sialidase for Sialoglycoconjugate Synthesis

Bacterial sialidases show different hydrolysis activities, kinetics, types of regioselectivity, and affinity toward various sialoglycoconjugates as a substrate. Many bacterial enzymes are capable of the hydrolysis of a broad range of sialoglycan substrates with either the α(2,3)-, α(2,6)-, or the α(2,8)-linked sialic acid. On the other hand, certain sialidase isoenzymes derived from gram-negative bacteria, *Salmonella typhimurium* LT2 and *Vibrio cholerae*, and gram-positive bacteria, *Clostridium chauvoei*, *Clostridium septicum*, *Clostridium sordellii*, and *Clostridium tertium*, have a relatively higher level of hydrolysis activity toward the α(2,3)-linked over the α(2,6)-linked sialic acid (Kim et al. 2011). These regioselective hydrolysis activities and stereoselective substrate specificities of the sialidases are useful characteristics for the chemoenzymatic synthesis of glycoconjugates using glycosidases and glycosyltransferases (Wang and Huang 2009; Chen and Varki 2010).

Generally, a chemical synthesis approach for glycoconjugates is known as challenging work. A synthetic procedure through several protection/de-protection steps for regio- and stereo-specific bond formations is very complicated. The purification and extraction steps occasionally require a considerable amount of time to recover intermediates. Moreover, the final products are usually obtained with low yields and low productivity levels. A chemical sialylation reaction for sialoglycoconjugates is more feasible than other types of syntheses of neutral glycans owing to the chemical structures of sialic acids, which include hindered and disfavored tertiary anomeric centers among nine carbon α-keto aldonic acids, a carboxyl group linked to the anomeric carbon, acetyl group and other similar characteristics (Chen and Varki 2010).

In contrast, a chemoenzymatic approach for sialoglycoconjugates is a sophisticated tool involving enzymatic reactions by regio- and stereo-selective sialidases or sialyltransferases. Enzymatic reactions have been considered as a promising means of selectively creating regio- and stereo-specific bond formations of glycosides under mild conditions without the need for elaborate protection or de-protection processes. However the use of sialyltransferase requires an expensive nucleotide-sugar, CMP-sialic acid, as a sialic acid donor. In contrast, sialidase can synthesize sialoglycoconjugates using less expensive sialic acid-linked glycosides through condensation and trans-glycosylation reactions (Ajisaka et al. 1994; Crout and Vic 1998; Schmidt et al. 2000). In comparison with sialyltransferase-catalyzed reaction, sialidase-catalyzed trans-sialylation has several advantages, including the use of various ready-made or natural sialic acid-linked substrates as a donor substrate, relaxed substrate specificity for acceptors, easy access for many bacterial sialidases, and flexibility of the enzyme reaction conditions for enhanced productivity. This process is also feasible with the addition of other co-organic solvents as well as alternative reaction media (Wang and Huang 2009).

13.4.1 Trans-sialylation by NanH Sialidase for Sialyl-Linkage Formation

C. diphtheriae KCTC3075 NanH is a secreted sialidase which is able to transfer sialic acid to an asialoglycoconjugate and to hydrolyze α(2,3)- and α(2,6)-linked sialic acids (Kim et al. 2010a). Regarding the sialidase hydrolysis activity toward various substrates, NanH shows the capability to cleave α(2,3)- and α(2,6)-linked sialic acids (Table 13.1). The relative hydrolysis activity of the enzyme showed the highest value toward sialyl-α(2,6)-lactose [Neu5Ac-α(2,6)-Gal-β(1,4)-Glc], with more preferable cleavage for the α(2,6)-linkage, and comparable activity for the α(2,3)-linked sialyllactose [Neu5Ac-α(2,3)-Gal-β(1,4)-Glc]. NanH also revealed the higher affinity and hydrolysis activity toward natural sialic acid substrates as compared to synthetic substrates. It is also remarkable that the relative hydrolysis activity toward sialic acid linked to glycoproteins is much lower than that by sialo-oligosaccharides.

NanH sialidase is a hydrolase cleaving the α(2,3)- linkage and α(2,6)-linkage of sialic acid conjugated to oligosaccharides and glycoproteins. However, this enzyme can also catalyze the formation of sialic acid-linkage through trans-sialylation as a reverse reaction under appropriate conditions. *C. diphtheriae* KCTC3075 NanH is an exo-α-sialidase which catalyzes the transfer of the terminal sialic acid unit to an asialoglycoconjugate acceptor (Table 13.2). In a test of trans-sialylation activity using both α(2,3)- or α(2,6)-linked sialoglycoconjugates as a sialic acid donor and 4-methylumbelliferyl-α-D-galactopyranoside (MU-Gal) as a sialic acid acceptor, *C. diphtheriae* sialidase showed trans-sialylation activity toward all of the donor

Table 13.1 Hydrolysis activity of *C. diphtheriae* NanH sialidase toward sialoglycoconjugate substrates. (*Source*: From Kim et al. 2010a. With permission)

Sialidase substrates	Representative glycan structure	Relative enzyme activity[a](%)	K_m(mM)	V_{max}(U mg^{-1})
Natural sialic acid				
Sialyl-α2,3-lactose	Neu5Acα2,3Galβ1,4Glc	89.8±7.1	5.5±1.7	41.0±6.2
Sialyl-α2,6-lactose	Neu5Acα2,6Galβ1,4Glc	100.0±3.8	7.0±2.3	57.0±5.3
Unnatural sialic acid				
*p*NP-α-sialoside	–	65.5±7.9	7.3±2.0	11.8±1.8
MU-α-sialoside	–	48.9±5.0	7.9±2.2	11.4±1.8
Glycoprotein				
Fetuin	Neu5Acα2,3Galβ1,4GlcNAc Neu5Acα2,3Galβ1,3GalNAc Neu5Acα2,3Galβ1,3(Neu5Acα2,6) GalNAc	17.2±2.4	N.D.	N.D.
α1-Acid glycoprotein	(Neu5Acα2,3(6)) Galβ1,4GlcNAc	24.1±1.3	N.D.	N.D.
Transferrin	Neu5Acα2,6Galβ1,4GlcNAc	15.5±1.7	N.D.	N.D.

[a] The relative hydrolysis activities were determined for *C. diphtheriae* NanH with 1 mM each of sialosubstrate. The value for sialyl-α2,6-lactose obtained after 1 h incubation at 37°C was set at 100%.
ND not determined

Table 13.2 Trans-sialylation activity of *C. diphtheriae* NanH sialidase toward various sialoglycoconjugate donor substrates. (*Source:* From Kim et al. 2010a. With permission)

Donor substrates[a]	Relative enzyme activity (%)	K_m (mM)	V_{max}(U mg^{-1})
Natural sialic acid			
Sialyl-α2,3-lactose; Neu5Acα2,3Galβ1,4Glc	100.0±5.6	12.2±1.9	1.6±0.2
Sialyl-α2,6-lactose; Neu5Acα2,6Galβ1,4Glc	99.2±8.5	15.4±3.8	1.7±0.2
Unnatural sialic acid			
pNP-α-sialoside	90.0±7.0	N.D	N.D
MU-α-sialoside	68.2±2.8	N.D	N.D
Glycoprotein			
Fetuin	61.2±5.3	N.D	N.D
Transferrin	47.5±12.2	N.D	N.D

[a] MU-α-D-galactopyranoside (MU-Gal) was used for the sialic acid acceptor sugar.
ND not determined

substrates tested (Table 13.2). Interestingly, the enzyme displays relatively equivalent activity levels toward both sialyl-α(2,3)-lactose [Neu5Ac-α(2,3)-Gal-β(1,4)-Glc] and sialyl-α(2,6)-lactose [Neu5Ac-α(2,6)-Gal-β(1,4)-Glc], at 1.6±0.2 U mg^{-1} and 1.7±0.2 U mg^{-1} towards the α(2,3)- and α(2,6)-isomers, respectively. In comparison with the level of hydrolysis activity toward various sialoglycoconjugates,

C. diphtheriae NanH sialidase can also transfer sialic acids of unnatural substrates and glycoproteins to MU-Gal. Interestingly, the enzyme showed relatively high levels of activity toward two unnatural sialic acid donors, *p*NP-α-sialoside and MU-α-sialoside, whereas it showed relatively low detectable levels of activity towards the glycoprotein substrates fetuin and transferrin compared to other sialoglycoconjugate donor substrates when tested.

13.4.2 Synthesis of Sialylated Glycoproteins by Trans-sialylation

Sialic acid in the *N*-linked glycan of glycoproteins is observed at the terminal position. The terminal sialic acid is an important factor which determines the quality of a therapeutic glycoprotein. It influences the *in vivo* half-life of the glycoprotein by protecting the protein from clearance by the hepatic asialoglycoprotein receptor (Bork et al. 2009; Kim et al. 2011). To enhance the sialylation of recombinant therapeutic proteins, *in vivo* and *in vitro* modification of the glycans have been extensively developed over the last few decades (Bork et al. 2009). For example, as *in vivo* approaches, several strategies have been developed to increase the metabolic flux in the biosynthetic pathway for an activated sialic acid, CMP-Neu5Ac, through the introduction of genes encoding the CMP-sialic acid transporter, control of the UDP-*N*-acetylglucosamine 2-epimerase/*N*-acetylmannosamine kinase (GNE) activity to reduce feedback inhibition, and the deletion of genes encoding endogenous sialidases. Other efforts have sought to synthesize α(2,3)- or α(2,6)-conjugated sialic acid formation by the heterologous expression of a gene encoding a linkage specific-sialyltransferase (Bork et al. 2009). The *in vitro* alternative method for an enhancement of sialylation in the glycoprotein is enzymatic sialylation using a sialyltransferase or a sialidase (Raju et al. 2001; Bork et al. 2009; Kim et al. 2010b).

One good example for *in vitro* sialylation using sialyltransferase and CMP-Neu5Ac is Etanercept®, a tetra *N*-glycosylated recombinant fusion glycoprotein, which is the homodimeric human type-2 tumor necrosis factor receptor fused to the hinge and Fc regions of the human IgG1 heavy chain (Raju et al. 2001; Bork et al. 2009). The terminal sialylation content of the protein within the heterologous glycan produced in a mammalian cell was increased to approximately 20-23 % through a two-step enzyme reaction using β(1,4)-galactosyltransferase and α(2,3)-sialyltransferase (Raju et al. 2001).

Trans-sialylation of a therapeutic glycoprotein using sialidase or trans-sialidase is one type of reaction that mimics the behavior of pathogenic microorganisms. Certain pathogenic strains of clinically isolated *C. diphtheriae* or *Trypanosoma* species are unable to synthesize sialic acid by themselves. Thus, these pathogens take out the sugar at the sialoglycoconjugates of infectious hosts *via* their surface-localized sialidase or trans-sialidase to decorate their cell surface-glycoconjugates to escape the host's immune system and to interact with the host cells (Scudder et al. 1993; Mattos-Guaraldi et al. 1998; Vimr et al. 2004). As a type of functional mimicry of these enzymes of pathogenic microorganisms, *Trypanosoma cruzi* trans-sialidase displayed on a yeast cell surface was applied to synthesize an α(2,3)-sialylated glycoconjugate using sialyl-α(2,3)-lactose as a sialic acid donor and a biantennary bigalactosylated complex *N*-glycan as an acceptor (Ryckaert et al. 2005).

In a parallel approach, *C. diphtheriae* NanH was engineered to display on a yeast cell surface (Kim et al. 2010b). In a trans-sialylation reaction with pyridylamino (PA)-labeled asialo-*N*-glycan as an acceptor and *p*NP-α-Neu5Ac as a sialic acid donor, *C. diphtheriae* NanH sialidase immobilized on the yeast surface transferred sialic acid to human-type asialo-*N*-glycans with a yield of less than 15 % (Fig. 13.3a): Asialo-biantennary *N*-glycan (PA-001), asialo-triantennary *N*-glycan (PA-002), and fucosylated asialo-triantennary *N*-glycan (PA-010) were sialylated at yields of 8.7 %, 13.7 %, and 9.0 %, respectively. Interestingly, the sialylation efficiency of triantennary *N*-glycan was slightly higher than that of the other biantennary *N*-glycans, with a conversion ratio approximately 13 % higher. For a further evaluation of the trans-sialylation activity of the immobilized *C. diphtheriae* NanH sialidase toward an asialoglycoprotein, the enzyme activity was tested using asialofetuin as a model acceptor glycoprotein and *p*NP-α-Neu5Ac as a sialic acid donor. After an enzyme reaction, the syntheses of the sialic acid linkages in the sialylated products were detected by lectin blot analyses using two sialic acid-specific lectins, *Maackia amurensis* (MAA) and *Sambucus nigra* (SNA-1), recognizing the terminal α(2,3)- and α(2,6)-sialic acids linked to galactose (Gal), respectively. The lectin blot analysis clearly revealed that the enzyme is able to transfer sialic acid to the glycan of asialofetuin with the α(2,6)- as well as the α(2,3)-linkage, although not all of the protein was completely sialylated (Fig. 13.3b). In a negative control, no protein bands of unreacted asialofetuin were observed on the lectin blot analysis. However, the sialylated fetuin, detected as a single band, implied that not all of the asialofetuin or not all of asialoglycans on the protein in the reaction mixture could be completely sialylated. Thus, only a small portion of the sialylated protein was detected in the assay. The glycoproteins sialylated by *C. diphtheriae* NanH sialidase could be also confirmed by a mobility shifting assay using an isoelectric focusing (IEF) gel and a lectin blot analysis (Fig. 13.3c). The sialylated fetuin was found to have a ladder pattern (lane 2) in the lectin blot, in which its mobility was shifted to a negative charge due to the change in the pI value of proteins harboring the sialic acid. To improve the sialylation efficiency in the enzyme reaction, 30 % (v/v) of dimethyl sulfoxide as an organic co-solvent was added to the reaction mixture. However, the protein mobility was not changed (lane 1). The sialylation efficiency of an asialoglycoprotein by the immobilized *C. diphtheriae* NanH sialidase may be similar to that of the free asialo-*N*-glycans, with a conversion ratio of less than 15 % (Fig. 13.3a).

For an *in vitro* trans-sialylation reaction to synthesize sialoglycoconjugates, the NanH sialidase has huddles of a low production yield as well as hydrolysis of sialylated products by itself. This can be overcome by protein engineering through the mutagenesis of *C. diphtheriae* NanH sialidase for protection of the newly formed sialyl linkage against hydrolysis. Moreover, optimization of the trans-sialylation reaction conditions, such as the enzyme reaction temperature, acceptor/donor ratio, reaction time, and appropriate co-solvents for the enhancement of the transglycosylation process will lead to an improvement of the productivity and the yields for sialoglycoconjugate synthesis.

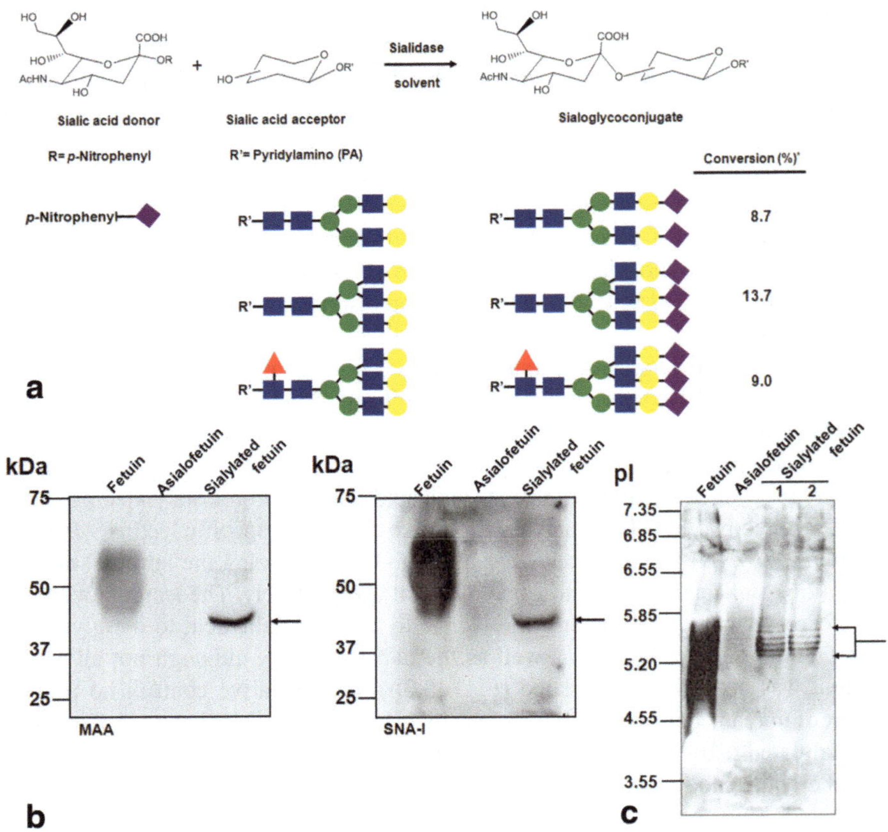

Fig. 13.3 Trans-sialylation activities of *C. diphtheriae* NanH sialidase immobilized on the surface of a yeast cell. **a** Sialylation of pyridylamino (PA)-labeled asialoglycans as a sialic acid acceptor using *p*NP-α-Neu5Ac as a sialic acid donor. *The conversion ratio (percent) was calculated by the following formula: Conversion (%) $= 100 \times [\text{product}]_t$ / $([\text{substrate}]_t + [\text{product}]t)$. **b** Lectin blot analysis of the sialylation of asialofetuin separated on 8 % SDS-PAGE. The sialylated products were detected by *M. amurensis* (MAA) and *S. nigra* (SNA-1) lectins. Fetuin and asialofetuin were used as a positive and a negative control, respectively. **c** Mobility shifting test of the sialylated fetuin by electrophoresis using IEF-gel (pH gradient 3–7) and a lectin blot analysis with MAA lectin. Lane 1, sialylated products reacted with 30 % (v/v) dimethyl sulfoxide (DMSO) as a co-solvent; lane 2, sialylated fetuin reacted without a co-solvent. The arrows indicate the sialylated products. (Reprinted with minor adaptation from Kim et al. 2010b; Kim et al. 2011. With permission)

13.4.3 Other Potential Applications

The identification of glycan structures which enhance the functional diversity and influence the biological activity in therapeutic glycoproteins or in clinical samples is an important research area (Bork et al. 2009; Mariño et al. 2010). Because naturally occurring glycoconjugates harbor regio- and stereo-specific bond formations, analyses of the glycan structures containing these complex linkages are worthy but difficult tasks. Nevertheless, the development of new analytical technologies (instrumentation-based chemical analyses) will lead to more precise, sensitive, reproducible, and robust analyses of these complex glycan structures (Mariño et al. 2010). Moreover, the glycan-linkage-specific exo-/endo-hydrolases in many bacterial enzymes, including α-sialidase, α-fucosidase, β-galactosidase, β-hexosaminidase, α-mannosidase and amidase (peptide N-glycosidase F, PNGase F), have been highlighted as useful tools for releasing a specific glycan or a glycan moiety from a glycoconjugate. Currently, sialidases derived from *S. pneumoniae*, *S. typhimurium,* and *A. ureafaciens* are mainly used in analyses of terminal sialic acids (Mariño et al. 2010). In addition to the applications of *C. diphtheriae* NanH sialidase for sialoglycoconjugate synthesis, as described previously, *vice versa,* the enzyme can be also used as a potential enzyme in an analysis of the glycan structure. This sialidase contains broad hydrolysis activities toward various sialoglyconjugates and is able to hydrolyze both α(2,3)- and α(2,6)-linked sialic acids (Table 13.1).

Another potential application of *C. diphtheriae* NanH sialidase and the homologous protein NanI may be as a target antigen for vaccinations. As bacterial sialidases are considered to be a virulence factor which recognizes sialic acid for host infection or adhesion, *C. diphtheriae* enzymes would be putative target proteins (Jedrzejas 2001; King et al. 2006 Lehmann et al. 2006; Li et al. 2011). Sialidase has been used as a vaccine target for virus influenza, bacterial pneumonia, and pathogenic protozoa (Tai 2006; Johansson and Brett 2007; Silva et al. 2009). Recently, an acne vaccine target with a surface sialidase of *Propionibacterium acnes*, a gram-positive bacterium associated with acne vulgaris, showed a successful immune response *in vivo* and *in vitro* model systems (Nakatsuji et al. 2008). Interestingly, *P. acnes* sialidase is also a cell-wall-anchored protein containing the sortase-cleavage signal LPXTG at its C-terminus, like *C. diphtheriae* NanH sialidase. Iron-mediated inhibition of diphtheria toxin production may be related to the expression of a surface-localized and secreted NanH sialidase, although the mechanism of these factors and how they influence each other remain vague (Warren and Spearing 1963; Mattos-Guaraldi et al. 1999; Moreira et al. 2003). Thus, this suggests the possibility of NanH and its homologous protein as alternative antigenic targets against the pathogenic bacterium *C. diphtheriae*. Moreover, a yeast system for the NanH sialidase immobilized on the cell surface, which was used for a trans-sialylation reaction, can be applied to the development of a novel sialidase vaccine as a natural adjuvant to improve immune responses and to increase the level of protein stability (Kim et al. 2010b; Jahns and Rehm 2012).

13.5 Conclusion

Bacterial sialidases are potential enzymes for trimming terminal α(2,3)-, α(2,6)-, and α(2,8)-sialic acids linked to glycoconjugates. These catalytic activities of sialidases can be applied to the synthesis of sialylated glycoconjugates and glycoproteins with a regio- and stereo-selective sialic acid-linkage *via* trans-sialylation. This chapter concentrates on the biochemical properties and the structural features of *C. diphtheriae* NanH sialidase and its homologous protein to synthesize sialoglycoconjugates through chemoenzymatic approaches. Although the cellular mechanisms and the biological functions of *C. diphtheriae* sialidases as a putative virulence factor are unidentified, the NanH enzyme displayed potential catalytic activities for hydrolysis and trans-sialylation toward natural and unnatural sialic acids conjugated to oligosaccharides as well as glycoproteins. In addition, the trans-sialylation activity for asialoglycoprotein harboring complex *N*-glycans showed that the sialidase can be utilized for the *in vitro* engineering of the *N*-glycan of glycoproteins. However, an improvement of the synthetic yield for sialoglycoconjugates through trans-glycosylation requires further investigation to increase the enzymatic activity of trans-sialylation via protein engineering, to optimize the reaction condition, and to develop a novel enzymatic process.

References

Ajisaka H, Fujimoto H, Isomura M (1994) Regioselective transglycosylation in the synthesis of oligosaccharide: comparison of β-galactosidases and sialidases of various origin. Carbohydr Res 259:103–115

Bennett-Lovsey RM, Herbert AD, Sternberg MJ, Kelley LA (2008) Exploring the extremes of sequence/structure space with ensemble fold recognition in the program Phyre. Proteins 70:611–625

Blumberg BS, Warren L (1961) The effect of sialidase on transferrins and other serum proteins. Biochim Biophys Acta 50:90–101

Bork K, Horstkorte R, Weidemann W (2009) Increasing the sialylation of therapeutic glycoproteins: the potential of the sialic acid biosynthetic pathway. J Pharm Sci 98:3499–3508

Cerdeño-Tárraga AM, Efstratiou A, Dover LG, Holden MT, Pallen M, Bentley SD, Besra GS, Churcher C, James KD, De Zoysa A, Chillingworth T, Cronin A, Dowd L, Feltwell T, Hamlin N, Holroyd S, Jagels K, Moule S, Quail MA, Rabbinowitsch E, Rutherford KM, Thomson NR, Unwin L, Whitehead S, Barrell BG, Parkhill J (2003) The complete genome sequence and analysis of *Corynebacterium diphtheriae* NCTC13129. Nucleic Acids Res 31:6516–6523

Chen X, Varki A (2010) Advances in the biology and chemistry of sialic acids. ACS Chem Biol 5:163–176

Corfield AP, Higa H, Paulson JC, Schauer R (1983) The specificity of viral and bacterial sialidases for α(2–3)- and α(2–6)-linked sialic acids in glycoproteins. Biochim Biophys Acta 744:121–126

Crout DHG, Vic G (1998) Glycosidases and glycosyl transferases in glycoside and oligosaccharide synthesis. Curr Opin Chem Biol 2:98–111

Delahay RM, Frankel G (2002) Coiled-coil proteins associated with type III secretion systems: a versatile domain revisited. Mol Microbiol 45:905–916

Iwamori M, Ohta Y, Uchida Y, Tsukada Y (1997) *Arthrobacter ureafaciens* sialidase isoenzymes, L, M1 and M2, cleave fucosyl GM1. Glycoconj J 14:67–73

Iwamori M, Kaido T, Iwamori Y, Ohta Y, Tsukamoto K, Kozaki S (2005) Involvement of the C-terminal tail of *Arthrobacter ureafaciens* sialidase isoenzyme M in cleavage of the internal sialic acid of ganglioside GM1. J Biochem 1(38):327–334

Jahns AC, Rehm BH (2012) Relevant uses surface proteins-display on self-organized biological structures. Microb Biotechnol 5:188–202

Jedrzejas MJ (2001) Pneumococcal virulence factors: structure and function. Microbiol Mol Biol Rev 65:187–207

Johansson BE, Brett IC (2007) Changing perspective on immunization against influenza. Vaccine 25:3062–3065

Kim S, Oh DB, Kwon O, Kang HA (2010a) Identification and functional characterization of the NanH extracellular sialidase from *Corynebacterium diphtheriae*. J Biochem 147:523–533

Kim S, Oh DB, Kwon O, Kang HA (2010b) Construction of an *in vitro* trans-sialylation system: surface display of *Corynebacterium diphtheriae* sialidase on *Saccharomyces cerevisiae*. Appl Microbiol Biotechnol 88:893–903

Kim S, Oh DB, Kang HA, Kwon O (2011) Features and applications of bacterial sialidases. Appl Microbiol Biotechnol 91:1–15

King SJ, Hippe KR, Weiser JN (2006) Deglycosylation of human glycoconjugates by the sequential activities of exoglycosidases expressed by *Streptococcus pneumoniae*. Mol Microbiol 59:961–974

Lehmann F, Tiralongo E, Tiralongo J (2006) Sialic acid-specific lectins: occurrence, specificity and function. Cell Mol Life Sci 63:1331–1354

Li J, Sayeed S, Robertson S, Chen J, McClane BA (2011) Sialidases affect the host cell adherence and epsilon toxin-induced cytotoxicity of *Clostridium perfringens* type D strain CN3718. PLoS Pathog 7:e1002429

Luo Y, Li SC, Li YT, Luo M (1999) The 1.8 Å structures of Leech intramolecular trans-sialidase complexes: evidence of its enzymatic mechanism. J Mol Biol 285:323–332

Magdesian MH, Giordano R, Ulrich H, Juliano MA, Juliano L, Schumacher RI, Colli W, Alves MJ (2001) Infection by *Trypanosoma cruzi*. Identification of a parasite ligand and its host cell receptor. J Biol Chem 276:19382–19389

Manco S, Hernon F, Yesilkaya H, Paton JC, Andrew PW, Kadioglu A (2006) Pneumococcal neuraminidases A and B both have essential roles during infection of the respiratory tract and sepsis. Infect Immun 74:4014–4020

Mariño K, Bones J, Kattla JJ, Rudd PM (2010) A systematic approach to protein glycosylation analysis: a path through the maze. Nat Chem Biol 6:713–723

Mattos-Guaraldi AL, Formiga LC, Andrade AF (1998) Trans-sialidase activity for sialic acid incorporation on *Corynebacterium diphtheriae*. FEMS Microbiol Lett 168:167–172

Mattos-Guaraldi AL, Cappelli EA, Previato JO, Formiga LC, Andrade AF (1999) Characterization of surface saccharides in two *Corynebacterium diphtheriae* strains. FEMS Microbiol Lett 170:159–166

Moreira L de O, Andrade AF, Vale MD, Souza SM, Hirata R Jr, Asad LM, Asad NR, Monteiro-Leal LH, Previato JO, Mattos-Guaraldi AL (2003) Effects of iron limitation on adherence and cell surface carbohydrates of *Corynebacterium diphtheriae* strains. Appl Environ Microbiol 69:5907–5913

Nakatsuji T, Liu YT, Huang CP, Gallo RL, Huang CM (2008) Vaccination targeting a surface sialidase of *P. acnes*: Implication for new treatment of acne vulgaris. PLoS one 3:e1551

Newstead SL, Potter JA, Wilson JC, Xu G, Chien CH, Watts AG, Withers SG, Taylor GL (2008) The structure of *Clostridium perfringens* NanI sialidase and its catalytic intermediates. J Biol Chem 283:9080–9088

Raju TS, Briggs JB, Chamow SM, Winkler ME, Jones AJ (2001) Glycoengineering of therapeutic glycoproteins: *in vitro* galactosylation and sialylation of glycoproteins with terminal *N*-acetylglucosamine and galactose residues. BioChemistry 40:8868–8876

Ryckaert S, Martens V, De Vusser K, Contreras R (2005) Development of a *S. cerevisiae* whole cell biocatalyst for *in vitro* sialylation of oligosaccharides. J Biotechnol 119:379–388

Schmidt D, Sauerbrei B, Thiem J (2000) Chemoenzymatic synthesis of sialyl oligosaccharides with sialidases employing transglycosylation methodology. J Org Chem 65:8518–8526

Scudder P, Doom JP, Chuenkova M, Manger ID, Pereira ME (1993) Enzymatic characterization of β-D-galactoside α2,3-trans-sialidase from *Trypanosoma cruzi*. J Biol Chem 268:9886–9891

Silva MS, Prazeres DM, Lança A, Atouguia J, Monteiro GA (2009) Trans-sialidase from *Trypanosoma brucei* as a potential target for DNA vaccine development against African trypanosomiasis. Parasitol Res 105:1223–1229

Tai SS (2006) *Streptococcus pneumoniae* protein vaccine candidates: properties, activities and animal studies. Crit Rev Microbiol 32:139–153

Tanaka H, Ito F, Iwasaki T (1994) Two sialidases which preferentially hydrolyze sialyl α2–8 linkage from *Bacteroides fragilis* SBT3182. J Biochem 115:318–321

Tonelli RR, Giordano RJ, Barbu EM, Torrecilhas AC, Kobayashi GS, Langley RR, Arap W, Pasqualini R, Colli W, Alves MJ (2010) Role of the gp85/trans-sialidases in *Trypanosoma cruzi* tissue tropism: preferential binding of a conserved peptide motif to the vasculature *in vivo*. PLoS Negl Trop Dis 4:e864

Uchiyama S, Carlin AF, Khosravi A, Weiman S, Banerjee A, Quach D, Hightower G, Mitchell TJ, Doran KS, Nizet V (2009) The surface-anchored NanA protein promotes pneumococcal brain endothelial cell invasion. J Exp Med 206:1845–1852

Varki A (1999) Sialic acids. In: Varki A, Cummings R, Esko J, Freeze H, Hart G, Marth J (eds) Essential of glycobiology. Cold Spring Harbor Laboratory Press, Cold Spring Harbor, New York, pp 195–209

Varki A (2007) Glycan-based interactions involving vertebrate sialic-acid-recognizing proteins. Nature 446:1023–1029

Vimr ER, Kathryn AK, Deszo EL, Steenbergen SM (2004) Diversity of microbial sialic acid metabolism. Microbiol Mol Biol Rev 68:132–153

Wang LX, Huang W (2009) Enzymatic transglycosylation for glycoconjugate synthesis. Curr Opin Chem Biol 13:592–600

Warren L, Spearing CW (1963) Sialidase (Neuraminidase) of *Corynebacterium diphtheriae*. J Bacteriol 86:950–955

Xu G, Potter JA, Russell RJM, Oggioni MR, Andrew PW, Taylor GL (2008) Crystal structure of the NanB sialidase from *Streptococcus pneumoniae*. J Mol Biol 384:436–449

Chapter 14
Molecular Genetic Tools for Research in *Corynebacterium diphtheriae*

Diana M. Oram

Abstract Research into the molecular mechanisms used by *Corynebacterium diphtheriae* to cause disease has been aided by a wide variety of innovative, recently developed, molecular, biochemical and genetic tools. Multiple compatible plasmid origins are now available and the methods used to introduce DNA to *C. diphtheriae* have been both improved and expanded to include conjugation. In the last decade, a technique to perform transposon mutagenesis in *C. diphtheriae* was developed as well as phage-based vectors that permit introduction of DNA at a known specific chromosomal site. Genetic systems that model *C. diphtheriae* gene regulation in *E. coli* have been invaluable tools and *in vitro* systems to characterize protein binding and to identify genes controlled by the diphtheria toxin repressor are available. Finally the recent expansion of sequencing and bioinformatics has yielded multiple genomic sequences for comparison both to each other and to the genomes of pathogens from other species. The vast expansion of tools available to characterize the pathogenic mechanisms of *C. diphtheriae* has been a boon to researchers; resulting in the identification of new virulence factors, including pili and a heme oxygenase, in a species that has been studied in laboratories for more than 100 years.

Keywords Conjugation · Lysogenization · SELEX · Transformation · Transposon

D. M. Oram (✉)
Department of Microbial Pathogenesis, School of Dentistry,
University of Maryland, Baltimore, MD 21201, USA

Laboratory of Respiratory and Special Pathogens, Division of Bacterial,
Parasitic, and Allergenic Products, Center for Biologics Evaluation and Research,
Food and Drug Administration, Bldg 29 Room 108, 8800 Rockville Pike,
Bethesda, MD 20892, USA
e-mail: Diana.Oram@fda.hhs.gov

A. Burkovski (ed.), *Corynebacterium diphtheriae and Related Toxigenic Species*,
DOI 10.1007/978-94-007-7624-1_14, © Springer Science+Business Media Dordrecht (outside the USA) 2014

14.1 Introduction

In 1883 Klebs recognized bacteria that were to become known as *Corynebacterium diphtheriae* in stained samples of pseudomembranes from the throats of patients with the respiratory disease diphtheria. The primary virulence factor of *C. diphtheriae*, diphtheria toxin was subsequently discovered in the late 1800s (Roux and Yersin 1888). Although the structure, function, mode of action, and virtually every other aspect of diphtheria toxin has been characterized, in detail, over the last 100+ years, details related to other virulence factors expressed by *C. diphtheriae* as well as the mechanisms and strategies used by this species to cause disease are decidedly lacking. In fact even the precise function of diphtheria toxin in promoting growth of *C. diphtheriae* is not clearly defined, though the toxin's effects on host cells are well-characterized (Oram and Holmes 2006). This knowledge gap has been caused, at least in part, by a lack of easily accessible molecular and genetic tools for use in *C. diphtheriae*.

In Gram negative bacteria, particularly those related to *Escherichia coli*, there are a wealth of techniques available to analyze gene structure and function. In contrast, development of the tools to perform these same analyses in Gram positive bacterial species has lagged decades behind. Analyses of the molecular basis of sporulation in the Gram positive bacterial species *Bacillus subtilis* spurred the development of molecular tools for this species but tools for use in species more distantly related to either *E. coli* and *B. subtilis* (including *C. diphtheriae*) were still rare at the beginning of this century. This trend has been reversed in recent years with the development of molecular tools for use not only in *C. diphtheriae* but in other Gram positive *Actinomycetales* including those in the genus *Mycobacterium*.

Several recent advances have resulted in tools to dissect gene regulation and pathogenesis in *C. diphtheriae* on a molecular level. The genome of *C. diphtheriae* strain NCTC13129 was sequenced completely in 2003, allowing a comparison between this genome and that of nonpathogenic corynebacteria as well as other related bacterial pathogens (Cerdeno-Tarraga et al. 2003). Then in 2012, the genomes of 12 additional *C. diphtheriae* strains were sequenced and annotated thereby permitting in-depth analyses of both strain-specific genes and those conserved across multiple strains (Trost et al. 2012). Plasmids capable of replicating in both *C. diphtheriae* and *E.coli* were isolated in the last 15 years as well as improved methods for transformation. In the last decade, methods for delivering DNA to the cytoplasm using conjugation and for isolating transposon insertion mutants were developed. In 2007, a bacteriophage-based integration system was developed that permits integration of exogenous DNA into a specific site in the chromosome of *C. diphtheriae*. These advances have ushered in a new molecular phase in the characterization of an old infectious disease foe and have resulted in an exponential expansion of our knowledge of the mechanisms used by *C. diphtheriae* to cause disease.

14.2 Early Genetic Analyses

The Power of Phage The first steps in using genetics to characterize the pathogenesis of *C. diphtheriae* occurred in the 1950s when Freeman reported that nontoxinogenic strains became toxinogenic after infection with specific corynephage (Freeman 1951). Characterization of phage host interactions from this decade were instrumental in demonstrating that corynephage can be either tox+ or tox− and that lysogenization of a nontoxinogenic strain of *C. diphtheriae* by a tox+ phage results in the ability to make diphtheria toxin. Conversely, curing a strain of the tox+ prophage results in loss of the ability to produce diphtheria toxin; thereby establishing that toxinogenicity in *C. diphtheriae* is determined by phage conversion (Groman 1953a, b; Barksdale and Pappenheimer 1954). These analyses were the first to show that the ability of the bacterium to make diphtheria toxin changes as a consequence of infection by a particular phage (reviewed in (Groman 1984; Holmes 2000)).

Further insights into both the location and regulation of the gene encoding diphtheria toxin came from genetic mapping and phage crosses. Analysis of recombination between toxigenic and nontoxigenic corynephages, demonstrated that the tox+ and tox− phenotypes segregated as alleles at a single genetic locus (Holmes and Barksdale 1969). Furthermore strains of *C. diphtheriae* can produce diphtheria toxin, not only, when a prophage is present in the genome, but also when a toxigenic phage is vegetatively replicating in the bacterium (Matsuda and Barksdale 1966; Gill et al. 1972). These findings helped demonstrate that the structural gene for diphtheria toxin is not coordinately regulated with other phage genes, such as those required for lysogeny. Two attachment sites for phage insertion were also identified in the *C. diphtheriae* chromosome (Rappuoli et al. 1983). The prophage genetic map was shown to be circularly permuted with respect to the vegetative genetic map. The gene encoding diphtheria toxin, *tox,* which occupies a central position in the vegetative map, was shown to be located at one end of the prophage map immediately adjacent to the attachment site (Holmes 1976; Laird and Groman 1976). The fact that *tox* is located at the end of the prophage genetic map suggests that it was incorporated into a nontoxinogenic ancestor phage during a horizontal gene transfer event.

Analyzing the Impact of Iron on Diphtheria Toxin Production Production of diphtheria toxin by toxinogenic strains of *C. diphtheriae* is affected dramatically by the content of the culture medium. The best-studied effect is that of iron. Early work demonstrated that diphtheria toxin production is greatest when *C. diphtheriae* is grown under iron-limiting conditions and is inhibited in high-iron conditions (Pappenheimer and Johnson 1936; Mueller 1940). Studies of the molecular mechanisms responsible for this iron-dependent effect began in the 1970s when mutants were isolated that were insensitive to the effects of iron (Kanei et al. 1977). Some of the mutations responsible for this phenotype were mapped to the bacterial chromosome, rather than the genome of the phage, indicating a role for a chromosomally encoded factor in controlling expression of diphtheria toxin. In addition, mutant phages were also isolated that enabled *C. diphtheriae* to produce diphtheria toxin under high-iron conditions and these mutations were cis-dominant (Murphy et al. 1976; Welkos and Holmes 1981). These data supported the hypothesis that the chromosome of *C. diphtheriae* encodes a repressor that can utilize iron as a co-repressor to inhibit transcription of *tox* under high-iron growth conditions.

The structural gene for diphtheria toxin, *tox* was cloned in the 1980s (Greenfield et al. 1983; Ratti et al. 1983) and two groups independently cloned the gene *dtxR* encoding the repressor by screening libraries of genes from *C. diphtheriae* in *E. coli* for iron-dependent inhibition of a reporter gene expressed under control of Ptox (Boyd et al. 1990; Schmitt and Holmes 1991). The diphtheria toxin repressor DtxR is the founding member of a large group of metal-dependent transcriptional regulators found in a wide variety of both pathogenic and nonpathogenic bacteria (reviewed in (Holmes 2000; Hantke 2001).

14.3 The Beginning of Genetic Engineering and Modeling Regulation in *E. coli*

Transformation and Plasmids Direct transformation of *C. diphtheriae* with DNA is significantly hampered by both the relatively impermeable Coryneform cell wall, as well as host DNA restriction barriers (Puech et al. 2001; Tauch et al. 2002). Nevertheless, several recent reports have significantly expanded the plasmids and systems available for genetic manipulation of *C. diphtheriae*. Several groups have reported methods to improve the DNA transformation efficacy into *C. diphtheriae*. These methods include altering the growth medium of *C. diphtheriae* prior to transformation in an effort to weaken the cell wall, isolating DNA from other coryneform species rather than from *E. coli* immediately prior to transformation and utilizing a heat-shock to inhibit the activity of restriction systems (Schmitt 1997b; Oram et al. 2002; Tauch et al. 2002). In 2003, and as discussed in more detail below, conjugation was used for the first time to introduce DNA to *C. diphtheriae* (Ton-That and Schneewind 2003). These advances have certainly made genetic analyses of *C. diphtheriae* easier.

Several plasmid origins which function to support replication in *C. diphtheriae* have been isolated and characterized. One of these is from pNG2, a plasmid that is capable of replicating to high copy number in *E. coli* while maintaining a low copy number in *C. diphtheriae* (Serwold-Davis et al. 1987; Tauch et al. 2003). A second origin comes from the well-characterized (Sonnen et al. 1990) and fully sequenced (Neŝvera et al. 1997) pGA1. In fact, this origin has been shown to be compatible with pNG2 permitting these two plasmids to be maintained in the same *C. diphtheriae* cell (Bibb et al. 2007). Finally two other origins originally isolated from cryptic plasmids in *C. glutamicum,* namely those from pBL1 and pHM1519 (Wohlleben et al. 1993) have been shown to support replication in *C. diphtheriae* (Tauch et al. 2002). The final three replication origins from pGA1, pBL1 and pHM1519 do not support stable replication in *E. coli* but the work of A. Tauch and colleagues resulted in multiple shuttle vectors that contain both *C. diphtheriae* specific origins and replication systems that function in *E. coli* and thereby greatly expanding the repertoire of plasmids available for use in *C. diphtheriae.*

Modeling DtxR-Dependent Gene Regulation There are a large number of vectors and genetic manipulation techniques available for studying gene regulation in *E. coli.* In addition, assays performed *in vitro* can provide key information, such as conditions required for protein activity. Some of these systems have been adapted to model gene regulation from *C. diphtheriae* and to subsequently identify DtxR-regulated genes. Transcriptional fusions, where a promoterless reporter gene is fused to a promoter of interest, are extremely useful in determining factors that impact gene expression. A transcription fusion system that models promoter regulation by DtxR in *E. coli* was developed early in the 1990s (Schmitt and Holmes 1991). In this system DtxR is expressed in *E. coli* on a plasmid and, on a separate compatible plasmid, DNA containing a putative DtxR-regulated promoter is cloned upstream of a gene *lacZ* encoding β-galactosidase. The expression of *lacZ* is dependent on the activity of the putative promoter so the effects of DtxR, iron or other factors can them be determined by assaying for β-galactosidase activity in an *E. coli* strain that lacks a functional chromosomal copy of *lacZ.*

Transcriptional Fusions Several groups have used transcriptional fusion systems in *E. coli* to identify DtxR-regulated promoters resulting in the identification of at least seven *C. diphtheriae* promoters; the *tox* promoter, iron regulated promoters 1 through 5, and the promoter for *hmuO* (Schmitt and Holmes 1994; Lee et al. 1997; Schmitt 1997a). *C. diphtheriae* strains do not express endogenous β-galactosidase activity and therefore the transcriptional fusion technology was easily adapted for use in *C. diphtheriae* where a promoterless copy of *lacZ* is contained on a replicating plasmid and the need to express DtxR on another compatible plasmid is eliminated (Schmitt 1997a; Oram et al. 2006). Using the promoter fusion vector in *C. diphtheriae* permits the characterization of the impact of transcriptional regulators on promoter activity in a native host environment rather than in *E. coli.* An example of the utility of the *C. diphtheriae* transcriptional fusion systems can be seen in the recent identification of multiple Zur and zinc- dependent promoters (Smith et al. 2009).

SELEX or Systematic Evolution of Ligands by Exponential Enrichment An *in vitro* protein-DNA binding assay has also been used to identify promoters regulated by DtxR (Qian et al. 2002). This system is based on SELEX which was first used to identify Fur-regulated promoters from *Pseudomonas aeruginosa* (Ochsner and Vasil 1996). The success of this method depends upon the production and purification of DtxR in large quantities as well as the availability of an antibody that recognizes DtxR. DNA fragments that bind to DtxR can then be enriched exponentially by purifying DNA-DtxR complexes using anti-DtxR antibodies. Finally PCR is used to amplify the DNA fragments isolated from the complexes and the purification cycle is repeated several times. Iron regulated promoter 6, IRP6 was identified using SELEX as a promoter that is regulated by DtxR in *C. diphtheriae* (Qian et al. 2002).

DRTA or Diphtheria Toxin Repressor Titration Assay In 2003, the Schmitt laboratory developed a repressor titration assay to identify DtxR-regulated in genes in *C. diphtheriae* (Kunkle and Schmitt 2003). This clever system, DRTA is based on the same principals as the widely used ferric uptake repressor titration assay or FURTA (Stojiljkovic et al. 1994). The basic concept of both systems is that the presence of multiple copies of a repressor binding site will result it sequestration of a limited amount of the cognate repressor protein and thereby result in de-repression (or activity) of a second repressed promoter that is present in low copy. Therefore DRTA uses the low copy plasmid pDRTA to both produce DtxR and as a transcriptional reporter in *E. coli*. The transcriptional reporter consists of the best-characterized DtxR-regulated promoter P*tox* controlling transcription of a full-length copy of *lacZ*. Transcription from P*tox* results in the production of β-galactosidase which can be easily detected and quantified. The second component of this system is a high copy plasmid containing DNA fragments that encode suspected DtxR binding sites. Kunkle et al. clearly showed that when known DtxR binding sites were included on the high-copy plasmid, de-repression of the *tox* promoter resulted in blue colonies on agar plates containing XGAL. They then went on to use DRTA to identify 10 previously uncharacterized DtxR binding sites in *C. diphtheriae* (Kunkle and Schmitt 2003).

14.4 Genetic Manipulation of the *C. diphtheriae* Chromosome

Homologous Recombination Exploitation of the host cell homologous recombination system to construction defined mutations in the chromosome is routinely used in *E. coli* and a wide variety of other bacterial species but in *C. diphtheriae* this technique was not used successfully until 2001 when a targeted allelic exchange method was developed (Schmitt and Drazek 2001). This system was the first method to construct defined mutations in both *C. diphtheriae* and *C. ulcerans* and it utilizes a single crossover event between an internal portion of a cloned gene and the chromosomal copy of the gene. The success of the method in *C. diphtheriae* was the result of several insightful modifications to a technique first used to integrate plasmid vector sequences and thereby disrupt gene function in the chromosome of *C. glutamicum* (Reyes et al. 1991).

Previous attempts to transform *C. diphtheriae* with a non-replicating fragment of DNA containing sequences with homology to the genome did not result in detectable recombination between the transformed sequences and chromosomal sequences. The potential explanations for the failure to detect recombination include a low level of endogenous homologous recombination in *C. diphtheriae*, inefficacy of transformation and high levels of host DNA restriction. Schmitt and Drazek solved the problem of restriction by isolating the recombination substrate DNA from *C. diphtheriae* so that it would be modified by endogenous enzymes, and therefore not be recognized by restriction systems. To do this they first constructed in *E. coli* a plasmid that included a DNA substrate designed to inactivate *hmuT* when it recombined with the chromosomal copy of the gene. Next they transformed *C. diphtheriae* with the replicating plasmid that included the *hmuT* inactivation substrate and re-isolated the plasmid DNA from *C. diphtheriae*. The recombination substrate plasmid was next digested with restriction enzymes that flanked the replication origin resulting in two DNA fragments one of which included the replication origin and one that included the *hmuT* recombination construct. The DNA fragment containing the *hmuT* recombination substrate was purified, ligated at low DNA concentration to promote intramolecular ligation (circularization), and re-transformed into *C. diphtheriae*. All of this careful manipulation resulted in activation of *hmuT* and the first directed mutation in the chromosome of *C. diphtheriae* (Schmitt and Drazek 2001).

Transposon Mutagenesis Transposon mutagenesis is a powerful tool used in a wide variety of bacterial species to determine the function of gene products and study gene regulation. Transposons and related elements are well studies in *E. coli* but the complexity of the transposition reaction has often made this technique difficult to adapt for use in other bacterial species. One exception is Tn*5*, which, following electroporation, has been shown to insert *via* transposition in the chromosome not only of its native host, *E. coli* but also into the chromosomes of an ever growing list of prokaryotic and eukaryotic species (Goryshin et al. 2000; Kirby 2007). The transposition of Tn*5* requires only the DNA ends of the transposon, the cognate transposase protein, and a DNA target both *in vitro* and *in vivo* (Goryshin and Reznikoff 1998).

In 2002 a Tn*5* based system was used to construct a bank of marked mutations in the chromosome of *C. diphtheriae* (Oram et al. 2002). This was accomplished by using electroporation to deliver preformed transposition intermediates consisting of a transposable DNA molecule bound by an active transposase protein (EZ-Tn5™ Epicentre Madison, WI USA). The DNA molecule included a gene encoding resistance to an antibiotic as well as the binding sites and associated cleavage sites for the Tn*5* transposase. Since Tn5 does not require specific host factors and the activity of the transposase is activated by divalent cations found in the host cell (primarily magnesium), once the complex enters the cytoplasm the transposae mediates transposition of the DNA molecule into a site in the chromosome. By selecting for resistance to the antibiotic, cells with chromosomal insertions of the input DNA were selected. Transposition of Tn*5* in *C. diphtheriae* had no detectable target site specificity. Oram et al. next used this technique to isolate two independent insertion mutants with altered expression of DtxR. The Tn*5* insertion site of one of these two mutants occurred within *dtxR* and no detectable DtxR was produced in this strain. This work demonstrated, for the first time, that *dtxR* is not an essential gene (Oram et al. 2002).

Conjugation and Counter-Selection In 2003, Ton-That and Schneewind described two advances to genetic engineering in *C. diphtheriae*; first they showed that a broad-host range conjugative transfer system could be used to efficiently transfer

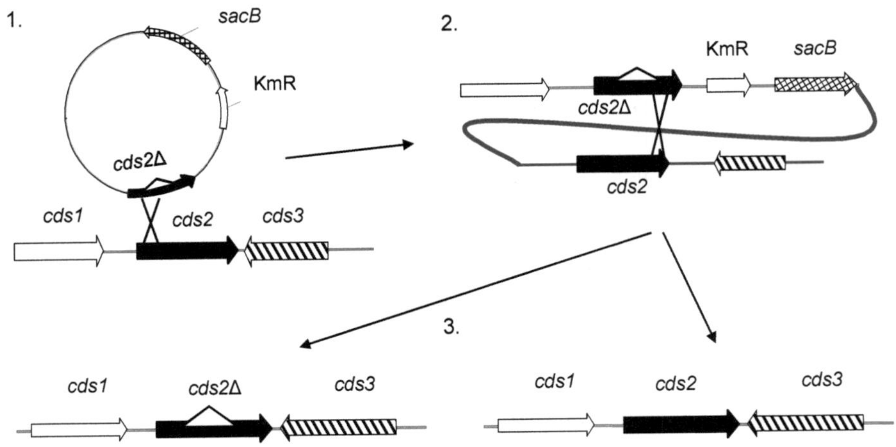

Fig. 14.1 Construction of an unmarked gene deletion in the chromosome [A non-replicating plasmid introduced by conjugation is shown as a circle that includes *sacB* and a gene encoding resistance to an antibiotic (KmR). Included on this plasmid is a copy of a gene of interest with and internal deletion (cds2D) (Part 1). Homologous recombination between the regions of identity on the plasmid and chromosome results in insertion of the plasmid and a strain that is resistance to antibiotic and sensitive to sucrose (Part 2). A second recombination event between the regions of identity on the inserted plasmid and the chromosome results in excision of the plasmid and a strain that is sensitive to antibiotic and resistant to sucrose. The second recombination event can generate either of two gene arrangements on the host chromosome (Part 3).]

DNA into *C. diphtheriae* thus providing the first practical alternative to electroporation, and second, they demonstrated that sucrose could be used to select against *C. diphtheriae* cells that expressed the *B. subtilis sacB* gene (Ton-That and Schneewind 2003). RP4 family conjugative systems are well-characterized and have been widely used to introduce DNA to a variety of bacterial species, with perhaps the best known example being their use in the environmentally ubiquitous, opportunistic human pathogen *P. aeruginosa* (Suh et al. 2004; Alvarez-Martinez and Christie 2009). Counter-section or the ability to select against a cell expressing a particular gene rather than for the presence of a gene, as is commonly done with genes encoding antibiotic resistance, is a powerful tool in constructing mutants that do not include exogenous DNA (Sander et al. 1995; Skorupski and Taylor 1996; Pelicic et al. 1997). The *B. subtilis* gene for levansucrase, *sacB* has been used as a counter-selectable marker in a variety of Gram negative as well as some Gram positive bacteria (Gay et al. 1985; Jager et al. 1992). The introduction of these two methods made the construction of unmarked mutations in the chromosome of *C. diphtheriae* possible (Fig. 14.1).

The introduction of DNA using conjugation from an *E. coli* host, where the trans-acting conjugal functions are engineered into the *E. coli* chromosome and the cis-acting factors, in this case the origin of transfer *ori*T, is encoded on the transferred plasmid, provides several advantages over other methods of DNA transfer. For one, conjugal DNA transfer utilizes a single-stranded intermediate which has two practical advantages for genetic engineering. First single stranded DNA is not a substrate for restriction and second single stranded DNA is an excellent substrate for the enzymes required for homologous recombination. Another advantage of introducing DNA using a specialized mating *E. coli* donor strain is that the transferred DNA lacks the trans-acting conjugation factors and is therefore unable to mediate further transfer events. Additionally, using the RP4 conjugation system to introduce DNA into *C. diphtheriae* is straight-forward and uses a high-transfer-efficiency mating method. Mating requires nothing other than mixing together the donor and recipient strains under mating conditions and then providing a selection for separating transconjugants from the parent strains. By taking advantage of the intrinsic resistance of *C. diphtheriae* to the gyrase inhibitor nalixidic acid and combining this with a selection for a marker on the transferred DNA, such as resistance to kanamycin, transconjugants are easily selected.

The second advance introduced by Ton-That and Schneewind was the use of *sacB* as a counter-selectable marker to facilitate detection of allelic exchange following introduction of a non-replicating plasmid using conjugation. By including an antibiotic resistance gene on the plasmid they were able to select for insertion of the incoming plasmid by way of homologous recombination to yield a Campbell-type insertion. A second recombination event which deleted the plasmid sequences was then selected using resistance to sucrose; the plasmid contains regions with homology to the *C. diphtheriae* chromosome, a gene encoding antibiotic resistance, and *sacB*. As *C. diphtheriae* cannot metabolize sucrose, expression of the plasmid encoded *sacB* leads to accumulation of toxic byproducts, thereby precipitating detection of recombinants that have lost the plasmid. These recombinants are then screened for those with the desired chromosomal gene arrangement using DNA analysis techniques such as PCR (Fig. 14.1). Since its description, this technique has been used to construct multiple independent directed mutations in the chromosome of *C. diphtheriae* (Ton-That et al. 2004; Smith et al. 2009; Bibb and Schmitt 2010).

Site Specific Integration In many bacterial species the availability of genome sequences and the characterization of bacteriophage lifecycles has resulted in the development of vector systems for the targeted insertion of DNA in the chromosome (Lee et al. 1991; Moreau et al. 1999; Bibb and Hatfull 2002; Lauer et al. 2002). These systems exploit the integrase enzyme and *att*P site of temperate phages as well as the attachment site in the bacterial chromosome *att*B (Groth and Calos 2004). An integrase-dependent, site-specific recombination reaction between a vector-borne *attP* and the chromosomal *attB* generates the recombinant *attL* and *attR* sites and causes integration of the vector (which can include virtually any foreign DNA sequence) into the chromosome. In 2007 Oram et al. developed an integrating vector system that exploits the *attP* site and integrase protein of diphtheria toxin encoding phage β for use in *C. diphtheriae* (Oram et al. 2007) (Fig. 14.2).

The phage based insertion vector developed for *C. diphtheriae* utilizes the integrase gene *int*/dip0182 from strain NCTC13129 (Cerdeno-Tarraga et al. 2003) and the *att*P from phage β (Barksdale and Pappenheimer 1954). In addition, vectors that

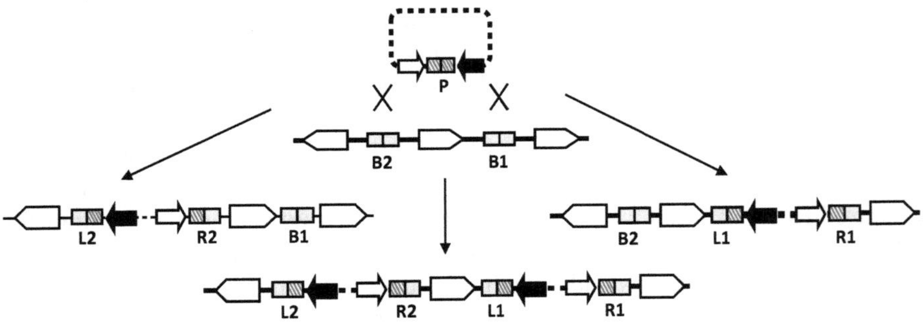

Fig. 14.2 Site-specific integration mediated by phage insertion vectors [Shown as a dotted line is integrating plasmid vector that includes *att*P (striped boxes) and *int* (black arrow). Additional genes (white arrow) may be included but are not essential. The region of a (non-lysogen) *C. diphtheriae* host chromosome that includes *att*B1 and *att*B2 (grey boxes) is shown below the plasmid. Maps of the chromosomes of strains in which insertions occur at *att*B1, *att*B2 or both attachment sites are shown at the bottom. The *att*L and *att*R sites that result following site-specific integration mediated by Int between *att*P and *att*B are labeled. Figure is not drawn to scale.]

include multiple cloning sites and a mobilizable origin of transfer were constructed to facilitate gene cloning and transfer to *C. diphtheriae*. The site-specific insertion of the β-phage-based vectors into the *C. diphtheriae* chromosome provides several advantages over replicating vectors including obviating problems arising from the multiplicity and variations in copy number, and the requirement for some form of selection to maintain an episome. The chromosomally inserted phage vector was stable without selection for antibiotics at least in part because excision requires the phage encoded protein Xis that is not included in the vector sequences.

Another attractive feature of using the phage site-specific recombination system is that *C. diphtheriae* carries two closely-spaced *attB* sites designated *attB*1 and *attB*2 (Rappuoli and Ratti 1984), which allows for single β-phage lysogens to still serve as recipients for the integrating vector (Fig. 14.2). For example the acquisition of *tox* (DIP0222) in strain NCTC13129 (Cerdeno-Tarraga et al. 2003) likely occurred from the integration of a 36.5 kb β-like phage at the *attB*1 site in a progenitor strain, with the concomitant formation of the *attL*1 and *attR*1 sites flanking the prophage sequence but because of the presence of the *attB*2 site, the phage integration vectors were still able to insert site-specifically in NCTC13129 (Oram et al. 2007). To demonstrate the utility of the integrating vector *dtxR* was transferred in single copy into a previously constructed Δ*dtxR* strain, with a consequent restoration of wild-type DtxR phenotypes. Finally, since both *C. glutamicum* and *C. ulcerans* possess the *attB* site(s) (Cianciotto et al. 1990), Oram et al. demonstrated that these two Coryneform species could also be transformed with the β phage-based vectors, thus establishing the utility of this vector system for species other than *C. diphtheriae*. In fact *C. glutamicum* is not a permissible host for phage β and this observation suggests that other bacterial species whose chromosomes include sequences similar to *att*B will be able to act as recipients of these phage-based vectors, even if they are not permissible host for β-phage replication.

14.5 Conclusions and Future Directions

The hard work of multiple researchers over the last 20 years has resulted in a variety of molecular and genetic tools to analyze gene expression in *C. diphtheriae*. These tools are now being used to characterize the virulence factors and pathogenic mechanisms used by *C. diphtheriae* to cause not only respiratory disease, but infections of the skin, heart, bloodstream and other sites. The reported incidence of extra-respiratory diseases caused by both toxigenic and nontoxigenic *C. diphtheriae* has increased; therefore there is both more interest and a greater need to identify therapeutics, other than the standard inactivated diphtheria toxin (toxoid) vaccine, to combat infections caused by *C. diphtheriae*. The availability of multiple genome sequences allows researchers to mine these genomes of traits associated with particular disease pathology. Once these genomic traits are identified molecular biology techniques can be used to test the hypotheses that the traits are involved in pathogenesis. In this way the diverse collection of genetic tools available for *C. diphtheriae* will facilitate the identification of new targets for antimicrobial development.

References

Alvarez-Martinez CE, Christie PJ (2009) Biological diversity of prokaryotic type IV secretion systems. Microbiol Mol Biol Rev 73:775–808

Barksdale WL, Pappenheimer AM Jr (1954) Phage-host relationships in nontoxigenic and toxigenic diphtheria bacilli. J Bacteriol 67:220–232

Bibb LA, Hatfull GF (2002) Integration and excision of the *Mycobacterium tuberculosis* prophage-like element, φRv1. Mol Microbiol 45:1515–1526

Bibb LA, Schmitt MP (2010) The ABC transporter HrtAB confers resistance to hemin toxicity and is regulated in a hemin-dependent manner by the ChrAS two-component system in *Corynebacterium diphtheriae*. J Bacteriol 192:4606–4617

Bibb LA, Kunkle CA, Schmitt MP (2007) The ChrA-ChrS and HrrA-HrrS signal transduction systems are required for activation of the *hmuO* promoter and repression of the *hemA* promoter in *Corynebacterium diphtheriae*. Infect Immun 75:2421–2431

Boyd J, Oza MN, Murphy JR (1990) Molecular cloning and DNA sequence analysis of a diphtheria *tox* iron-dependent regulatory element (*dtxR*) from *Corynebacterium diphtheriae*. Proc Natl Acad Sci U S A 87:5968–5972

Cerdeño-Tárraga AM, Efstratiou A, Dover LG, Holden MT, Pallen M, Bentley SD, Besra GS, Churcher C, James KD, De Zoysa A, Chillingworth T, Cronin A, Dowd L, Feltwell T, Hamlin N, Holroyd S, Jagels K, Moule S, Quail MA, Rabbinowitsch E, Rutherford KM, Thomson NR, Unwin L, Whitehead S, Barrell BG, Parkhill J (2003) The complete genome sequence and analysis of *Corynebacterium diphtheriae* NCTC13129. Nucleic Acids Res 31:6516–6523

Cianciotto N, Serwold-Davis T, Groman N, Ratti G, Rappuoli R (1990) DNA sequence homology between *attB*-related sites of *Corynebacterium diphtheriae*, *Corynebacterium ulcerans*, *Corynebacterium glutamicum*, and the *attP* site of γ-corynephage. FEMS Microbiol Lett 66:299–301

Freeman VJ (1951) Studies on the virulence of bacteriophage-infected strains of *Corynebacterium diphtheriae*. J Bacteriol 61:675–688

Gay P, Le Coq D, Steinmetz M, Berkelman T, Kado CI (1985) Positive selection procedure for entrapment of insertion sequence elements in gram-negative bacteria. J Bacteriol 164:918–921
Gill DM, Uchida T, Singer RA (1972) Expression of diphtheria toxin genes carried by integrated and nonintegrated phage β. Virology 50:664–668
Goryshin IY, Reznikoff WS (1998) Tn*5* *in vitro* transposition. J Biol Chem 273:7367–7374
Goryshin IY, Jendrisak J, Hoffman LM, Meis R, Reznikoff WS (2000) Insertional transposon mutagenesis by electroporation of released Tn*5* transposition complexes. Nat Biotechnol 18:97–100
Greenfield L, Bjorn MJ, Horn G, Fong D, Buck GA, Collier RJ, Kaplan DA (1983) Nucleotide sequence of the structural gene for diphtheria toxin carried by corynebacteriophage β. Proc Natl Acad Sci U S A 80:6853–6857
Groman NB (1953a) Evidence for the induced nature of the change from nontoxigenicity to toxigenicity in *Corynebacterium diphtheriae* as a result of exposure to specific bacteriophage. J Bacteriol 66:184–191
Groman NB (1953b) The relation of bacteriophage to the change of *Corynebacterium diphtheriae* from avirulence to virulence. Science 117:297–299
Groman NB (1984) Conversion by corynephages and its role in the natural history of diphtheria. J Hyg 93:405–417
Groth AC, Calos MP (2004) Phage integrases: biology and applications. J Mol Biol 335:667–678
Hantke K (2001) Iron and metal regulation in bacteria. Curr Opin Microbiol 4:172–177
Holmes RK (1976) Characterization and genetic mapping of nontoxinogenic (*tox*) mutants of corynebacteriophage β. J Virol 19:195–207
Holmes RK (2000) Biology and molecular epidemiology of diphtheria toxin and the *tox* gene. J Infect Dis 181(Suppl 1):156–167
Holmes RK, Barksdale L (1969) Genetic analysis of tox+ and tox– bacteriophages of *Corynebacterium diphtheriae*. J Virol 3:586–598
Jäger W, Schäfer A, Pühler A, Labes G, Wohlleben W (1992) Expression of the *Bacillus subtilis sacB* gene leads to sucrose sensitivity in the gram-positive bacterium *Corynebacterium glutamicum* but not in *Streptomyces lividans*. J Bacteriol 174:5462–5465
Kanei C, Uchida T, Yoneda M (1977) Isolation from *Corynebacterium diphtheriae* C7(*β*) of bacterial mutants that produce toxin in medium with excess iron. Infect Immun 18:203–209
Kirby JR (2007) In vivo mutagenesis using EZ-Tn*5*. Methods Enzymol 421:17–21
Kunkle CA, Schmitt MP (2003) Analysis of the *Corynebacterium diphtheriae* DtxR regulon: identification of a putative siderophore synthesis and transport system that is similar to the *Yersinia* high-pathogenicity island-encoded yersiniabactin synthesis and uptake system. J Bacteriol 185:6826–6840
Laird W, Groman N (1976) Prophage map of converting corynebacteriophage β. J Virol 19:208–219
Lauer P, Chow MY, Loessner MJ, Portnoy DA, Calendar R (2002) Construction, characterization, and use of two *Listeria monocytogenes* site-specific phage integration vectors. J Bacteriol 184:4177–4186
Lee MH, Pascopella L, Jacobs WR Jr, Hatfull GF (1991) Site-specific integration of mycobacteriophage L5: integration-proficient vectors for *Mycobacterium smegmatis*, *Mycobacterium tuberculosis*, and bacille Calmette-Guérin. Proc Natl Acad Sci U S A 88:3111–3115
Lee JH, Wang T, Ault K, Liu J, Schmitt MP, Holmes RK (1997) Identification and characterization of three new promoter/operators from *Corynebacterium diphtheriae* that are regulated by the diphtheria toxin repressor (DtxR) and iron. Infect Immun 65:4273–4280
Matsuda M, Barksdale L (1966) Phage-directed synthesis of diphtherial toxin in non-toxinogenic *Corynebacterium diphtheriae*. Nature 210:911–913
Moreau S, Blanco C, Trautwetter A (1999) Site-specific integration of corynephage Ø16: construction of an integration vector. Microbiology 145:539–548
Mueller JH (1940) Nutrition of the diphtheria bacillus. Bacteriol Rev 4:97–134

Murphy JR, Skiver J, McBride G (1976) Isolation and partial characterization of a corynebacteriophage *β*, *tox* operator constitutive-like mutant lysogen of *Corynebacterium diphtheriae*. J Virol 18:235–244

Neŝvera J, Pátek M, Hochmannová J, Abrhamová Z, Bečvářová V, Jelínková M, Vohradský J (1997) Plasmid pGA1 from *Corynebacterium glutamicum* codes for a gene product that positively influences plasmid copy number. J Bacteriol 179:1525–1532

Ochsner UA, Vasil ML (1996) Gene repression by the ferric uptake regulator in *Pseudomonas aeruginosa*: cycle selection of iron-regulated genes. Proc Natl Acad Sci U S A 93:4409–4414

Oram DM, Holmes RK (2006) Diphtheria Toxin. In: Alouf JE, Popoff MR (eds) The comprehensive sourcebook of bacterial protein toxins, 3rd edn. Academic Press 245–256

Oram DM, Avdalovic A, Holmes RK (2002) Construction and characterization of transposon insertion mutations in *Corynebacterium diphtheriae* that affect expression of the diphtheria toxin repressor (DtxR). J Bacteriol 184:5723–5732

Oram DM, Jacobson AD, Holmes RK (2006) Transcription of the contiguous *sigB*, *dtxR*, and *galE* genes in *Corynebacterium diphtheriae*: evidence for multiple transcripts and regulation by environmental factors. J Bacteriol 188:2959–2973

Oram M, Woolston JE, Jacobson AD, Holmes RK, Oram DM (2007) Bacteriophage-based vectors for site-specific insertion of DNA in the chromosome of Corynebacteria. Gene 391:53–62

Pappenheimer AM Jr, Johnson S (1936) Studies on diphtheria toxin production. I. The effect of iron and copper. Brit J Exp Pathol 17:335–341

Pelicic V, Jackson M, Reyrat JM, Jacobs WR Jr, Gicquel B, Guilhot C (1997) Efficient allelic exchange and transposon mutagenesis in *Mycobacterium tuberculosis*. Proc Nat l Acad Sci U S A 94:10955–10960

Puech V, Chami M, Lemassu A, Lanéelle MA, Schiffler B, Gounon P, Bayan N, Benz R, Daffé M (2001) Structure of the cell envelope of corynebacteria: importance of the non-covalently bound lipids in the formation of the cell wall permeability barrier and fracture plane. Microbiology 147:1365–1382

Qian Y, Lee JH, Holmes RK (2002) Identification of a DtxR-regulated operon that is essential for siderophore-dependent iron uptake in *Corynebacterium diphtheriae*. J Bacteriol 184:4846–4856

Rappuoli R, Ratti G (1984) Physical map of the chromosomal region of *Corynebacterium diphtheriae* containing corynephage attachment sites *attB*1 and *attB*2. J Bacteriol 158:325–330

Rappuoli R, Michel JL, Murphy JR (1983) Integration of corynebacteriophages β^{tox+}, ω^{tox+}, and γ^{tox-} into two attachment sites on the *Corynebacterium diphtheriae* chromosome. J Bacteriol 153:1202–1210

Ratti G, Rappuoli R, Giannini G (1983) The complete nucleotide sequence of the gene coding for diphtheria toxin in the corynephage omega (tox+) genome. Nucleic Acids Res 11:6589–6595

Reyes O, Guyonvarch A, Bonamy C, Salti V, David F, Leblon G (1991) 'Integron'-bearing vectors: a method suitable for stable chromosomal integration in highly restrictive corynebacteria. Gene 107:61–68

Roux E Jr, Yersin A (1888) Contribution a l'etude de la diphtherie. Ann Inst Pasteur 2:620–629

Sander P, Meier A, Böttger EC (1995) *rpsL+*: a dominant selectable marker for gene replacement in mycobacteria. Mol Microbiol 16:991–1000

Schmitt MP (1997a) Transcription of the *Corynebacterium diphtheriae hmuO* gene is regulated by iron and heme. Infect Immun 65:4634–4641

Schmitt MP (1997b) Utilization of host iron sources by *Corynebacterium diphtheriae*: identification of a gene whose product is homologous to eukaryotic heme oxygenases and is required for acquisition of iron from heme and hemoglobin. J Bacteriol 179:838–845

Schmitt MP, Drazek ES (2001) Construction and consequences of directed mutations affecting the hemin receptor in pathogenic *Corynebacterium* species. J Bacteriol 183:1476–1481

Schmitt MP, Holmes RK (1991) Iron-dependent regulation of diphtheria toxin and siderophore expression by the cloned *Corynebacterium diphtheriae* repressor gene *dtxR* in *C. diphtheriae* C7 strains. Infect Immun 59:1899–1904

Schmitt MP, Holmes RK (1994) Cloning, sequence, and footprint analysis of two promoter/operators from *Corynebacterium diphtheriae* that are regulated by the diphtheria toxin repressor (DtxR) and iron. J Bacteriol 176:1141–1149

Serwold-Davis TM, Groman N, Rabin M (1987) Transformation of *Corynebacterium diphtheriae, Corynebacterium ulcerans, Corynebacterium glutamicum*, and *Escherichia coli* with the *C. diphtheriae* plasmid pNG2. Proc Natl Acad Sci U S A 84:4964–4968

Skorupski K, Taylor RK (1996) Positive selection vectors for allelic exchange. Gene 169:47–52

Smith KF, Bibb LA, Schmitt MP, Oram DM (2009) Regulation and activity of a zinc uptake regulator, Zur, in *Corynebacterium diphtheriae*. J Bacteriol 191:1595–1603

Sonnen H, Schneider J, Kutzner HJ (1990) Corynephage Cog, a virulent bacteriophage of *Corynebacterium glutamicum*, and its relationship to ØGA1, an inducible phage particle from *Brevibacterium flavum*. J Gen Virol 71:1629–1633

Stojiljkovic I, Bäumler AJ, Hantke K (1994) Fur regulon in Gram-negative bacteria. Identification and characterization of new iron-regulated *Escherichia coli* genes by a fur titration assay. J Mol Biol 236:531–545

Suh SJ, Silo-Suh LA, Ohman DE (2004) Development of tools for the genetic manipulation of *Pseudomonas aeruginosa*. J Microbiol Methods 58:203–212

Tauch A, Kirchner O, Löffler B, Götker S, Pühler A, Kalinowski J (2002) Efficient electrotransformation of *Corynebacterium diphtheriae* with a mini-replicon derived from the *Corynebacterium glutamicum* plasmid pGA1. Curr Microbiol 45:362–367

Tauch A, Bischoff N, Brune I, Kalinowski J (2003) Insights into the genetic organization of the *Corynebacterium diphtheriae* erythromycin resistance plasmid pNG2 deduced from its complete nucleotide sequence. Plasmid 49:63–74

Ton-That H, Schneewind O (2003) Assembly of pili on the surface of *Corynebacterium diphtheriae*. Mol Microbiol 50:1429–1438

Ton-That H, Marraffini LA, Schneewind O (2004) Sortases and pilin elements involved in pilus assembly of *Corynebacterium diphtheriae*. Mol Microbiol 53:251–261

Trost E, Blom J, Soares Sde C, Huang IH, Al-Dilaimi A, Schröder J, Jaenicke S, Dorella FA, Rocha FS, Miyoshi A, Azevedo V, Schneider MP, Silva A, Camello TC, Sabbadini PS, Santos CS, Santos LS, Hirata R Jr, Mattos-Guaraldi AL, Efstratiou A, Schmitt MP, Ton-That H, Tauch A (2012) Pangenomic study of *Corynebacterium diphtheriae* that provides insights into the genomic diversity of pathogenic isolates from cases of classical diphtheria, endocarditis, and pneumonia. J Bacteriol 194:3199–3215

Welkos SL, Holmes RK (1981) Regulation of toxinogenesis in *Corynebacterium diphtheriae*. I. Mutations in bacteriophage β that alter the effects of iron on toxin production. J Virol 37:936–945

Wohlleben W, Muth G, Kalinowski J (1993) Genetic engineering of gram-positive bacteria. In: Rehm H-J, Reed G, Pühler A, Stadler P (eds) Biotechnology, vol 2. VCH, Weinheim, pp 455–450

Chapter 15
Diphtheria Toxin Based Molecules as Therapeutic Approaches

Ingo Schubert

Abstract Malignant diseases are still one of the main causes of death and due to this high mortality rate new therapy approaches are needed. During the last decades cancer specific antigens on the cell surface were identified. This fundamental discovery was the important step to develop ligand-directed-toxins and antibody-drug-conjugates. They consist of an antigen binding domain plus an effector moiety like bacterial or plant toxins, respectively. These kinds of new molecules achieved a higher selectivity and efficacy, eliminated side effects and ensured better drug delivery as the standard approaches in cancer therapy. In consequence, both types of molecules are promising candidates for further development to extend the existing chemotherapeutic and radioactive armamentarium. In this review, the existing approaches to apply diphtheria toxin in this respect are summarized.

Keywords Antibody-drug-conjugates · Ligand-directed-toxins · Lymphoma

15.1 Introduction

The 'magic bullet' concept introduced by Paul Ehrlich, inspired scientists to develop agents with the ability to bind target cells with a higher selectivity. Ligand-directed-toxins (LDTs) and antibody-drug-conjugates (ADCs) are fusion proteins consisting of a binding domain (antibody, antibody derived binding head, cytokine or chemokine), which is specific for a target cell, and a toxic component. The binding heads of these molecules allow to target the malignant cell with an increased efficacy. After binding to the target cell, these molecules are internalized by endocytosis and induce cell death. Over time, several different toxins have been used in LDTs and ADCs, such as *Pseudomonas* Exotoxin A (ETA), diphtheria toxin (DT), anthrax, ricin and saporin (Blythman et al. 1981; Frankel et al. 2002; Thorpe et al. 1985;

I. Schubert (✉)
Department of Biology, Chair of Microbiology, University of Erlangen-Nuremberg,
Staudtstr. 5, 91058, Erlangen, Germany
e-mail: Ingo.A.Schubert@gmail.com

A. Burkovski (ed.), *Corynebacterium diphtheriae and Related Toxigenic Species*,
DOI 10.1007/978-94-007-7624-1_15, © Springer Science+Business Media Dordrecht
(outside the USA) 2014

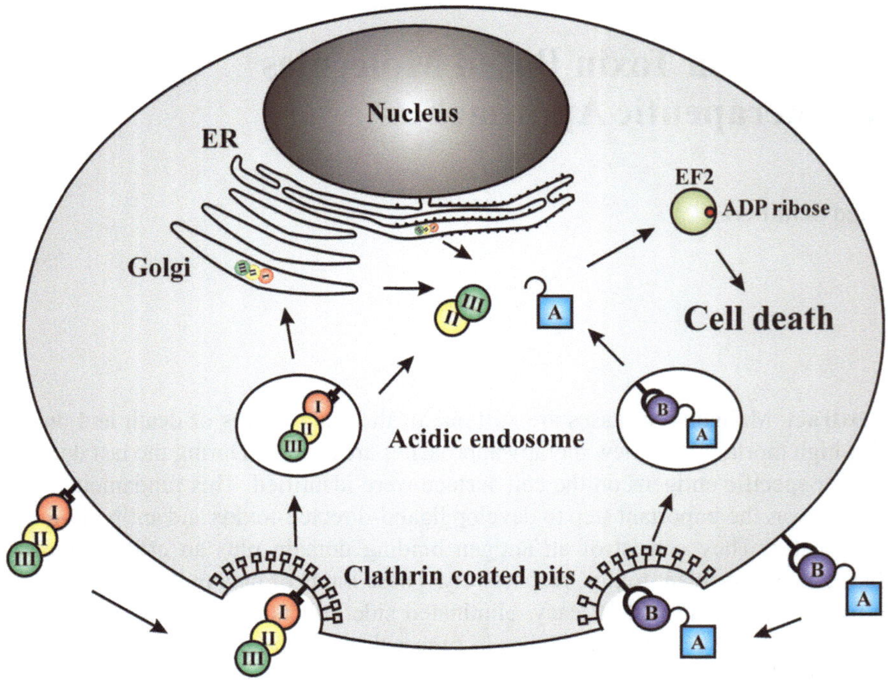

Fig. 15.1 Mechanism of action of *Pseudomonas* Exotoxin A (ETA) and diphtheria toxin (DT). After binding to a receptor, the toxin is internalised through clathrin coated pits into an endosome. Unfolding of the protein, cleavage by furin and the release of the catalytic domain into the cytosol is induced by low pH. Both toxins inhibit protein synthesis by ADP ribosylation of the elongation factor 2 which mediates cell death. *I* binding domain of ETA; *II* furin cleavage site of ETA; *III* catalytic domain of ETA; *B* binding moiety of DT; *A* enzymatic domain of DT; *ER* endoplasmatic reticulum; *EF2* elongation factor 2

Wels et al. 1992; Youle et al. 1986). The first generation of immunotoxins, developed 35 years ago, consisted of a toxic component chemically-linked to an antibody (ab). They showed tumour regression in some lymphoma patients, but, due to their size, they were ineffective in infiltrating solid tumours (Ghetie and Vitetta 2001).

After binding the target cell, ADCs and LDTs enter the cell via receptor-mediated endocytosis. Unfolding of *Pseudomonas* endotoxin A and diphtheria toxin is induced by the low pH in the endosome. Then, the toxins are proteolytically cleaved and the catalytic domains are released into the cytosol (Gonzalez and Wisnieski 1988), where they inhibit protein biosynthesis, leading to death like the unmodified toxins (Fig. 15.1).

15.1.1 Targeted Therapy Using Diphtheria Derived Ligand-Directed-Toxins (LDT) and Antibody-Drug Conjugates (ADCs)

During the last three decades, different kinds of ADCs and LDTs have been constructed on the basis of the toxic moiety of *Corynebacterium diphtheriae*. Most ligand-directed-toxins contain a cytokine as binding domain for the target cell (Table 15.1) and an antibody or an antibody-derivative (e.g. single chain Fragment variable; scFv) mediates binding of the ADCs (Table 15.2). Diphtheria toxin is a protein 535 amino acids in length with three domains. The enzymatic A domain, the binding B domain and the transmembrane domain which is located in the centre of the molecule. To improve of specificity of the ADCs and LDTs, they were designed with recombinant toxic components in which the binding domain was truncated or site-specifically mutated. A full-length version of diphtheria toxin would contain mutations at position 390 and 525 which inhibit its binding ability (Greenfield et al. 1987). Truncated versions in which the B domain had been deleted include DAB_{486}, DT388 and DAB_{389} (Chaudhary et al. 1991; Williams et al. 1987; Williams et al. 1990). Applications of combinations of different targeting moieties and toxin molecules are described below.

15.2 Ligand-Directed-Toxins (LDT)

A number of different ligands has been used to direct toxin molecules to target cells, including interleukins, growth factors and other peptides.

15.2.1 Binding Moiety Interleukin

DAB_{389}-IL6 Jean and co-workers replaced the portion of the diphtheria toxin which encodes the receptor binding domain with human IL-6. The resulting fusion protein DAB_{389}-IL-6 is selectively cytotoxic for eukaryotic cells bearing the interleukin 6 receptor, while cells, which are devoid of the IL-6 receptor, are resistant to the action of this recombinant toxin (Jean and Murphy 1991).

DAB_{389}-IL4 IL-4 is a cytokine which affects the growth and differentiation of various cell types. One of the common AIDS symptoms is the so-called AIDS-related Kaposi's sarcoma (KS). These KS infected cells express high affinity IL-4 binding sites. Recombinant human IL-4 minimally inhibited Kaposi's sarcoma cell growth and expression of IL-6. Investigators studied the effects of a fusion toxin DAB_{389}IL4 which mediates cellular toxicity only on cells expressing the IL-4 receptor (IL-4R). DAB_{389}IL4 inhibited protein synthesis in Kaposi's sarcoma cells at low concentrations and this effect was neutralized with an IL-4 specific antibody. They concluded

Table 15.1 Ligand-directed-toxins

Name	Binding moiety	Indication	Reference
DAB_{389}-IL-6	Interleukin 6 (IL-6)	Myeloma	(Jean and Murphy 1991)
DAB_{389}IL4	Interleukin 4 (IL-4)	AIDS-related Kaposi sarcoma	(Cai et al. 1997)
DAB_{389} sIL15	Simian interleukin 15 (sIL-15)	IL-2 and IL-15 dependent diseases	(vanderSpek et al. 1995)
DT_{388}IL3	Interleukin 3 (IL-3)	Acute myeloid leukemia (AML)	(Urieto et al. 2004)
DT_{390}IL3	Interleukin 3 (IL-3)	*ex vivo* purging of leukemia cells; autologous bone marrow transplantation	(Vallera et al. 1999)
DAB_{389}IL7	Interleukin 7 (IL-7)	Hematologic malignancies	(Sweeney et al. 1998)
DT_{390}IL18	Interleukin 18 (IL-18)	Autoimmune encephalomyelitis	(Jia et al. 2008)
DAB_{389} IL2 (Denileukin diftitox)	Interleukin 2 (IL-2)	cutaneous T cell lymphoma (CTCL)	(Manoukian and Hagemeister 2009)
DAB_{389}GRP	Gastrin-releasing peptide (GRP)	small cell lung cancer	(vanderSpek et al. 1997)
DAB_{389}-NT4	Rat neurotrophin-4/5	NT-4/5 nerve cell depletion	(Negro and Skaper 1997)
DAB_{389}EGF	Epidermal growth factor (EGF)	Several malignancies	(Shaw et al. 1991)
DT_{389}-bFGF	Basic fibroblast growth factor (bFGF)	Several malignancies	(Zhang et al. 2006)
DTGM	Granulocyte–macrophage colony-stimulating factor	Acute myeloid leukemia (AML)	(Frankel et al. 1999)
VEGF-DT385	Vascular endothelial growth factor (VEGF)	Several malignancies	(Olson et al. 1997)
DTAT	Aminoterminal (AT) fragment of the urokinase-type plasminogen activator	Glioblastoma multiforme (GBM)	(Vallera et al. 2002)
DTAT13	Aminoterminal (AT) fragment of the urokinase-type plasminogen activator; interleukin 13 (IL-13)	Glioblastoma multiforme (GBM)	(Rustamzadeh et al. 2006)
DTEGF13	Epidermal growth factor (EGF); interleukin 13 (IL-13)	Glioblastoma	(Oh et al. 2009)

Table 15.2 Antibody-drug conjugates

Name	Specificity	Indication	Reference
DT_{388}-anti-Tac(Fv)	Interleukin 2 receptor (IL-2R)	Chronic lymphocytic leukemia (CLL)	(Kreitman et al. 1992)
DTM1-E6scFv-PE40	Transferrin receptor (TfnR)	Several malignancies	(Nicholls et al. 1993)
A-dmDT390-bisFv(UCHT1)	CD3ε	T cell malignancies	(Woo et al. 2008)
A-dmDT390-scfbDb(C207)	Monkey CD3	T cell depletion in monkeys	(Kim et al. 2007)
A-dmDT390biscFv(2-6-15)	Porcine CD3	T cell depletion	(Wang et al. 2011)
DT2219	CD22/CD19	B cell leukemia and lymphoma	(Vallera et al. 2005)

that the expression of a functional IL-4R could serve as a target for novel therapy with agents such as DAB_{389}IL4 (Cai et al. 1997).

DAB_{389}-sIL15 The fusion toxin DAB_{389}-sIL15 is directed against simian interleukin 15 (sIL-15). The sIL-15 domain of the molecule binds to the IL-2/IL-15 receptor and is involved in signal transduction leading to DNA and protein synthesis. This recombinant molecule is internalized by receptor-mediated endocytosis and the catalytic domain of diphtheria toxin is delivered to the cell cytosol, where it kills the target cell. DAB_{389}-sIL15 could be a potential therapeutic agent in cases where patients have proven refractory to treatment with IL-2 specific toxins (vanderSpek et al. 1995).

DT_{388}-IL3 Acute myeloid leukemia (AML) is the most common leukemia in adults and the second most common leukemia in children. Interleukin 3 (IL-3) is needed for proliferation and differentiation of multi-potential and committed myeloid and lymphoid progenitors. A major advantage of therapies with IL-3 specific agents is, that the receptor is absent from mature myeloid cells whereas myeloid leukemia progenitors over-express the IL-3 receptor. A recombinant toxin consisting of truncated diphtheria toxin and the human IL-3 showed potent and selective killing of IL-3 receptor expressing AML cell lines and patient leukemic progenitors (Alexander et al. 2001). Therapy of SCID mice bearing human leukemia cells revealed *in vivo* efficacy (Black et al. 2003). The safety of DT_{388}-IL3 was tested in rodents and monkeys and was remarkable (Urieto et al. 2004).

DT_{390}-IL3 Another LDT directed against IL-3 is the molecule DT_{390}-IL3 with murine IL-3 as binding moiety. DT_{390}-IL3 was applied as an anti-leukaemic drug to mice. However, it was already toxic when administrated at low doses. Due to the high toxicity they alternatively tried to use this recombinant molecule *ex vivo* for purging of contaminated leukemic cells from bone marrow grafts in an autologous transplantation assay. Mice given treated bone marrow survive over 100 days while mice given untreated cells did not survive. So DT_{390}-IL3 may rather prove to be a useful agent for *ex vivo* purging of bone marrow grafts (Vallera et al. 1999).

DAB_{389}-IL7 Interleukin 7 (IL-7) is important for cell growth in haematological malignancies such as acute lymphoid leukaemia (ALL), chronic lymphoid leukaemia (CLL), AML, and Sezary syndrome. A recombinant fusion protein, DAB_{389}-IL-7, composed of the toxic domain of diphtheria toxin fused to IL-7, was constructed and tested. This fusion protein selectively inhibited protein synthesis in IL-7 receptor (IL-7R) positive cells. Due to the high expression of IL-7R on a variety of haematopoietic neoplasms, DAB_{389}-IL-7 may serve as a therapeutic agent for patients with IL-7R positive leukaemia and lymphoma (Sweeney et al. 1998).

DT_{390}-IL18 Antigen-presenting cells (APCs) expressing interleukin-18 receptor (IL-18R) were shown to be crucial for establishing and maintaing experimental autoimmune encephalomyelitis (EAE) in an animal model for multiple sclerosis. This LDT showed good results *in vivo*; EAE mice treated with the agent showed a delayed manifestation of EAE and decreased symptoms (e.g. infiltration of inflammatory cells into the brain) in comparison to the control group (Jia et al. 2008).

DAB_{389}IL2 DAB_{389}IL2, also known as denileukin diftitox, is the first commercially available recombinant fusion protein which is capable of delivering a cytotoxic agent directly to specific intracellular targets. Denileukin diftitox is composed of the full-length sequence of interleukin 2 (IL-2) and a modified cytotoxic DT. It binds to cells expressing the interleukin 2 receptor (IL-2R) and inhibits protein synthesis due to the diphtheria toxin fragment. In a mouse model with an IL-2R-expressing malignancy, denileukin diftitox prolonged survival of the treated group compared with controls. During the first in man study with patients bearing IL-2R-positive cutaneous T cell lymphoma (CTCL), the overall response rate was 36 % after administration of 18 μg/kg/day for 5 days every 3 weeks (Figgitt et al. 2000). Denileukin diftitox is FDA-approved only for CTCL, but other applications might be beneficial as well. Several clinical studies are in progress for different indications, but data are still pending (Manoukian and Hagemeister 2009).

Different groups showed a correlation between the presence of regulatory T-cells (Treg) and the efficacy of tumour vaccination, as Treg cells have been shown to inhibit anti-tumour immune responses. It was shown that the depletion of Treg using the IL-2 fusion protein denileukin diftitox decreased Treg function and increased antigen-specific T-cell response to a cancer vaccine in mice (Hobeika et al. 2012). *In vitro* studies demonstrated that canine Treg cells are a target of denileukin diftitox. Suppression of T-cell proliferation due to Treg was corrected by addition of denileukin diftitox. Application of the toxin resulted in a depletion of Treg which was followed by an increase in the immune response *in vitro* (Knueppel et al. 2012).

15.2.2 Growth Factors and Other Peptides as Binding Domain

DAB_{389}GRP Gastrin-releasing peptide (GRP) is a bombesin-like peptide involved in the regulation of a large number of different cellular processes including e.g. exocrine secretion and smooth muscle contraction. GRP also functions as autocrine

growth factor on a number of neoplastic tissues, including small cell lung cancer cells (SCLC), a property which has been exploited for toxin delivery. The fusion protein DAB_{389}GRP showed receptor specific cytotoxicity which could be inhibited by an excess of GRP and anti-GRP antibody, respectively. A number of SCLC cell lines were tested and the LDT showed inhibition of protein synthesis in all of them (vanderSpek et al. 1997).

DAB_{389}-NT4 Neutrophins are a family of structurally related proteins which play a role in the development of the nervous system. The DAB_{389}NT4 is a recombinant molecule with a truncated version of diphtheria toxin serving as effector and the rat neutrophin-4/5 (NT-4/5) as binding domain. This fusion protein was cytotoxic for different neural cell lines. To study the role of neutrophins, gene knockouts can be used, but in a conventional knockout the protein is absent in all tissues during embryonic development. So DAB_{389}NT4 could be a useful agent for the selective depletion of NT-4/5 positive nerve cells *in vivo* (Negro and Skaper 1997).

DAB_{389}EGF A characteristic of several malignancies including those of the breast, bladder, prostate, lung, and neuroglia is the expression of the receptor for the epidermal growth factor (EGF). The fusion protein DAB_{389}EGF was constructed for treatment of these malignancies. The toxin binds specifically to the EGF receptor and inhibits protein synthesis in different EGF receptor expressing tumour cell lines. Studies of the kinetics of poisoning showed a fast effect which resulted in complete protein synthesis inhibition after a 15 min exposure of an EGF receptor positive cell line to DAB_{389}EGF. The ability of the toxin to effectively kill tumour cells at extremely low concentrations and the selectivity and rapid kinetics of cytotoxic action suggest that this molecule could be an efficient anti-tumour agent (Shaw et al. 1991).

DT_{389}-bFGF The high-affinity receptor for basic fibroblast growth factor (bFGF), fibroblast growth factor receptor (FGFR), is an important tumour-associated growth factor that contributes to proliferation and angiogenesis, and it is over-expressed in a number of tumour cell lines. A recombinant immunotoxin containing a truncated version of diphtheria toxin and human bFGF was designed, expressed and purified. The activity and potential anti-tumour effect of DT_{389}-bFGF was evaluated by testing the cytotoxicity on a human ovarian teratocarcinoma cell line with high-level expression of FGFR. The immunotoxin showed significant cytotoxicity and the effect could be blocked by an excess of bFGF and anti-bFGF antibodies, respectively. Additionally, cell lines with low expression of FGFR were found to be resistant to the recombinant molecule. FGFR could be an interesting target for tumour therapy and the FGFR targeting molecules might be a promising approach in the treatment of cancers (Zhang et al. 2006).

DTGM Frankel et al. constructed an AML-selective fusion toxin consisting of human granulocyte–macrophage colony-stimulating factor (GM-CSF) as binding domain fused to the catalytic domain of diphtheria toxin, the so-called DTGM. The receptor for GM-CSF is selectively expressed on mature and immature monocytes, granulocytes, or macrophages and on malignant myeloid leukemias (Cannistra et al. 1990).

Treatment failure of patients with AML is due to the development of multidrug resistant phenotypic blasts. The fusion toxin DTGM selectively inhibited protein synthesis and inducted apoptosis in receptor-positive cells (Burbage et al. 1997). Frankel and his group designed a high-level expression vector and a purification protocol for a phase I clinical trial with DTGM (Frankel et al. 1999).

VEGF-DT385 Tumour-derived vascular endothelial growth factor (VEGF) plays an important role in neovascularisation and in the development of tumour stroma. The VEGF receptors (VEGFR) are over-expressed on tumour cells of endothelial origin and nearly undetectable in the endothelium of normal tissues. This expression profile offers a selective advantage for targeting VEGFR positive tumours. VEGF-DT385 was selectively toxic to endothelial cell lines and inhibited neovascularisation. Athymic nude mice with established subcutaneous tumours were treated with daily injections of VEGF-DT385 or with free toxin. The conjugate-treated animals displayed a significant inhibition of tumour growth in comparison to the control animals treated with the toxin alone. Analysis of the tumours from VEGF-DT385 animals revealed haemorrhagic necrosis. In contrast, highly vascularised normal tissues from the same animals demonstrated no evidence of haemorrhage or tissue injury. This recombinant protein illustrates the selectively anti-tumour activity of a VEGF conjugate (Olson et al. 1997)

DTAT Patients with brain malignancies have a poor prognosis and new, selective therapies are urgently needed. The urokinase-type plasminogen activator (uPA) receptor (uPAR) is expressed on the surface of glioblastoma (GB) and could be a potential target for GB treatment. The recombinant fusion protein DTAT, which contains the catalytic portion of diphtheria toxin fused to the non-internalizing amino-terminal (AT) fragment of uPA, was highly potent and selective in killing uPAR-expressing glioblastoma cells. DTAT caused a statistically significant regression of cell-induced tumours in all mice in comparison with the control group. To investigate the molecule's potential toxicity, DTAT was given to tumour-free mice. DTAT had only small effects on kidney, liver, heart, lung and spleen. Due to its ability to target tumour cells and tumour vasculature simultaneously, combined with the lack of systemic effects, DTAT maybe a novel agent for glioblastoma therapy (Vallera et al. 2002).

15.2.3 Dual-Targeting Ligand-Directed-Toxins

DTAT13 DTAT13, another recombinant LDT consisting of truncated diphtheria toxin, an AT-fragment and a fragment of IL-13 was constructed to target receptors on glioblastoma cells. To determine the properties of a dual-targeting DTAT13, pharmacokinetic and bio-distribution experiments were performed. Binding analysis revealed that the IL-13 domain functioned independently of the AT and that the equilibrium constant for each binding domain was essentially the same as in the monovalent DTIL13 and DTAT molecules. Flow cytometry studies indicated that the bivalent DTAT13 bound better than the monovalent control molecules. DTAT13 inhibits the rate of protein synthesis in double positive cells faster than DTAT, but in a same order of magnitude as DT13. This study showed that DTAT13 has properties encompassing those of both DTIL13 and DTAT and warrants further consideration for clinical development (Rustamzadeh et al. 2006).

DTEGF13 A bispecific ligand-directed toxin (BLT) consisting of IL-13, EGF and a truncated version of DT was constructed to target glioblastoma cells. The corresponding molecule, DTEGF13, selectively killed a human glioblastoma cell line. This recombinant BLT has greater activity than the monospecific controls or their mixture. Aggressive brain tumours were established in nude rats with the U87 cell line genetically marked with a luciferase reporter gene. These were treated with two injections of the dual-targeting BLT, resulting in tumour eradication in 50% of the rats. In contrast a combination of the monospecific LDTs DTEGF and DTIL13 was not able to inhibit tumour growth. Anti-DT antibodies were not generated in normal immune-competent rats given the identical intracranial DTEGF13 therapy. These observations led to the conclusion that DTEGF13 is safe and could be an alternative drug for glioblastoma therapy (Oh et al. 2009).

15.3 Antibody-Drug Conjugates (ADCs)

15.3.1 Monospecific and Monovalent ADCs

DT_{388}-anti-Tac(Fv) The monoclonal antibody anti-Tac is specific for the IL-2 receptor, and in the fusion protein DT_{388}-anti-Tac(Fv), the variable domain of this antibody is fused to the C-terminus of a truncated diphtheria toxin. Peripheral blood mononuclear cells (PBMCs) from 14 chronic lymphoid leukaemia (CLL) patients were incubated with the recombinant ADC. It was active against eleven of the 14 samples tested, while an LDT consisting of the truncated diphtheria toxin and the human IL-2 as binding moiety was only cytotoxic for four of these samples. The patient samples, which were sensitive to the ADC, contained between 400 and 2,500 binding sites per cell. These data showed that DT_{388}-anti-Tac(Fv) is able to kill CLL cells even when there is only a low number of IL-2R (Kreitman et al. 1992)

DTM1-E6scFv-PE40 DTM1-E6scFv-PE40 is a tripartite protein, consisting of the diphtheria toxin mutant M1 (containing two mutations in the binding domain), the E6scFv directed against the human transferrin receptor (TfnR) followed by the *Pseudomonas* Exotoxin PE40 (lacking the binding domain).

The expression pattern of TfnR is limited in normal tissue, but it is widely distributed on cells from several malignancies. Therefore it could be a suitable target for therapy. DTM1-E6scFv-PE40 showed cytotoxicity to TfnR-positive cells. Although this molecule contains two bacterial toxin components, the cell death mediated was very similar to that mediated by the mono-toxin E6scFv-PE40. This would suggest that the main effect is due to the PE40 domain. However, this molecule showed that it is possible in principle to create a fusion protein with two effector domains (Nicholls et al. 1993).

15.3.2 Monospecific, Bivalent ADCs

A-dmDT390-bisFv(UCHT1) The bivalent ADC A-dmDT390-bisFv(UCHT1) for treatment of patients with T cell malignancies is a fusion protein composed of the catalytic domain of DT fused to two tandem scFvs. The scFv(UCHT1) is specific for human CD3ε and the fusion molecule was able to selectively kill CD3ε positive T cells. For evaluation of the maximum tolerated dose, pharmacokinetics and immunogenicity of the ADC, studies with rats and squirrel monkeys were performed by Woo and colleagues. The administration of the toxin did not affect liver function, renal function, the haemogram, nor did it produce serious organ histopathology in the animals tested. A-dmDT390-bisFv(UCHT1) had a plasma half life of 23 min on average in the two species. Due to these data, this ADC appears to be a promising drug for patients with CD3-positive T cell malignancies (Woo et al. 2008).

A-dmDT390-scfbDb(C207) T cell depletion in non-human primates is important for models of transplantation tolerance and autoimmune disease therapy. A-dmDT390-scfbDb(C207) consists of a truncated version of DT and a diabody directed against monkey CD3. The ADC format showed a 7-fold increased binding to T cells in comparison to the parental scFv(C207). It mediated 5- to 7-fold more bioactivity over the A-dmDT390bisFv(C207) and the diabody format was able to deplete T cells in monkeys (Kim et al. 2007), providing a putative animal model.

A-dmDT390-biscFv(2-6-15) Another molecule for T cell depletion is the ADC A-dmDT390-biscFv(2-6-15). This fusion protein is directed against porcine CD3 and it induces a profound but transient T-cell depletion which allows to investigate transplantation tolerance models and autoimmune disease therapies in pigs. A-dmDT390-biscFv(2-6-15) was tested in four animals and was able to deplete the T cell population without any detectable clinical toxicity. This monospecific, bivalent molecule could be utilized in experimental porcine models of transplantation tolerance, autoimmune disease therapy and treatment of T-cell leukaemia (Wang et al. 2011).

15.3.3 Dual-Targeting ADC

DT2219 A dual-targeting immunotoxin called DT2219, consisting of two scFvs recognizing CD19 and CD22 and the catalytic DT_{390} fragment as death effector, showed strong *in vivo* anti-leukaemic activity in a murine xenograft model. Treatment of transplanted mice with this agent resulted in long-term tumour-free survival, measured in a bioluminescent xenograft imaging model, in which the propagation of Raji Burkitt's lymphoma cells was tracked in real time (Vallera et al. 2005).

15.4 Conclusions

The various diphtheria toxin based molecules described above showed promising results *in vitro* and *in vivo*; they selectively killed the target cells and displayed only marginal or no clinical cytotoxicity in animal models. However, only one fusion protein, denileukin diftitox, has been approved by the FDA for use in CTCL up to now. Ongoing studies suggest the potential value in therapy of other diseases which express IL-2R, and investigators have demonstrated that it may be combined with other drugs to achieve synergistic effects (Manoukian and Hagemeister 2009). A newly designed phase II study showed that denileukin diftitox has significant clinical activity in unresectable stage IV melanoma patients due to T cell depletion and CD8-positive T cell expansion (Telang et al. 2011).

However, it is noticeable that many groups, which have worked with diphtheria toxin in the past and which have achieved remarkable results *in vivo* like Vallera and co-workers with DT2219, have now switched to *Pseudomonas* exotoxin as death domain. One reason could be to avoid immunogenicity and to achieve a higher potency of the agent (Vallera et al. 2010). In the late 90s, Brinkmann et al. investigated the effect of BCL-2 expression on the sensitivity of cancer cell lines to different immunotoxins. BCL-2, a mitochondrial membrane protein, has been shown to inhibit apoptosis (Sentman et al. 1991). Some malignant cells are relatively resistant to apoptosis due to overexpression of BCL-2 (Tsujimoto et al. 1985). ADCs with *Pseudomonas* exotoxin as effector domain were the only molecules which were able to induce cell death independent of the BCL-2 level (Brinkmann et al. 1997).

In summary, DT-derived LDTs and ADCs described here proved which molecule formats could lead to promising new therapies for several diseases. But, one has to keep in mind that it is important to identify and verify which combination of toxin and target is the best for each individual ailment.

References

Alexander RL, Ramage J, Kucera GL, Caligiuri MA, Frankel AE (2001) High affinity interleukin-3 receptor expression on blasts from patients with acute myelogenous leukemia correlates with cytotoxicity of a diphtheria toxin/IL-3 fusion protein. Leuk Res 25(10):875–881

Black JH, McCubrey JA, Willingham MC, Ramage J, Hogge DE, Frankel AE (2003) Diphtheria toxin-interleukin-3 fusion protein (DT(388)IL3) prolongs disease-free survival of leukemic immunocompromised mice. Leukemia 17(1):155–159

Blythman HE, Casellas P, Gros O, Gros P, Jansen FK, Paolucci F, Pau B, Vidal H (1981) Immunotoxins: hybrid molecules of monoclonal antibodies and a toxin subunit specifically kill tumour cells. Nature 290(5802):145–146

Brinkmann U, Mansfield E, Pastan I (1997) Effects of BCL-2 overexpression on the sensitivity of MCF-7 breast cancer cells to ricin, diphtheria and *Pseudomonas* toxin and immunotoxins. Apoptosis 2(2):192–198

Burbage C, Tagge EP, Harris B, Hall P, Fu T, Willingham MC, Frankel AE (1997) Ricin fusion toxin targeted to the human granulocyte-macrophage colony stimulating factor receptor is selectively toxic to acute myeloid leukemia cells. Leuk Res 21(7):681–690

Cai J, Zheng T, Murphy J, Waters CA, Lin GY, Gill PS (1997) IL-4R expression in AIDS-KS cells and response to rhIL-4 and IL-4 toxin (DAB389-IL-4). Invest New Drugs 15(4):279–287

Cannistra SA, Groshek P, Garlick R, Miller J, Griffin JD (1990) Regulation of surface expression of the granulocyte/macrophage colony-stimulating factor receptor in normal human myeloid cells. Proc Natl Acad Sci U S A 87(1):93–97

Chaudhary VK, FitzGerald DJ, Pastan I (1991) A proper amino terminus of diphtheria toxin is important for cytotoxicity. Biochem Biophys Res Commun 180(2):545–551

Figgitt DP, Lamb HM, Goa KL (2000) Denileukin diftitox. Am J Clin Dermatol 1(1):67–72; discussion 73

Frankel AE, Ramage J, Latimer A, Feely T, Delatte S, Hall P, Tagge E, Kreitman R, Willingham M (1999) High-level expression and purification of the recombinant diphtheria fusion toxin DTGM for PHASE I clinical trials. Protein Expr Purif 16(1):190–201

Frankel AE, Powell BL, Duesbery NS, Vande Woude GF, Leppla SH (2002) Anthrax fusion protein therapy of cancer. Curr Protein Pept Sci 3(4):399–407

Ghetie V, Vitetta ES (2001) Chemical construction of immunotoxins. Mol Biotechnol 18(3):251–268

Gonzalez JE, Wisnieski BJ (1988) An endosomal model for acid triggering of diphtheria toxin translocation. J Biol Chem 263(30):15257–15259

Greenfield L, Johnson VG, Youle RJ (1987) Mutations in diphtheria toxin separate binding from entry and amplify immunotoxin selectivity. Science 238(4826):536–539

Hobeika AC, Morse MA, Osada T, Peplinski S, Lyerly HK, Clay TM (2012) Depletion of human regulatory T cells. Methods Mol Biol 707:219–231

Jean LF, Murphy JR (1991) Diphtheria toxin receptor-binding domain substitution with interleukin 6: genetic construction and interleukin 6 receptor-specific action of a diphtheria toxin-related interleukin 6 fusion protein. Protein Eng 4(8):989–994

Jia J, Li H, Tai S, Lv M, Liao M, Yang Z, Zhang B, Zhou B, Zhang G, Zhang L (2008) Construction and preliminary investigation of a plasmid containing a novel immunotoxin DT390-IL-18 gene for the prevention of murine experimental autoimmune encephalomyelitis. DNA Cell Biol 27(5):279–285

Kim GB, Wang Z, Liu YY, Stavrou S, Mathias A, Goodwin KJ, Thomas JM, Neville DM (2007) A fold-back single-chain diabody format enhances the bioactivity of an anti-monkey CD3 recombinant diphtheria toxin-based immunotoxin. Protein Eng Des Sel 20(9):425–432

Knueppel A, Lange S, Altmann S, Sekora A, Knuebel G, Vogel H, Lindner I, Freund M, Junghanss C (2012) Upfront denileukin diftitox as *in vivo* regulatory T-cell depletion in order to enhance vaccination effects in a canine allogeneic hematopoietic stem cell transplantation model. Vet Immunol Immunopathol 145(1–2):233–240

Kreitman RJ, Chaudhary VK, Kozak RW, FitzGerald DJ, Waldman TA, Pastan I (1992) Recombinant toxins containing the variable domains of the anti-Tac monoclonal antibody to the interleukin-2 receptor kill malignant cells from patients with chronic lymphocytic leukemia. Blood 80(9):2344–2352

Manoukian G, Hagemeister F (2009) Denileukin diftitox: a novel immunotoxin. Expert Opin Biol Ther 9(11):1445–1451

Negro A, Skaper SD (1997) Synthesis and cytotoxic profile of a diphtheria toxin-neurotrophin-4 chimera. J Neurochem 68(2):554–563

Nicholls PJ, Johnson VG, Andrew SM, Hoogenboom HR, Raus JC, Youle RJ (1993) Characterization of single-chain antibody (sFv)-toxin fusion proteins produced *in vitro* in rabbit reticulocyte lysate. J Biol Chem 268(7):5302–5308

Oh S, Ohlfest JR, Todhunter DA, Vallera VD, Hall WA, Chen H, Vallera DA (2009) Intracranial elimination of human glioblastoma brain tumors in nude rats using the bispecific ligand-directed toxin, DTEGF13 and convection enhanced delivery. J Neurooncol 95(3):331–342

Olson TA, Mohanraj D, Roy S, Ramakrishnan S (1997) Targeting the tumor vasculature: inhibition of tumor growth by a vascular endothelial growth factor-toxin conjugate. Int J Cancer 73(6):865–870

Rustamzadeh E, Vallera DA, Todhunter DA, Low WC, Panoskaltsis-Mortari A, Hall WA (2006) Immunotoxin pharmacokinetics: a comparison of the anti-glioblastoma bi-specific fusion protein (DTAT13) to DTAT and DTIL13. J Neurooncol 77(3):257–266

Sentman CL, Shutter JR, Hockenbery D, Kanagawa O, Korsmeyer SJ (1991) bcl-2 inhibits multiple forms of apoptosis but not negative selection in thymocytes. Cell 67(5):879–888

Shaw JP, Akiyoshi DE, Arrigo DA, Rhoad AE, Sullivan B, Thomas J, Genbauffe FS, Bacha P, Nichols JC (1991) Cytotoxic properties of DAB486EGF and DAB389EGF, epidermal growth factor (EGF) receptor-targeted fusion toxins. J Biol Chem 266(31):21118–21124

Sweeney EB, Foss FM, Murphy JR, vanderSpek JC (1998) Interleukin 7 (IL-7) receptor-specific cell killing by DAB389 IL-7: a novel agent for the elimination of IL-7 receptor positive cells. Bioconjug Chem 9(2):201–207

Telang S, Rasku MA, Clem AL, Carter K, Klarer AC, Badger WR, Milam RA, Rai SN, Pan J, Gragg H, Clem BF, McMasters KM, Miller DM, Chesney J (2011) Phase II trial of the regulatory T cell-depleting agent, denileukin diftitox, in patients with unresectable stage IV melanoma. BMC Cancer 11(1):515

Thorpe PE, Brown AN, Bremner JA Jr, Foxwell BM, Stirpe F (1985) An immunotoxin composed of monoclonal anti-Thy 1.1 antibody and a ribosome-inactivating protein from Saponaria officinalis: potent antitumor effects *in vitro* and *in vivo*. J Natl Cancer Inst 75(1):151–159

Tsujimoto Y, Cossman J, Jaffe E, Croce CM (1985) Involvement of the *bcl-2* gene in human follicular lymphoma. Science 228(4706):1440–1443

Urieto JO, Liu T, Black JH, Cohen KA, Hall PD, Willingham MC, Pennell LK, Hogge DE, Kreitman RJ, Frankel AE (2004) Expression and purification of the recombinant diphtheria fusion toxin DT388IL3 for phase I clinical trials. Protein Expr Purif 33(1):123–133

Vallera DA, Seo SY, Panoskaltsis-Mortari A, Griffin JD, Blazar BR (1999) Targeting myeloid leukemia with a DT(390)-mIL-3 fusion immunotoxin: *ex vivo* and *in vivo* studies in mice. Protein Eng 12(9):779–785

Vallera DA, Li C, Jin N, Panoskaltsis-Mortari A, Hall WA (2002) Targeting urokinase-type plasminogen activator receptor on human glioblastoma tumors with diphtheria toxin fusion protein DTAT. J Natl Cancer Inst 94(8):597–606

Vallera DA, Todhunter DA, Kuroki DW, Shu Y, Sicheneder A, Chen H (2005) A bispecific recombinant immunotoxin, DT2219, targeting human CD19 and CD22 receptors in a mouse xenograft model of B-cell leukemia/lymphoma. Clin Cancer Res 11(10):3879–3888

Vallera DA, Oh S, Chen H, Shu Y, Frankel AE (2010) Bioengineering a unique deimmunized bispecific targeted toxin that simultaneously recognizes human CD22 and CD19 receptors in a mouse model of B-cell metastases. Mol Cancer Ther 9(6):1872–1883

vanderSpek JC, Sutherland J, Sampson E, Murphy JR (1995) Genetic construction and characterization of the diphtheria toxin-related interleukin 15 fusion protein DAB389 sIL-15. Protein Eng 8(12):1317–1321

vanderSpek JC, Sutherland JA, Zeng H, Battey JF, Jensen RT, Murphy JR (1997) Inhibition of protein synthesis in small cell lung cancer cells induced by the diphtheria toxin-related fusion protein DAB389 GRP. Cancer Res 57(2):290–294

Wang Z, Duran-Struuck R, Crepeau R, Matar A, Hanekamp I, Srinivasan S, Neville DM Jr, Sachs DH, Huang CA (2011) Development of a diphtheria toxin based antiporcine CD3 recombinant immunotoxin. Bioconjug Chem 22(10):2014–2020

Wels W, Harwerth IM, Mueller M, Groner B, Hynes NE (1992) Selective inhibition of tumor cell growth by a recombinant single-chain antibody-toxin specific for the erbB-2 receptor. Cancer Res 52(22):6310–6317

Williams DP, Parker K, Bacha P, Bishai W, Borowski M, Genbauffe F, Strom TB, Murphy JR (1987) Diphtheria toxin receptor binding domain substitution with interleukin-2: genetic construction and properties of a diphtheria toxin-related interleukin-2 fusion protein. Protein Eng 1(6):493–498

Williams DP, Snider CE, Strom TB, Murphy JR (1990) Structure/function analysis of interleukin-2-toxin (DAB486-IL-2). Fragment B sequences required for the delivery of fragment A to the cytosol of target cells. J Biol Chem 265(20):11885–11889

Woo JH, Bour SH, Dang T, Lee YJ, Park SK, Andreas E, Kang SH, Liu JS, Neville DM Jr, Frankel AE (2008) Preclinical studies in rats and squirrel monkeys for safety evaluation of the bivalent anti-human T cell immunotoxin, A-dmDT390-bisFv(UCHT1). Cancer Immunol Immunother 57(8):1225–1239

Youle RJ, Uckun FM, Vallera DA, Colombatti M (1986) Immunotoxins show rapid entry of diphtheria toxin but not ricin via the T3 antigen. J Immunol 136(1):93–98

Zhang H, Zhang S, Zhuang H, Lu F (2006) Cytotoxicity of a novel fibroblast growth factor receptor targeted immunotoxin on a human ovarian teratocarcinoma cell line. Cancer Biother Radiopharm 21(4):321–332

Index

A. Burkovski (ed.), *Corynebacterium diphtheriae and Related Toxigenic Species*, DOI 10.1007/978-94-007-7624-1, © Springer Science+Business Media Dordrecht (outside the USA) 2014

MIX
Papier aus verantwortungsvollen Quellen
Paper from responsible sources
FSC® C105338

If you have any concerns about our products,
you can contact us on
ProductSafety@springernature.com

In case Publisher is established outside the EU,
the EU authorized representative is:
Springer Nature Customer Service Center GmbH
Europaplatz 3, 69115 Heidelberg, Germany

Printed by Libri Plureos GmbH
in Hamburg, Germany